T0177985

Universitext

Universitext

Universitext is a series of textbooks that presents material from a wide variety of mathematical disciplines at master's level and beyond. The books, often well class-tested by their author, may have an informal, personal even experimental approach to their subject matter. Some of the most successful and established books in the series have evolved through several editions, always following the evolution of teaching curricula, into very polished texts.

Thus as research topics trickle down into graduate-level teaching, first textbooks written for new, cutting-edge courses may make their way into *Universitext.*

More information about this series at http://www.springer.com/series/223

Thomas Alazard • Claude Zuily

Tools and Problems in Partial Differential Equations

 Springer

Thomas Alazard
École Normale Supérieure Paris-Saclay
Université Paris-Saclay
Gif sur Yvette, France

Claude Zuily
Institut de Mathématique d'Orsay
Université Paris-Saclay
Orsay, France

ISSN 0172-5939   ISSN 2191-6675   (electronic)
Universitext
ISBN 978-3-030-50283-6   ISBN 978-3-030-50284-3   (eBook)
https://doi.org/10.1007/978-3-030-50284-3

Mathematics Subject Classification: 35, 76

This Springer imprint is published by the registered company Springer Nature Switzerland AG.
The registered company address is: Gewerbestrasse 11, 6330 Cham, Switzerland

To our families.

Introduction

The aim of this book is to present, through 65 fully solved long problems, various aspects of the current theory of partial differential equations (PDE). It is intended for graduate students who would like, through **practice**, to test their understanding of the theory.

The book is made of two parts. The first part (Chaps. 1–8) contains some theory (in the odd chapters) and the statements of the problems (in the even chapters). The second part contains the solutions (Chap. 9) and an appendix (Chap. 10), where we recall some basics of the classical analysis.

Even though the main purpose of this book is to present these problems and their solutions, for the reader's convenience, we have recalled some of the main theoretical results concerning each topic. This is why each chapter of problems is preceded by a short introduction recalling without proof the basic facts. However, some comments indicate where one may find the details of the proofs. This makes the book essentially self-contained for the reader.

Since the theory of partial differential equations is a very wide subject, it is by no means realistic to hope to describe all the topics in a single volume. Therefore, choices have to be made and we have chosen to focus on a few of them.

Let us now describe more precisely the contents of this book.

The first chapter is introductory. Some of the essential tools in functional analysis which are commonly used in PDE are recalled. This includes the main theorems concerning Fréchet, Banach, or Hilbert spaces. We have also recalled the main notions in distributions theory, including the analysis of the Fourier transform and the stationary phase formula.

The second chapter contains three problems on the subjects discussed in the previous one.

The third chapter begins with the description of some of the main function spaces used in PDE, namely Sobolev, Hölder, and Zygmund spaces (including their Littlewood–Paley characterization). Other tools, such as interpolation theory and paraproducts, are also presented.

In the fourth chapter, through 15 problems, we discuss various applications such as functional inequalities (product and composition rules in Sobolev spaces, Hardy inequality, etc.) and we introduce other function spaces such as spaces of analytic functions, uniformly local Sobolev spaces, weak Lebesgue spaces, space of bounded mean oscillation, etc.

The fifth chapter is concerned with the theory of microlocal analysis. We review the main notions about the pseudo-differential and paradifferential operators, the wave front set, and the microlocal defect measures and we describe some of the main results, such as continuity, Garding inequalities, and propagation of singularities.

In the sixth chapter, through 18 problems, we give some applications of these notions, in particular we explore the notions of symbols, hypoelliptic operators, smoothing effect, and Carleman inequalities.

The seventh chapter reviews the main classical partial differential equations that is the Laplace, wave, Schrödinger, heat, Burgers, Euler, Navier–Stokes equations, and their main properties are recalled.

The last chapter of the first part of the book contains 29 problems on the above equations. For instance, several problems about spectral theory for the Laplace equation, about linear and nonlinear wave and Schrödinger equations are stated. Moreover, other equations such as Monge–Ampère, the mean-curvature, kinetic, and Benjamin–Ono equations are discussed.

The second part of the book contains the detailed solutions to all the problems and a chapter gathering several fundamental results concerning the basics of classical analysis, such as Lebesgue integration, differential calculus, differential equations and holomorphic functions.

Acknowledgements

We would like to warmly thank the anonymous referees for their excellent work, which helped us to improve the presentation of this book.

However, it should be clear that the authors are solely responsible of any mistake remaining. A list of possible corrections will be available at http://talazard.perso.math.cnrs.fr.

After this book was proposed to Springer, we had several mail exchanges with Mr Rémi Lodh. We would like to warmly thank him for his suggestions, support, and efficiency.

April 2020   Thomas Alazard
                                                                                    Claude Zuily

Contents

Part I
Tools and Problems

Chapter 1
Elements of Functional Analysis and Distributions

The goal of this chapter is to recall, without proof, the main results in functional analysis: classical theorems about Fréchet, Hilbert, and Banach spaces, as well as fixed point theorems and an introduction to spectral theory. This is complemented by the main definitions in distributions theory, including results about the Fourier transform.

1.1 Fréchet Spaces

- **Seminorms.** Consider a vector space E on $\mathbf{C}$. A seminorm on E is a map $p : E \to [0, +\infty)$ such that, for all $x, y \in E$ and all $\lambda \in \mathbf{C}$,

$$(i) \quad p(\lambda x) = |\lambda| p(x), \quad (ii) \quad p(x + y) \le p(x) + p(y).$$

  A family $\mathcal{P}$ of seminorms on E is called *separating* if for every $x \in E$ there exists $p \in \mathcal{P}$ such that $p(x) \ne 0$.
- **Topology.** Let $\mathcal{P}$ be a family of separating seminorms on a vector space E. For $x_0 \in E, n \in \mathbf{N}, n \ge 1$ and $p \in \mathcal{P}$, set

$$V_{(p,n)}(x_0) = \left\{ x \in E : p(x - x_0) < \frac{1}{n} \right\}.$$

Let $\mathcal{V}_{x_0}$ be the collection of all finite intersections of the sets $V_{(p,n)}(x_0)$ and define a neighborhood of x_0 as a set which contains an element of $\mathcal{V}_{x_0}$. This defines a topology on E. Then $(E, \mathcal{P})$ is called a *locally convex space*.

© The Editor(s) (if applicable) and The Author(s), under exclusive license
to Springer Nature Switzerland AG 2020
T. Alazard, C. Zuily, *Tools and Problems in Partial Differential Equations*,
Universitext, https://doi.org/10.1007/978-3-030-50284-3_1

- A subset A of E is said to be *bounded* if,

$$\forall p \in \mathcal{P}, \quad \exists M > 0 : p(x) \leq M, \quad \forall x \in A.$$

- If $(E, \mathcal{P})$ and $(F, \mathcal{Q})$ are two locally convex spaces and $T : E \to F$ is a linear map then T is continuous if and only if,

$$\forall q \in \mathcal{Q}, \ \exists \mathcal{P}_0 \subset \mathcal{P} \text{ finite}, \ \exists C > 0 : q(Tx) \leq C \sum_{p \in \mathcal{P}_0} p(x), \forall x \in E.$$

- If $\mathcal{P} = (p_j)_{j \in \mathbf{N}}$ is countable then the topology defined above is *metrizable*, that is there exists a metric d on E which induces the same topology. Indeed for $x, y \in E$ we set,

$$d(x, y) = \sum_{j=0}^{+\infty} \frac{1}{2^j} \frac{p_j(x - y)}{1 + p_j(x - y)}.$$

If moreover the metric space (E, d) is complete then we say that E is a *Fréchet space*.

• **Examples**.

- A Banach space is a Fréchet space.
- Let Ω be an open subset of $\mathbf{R}^d$. Then $C^0(\Omega)$ denotes the space of continuous functions on Ω with complex values, $C^1(\Omega)$ denotes the space of differentiable functions whose partial derivatives belong to $C^0(\Omega)$, and, for $k \in \mathbf{N}, k \geq 2$,

$$C^k(\Omega) = \left\{ u \in C^{k-1}(\Omega) : \frac{\partial u}{\partial x_j} \in C^{k-1}(\Omega), 1 \leq j \leq d \right\}.$$

Recall that Ω can be written as $\Omega = \cup_{j=0}^{+\infty} K_j$, where the K_j's are compact and $K_j \subset K_{j+1}$. For $k \in \mathbf{N} \cup \{+\infty\}$, $u \in C^k(\Omega)$ and $j \in \mathbf{N}$ set,

$$p_j(u) = \sum_{|\alpha| \leq k} \sup_{x \in K_j} |\partial^\alpha u(x)| \qquad (k < +\infty),$$

$$p_j(u) = \sum_{|\alpha| \leq j} \sup_{x \in K_j} |\partial^\alpha u(x)| \qquad (k = +\infty).$$

Then $(C^k(\Omega), (p_j)_{j \in \mathbf{N}})$ is a Fréchet space. A sequence $(f_n)_{n \in \mathbf{N}} \subset C^k(\Omega)$ converges to f for this topology if and only if, for every $|\alpha| \leq k$, $(\partial^\alpha f_n)_{n \in \mathbf{N}}$ converges to $\partial^\alpha f$ uniformly on each compact subset of Ω.

1.2   Elements of Functional Analysis

In this section we recall several important results of functional analysis used in the study of partial differential equations. Unless expressly stated all the normed spaces considered below are normed spaces on $\mathbf{C}$.

1.2.1   Fixed Point Theorems

- **The Banach fixed point theorem.** Let (X, d) be a complete metric space and consider a map $f : X \to X$ satisfying the following condition: There exists $k \in \,]0, 1[$ such that,

$$d(f(x), f(y)) \le k d(x, y), \quad \forall x, y \in X.$$

  Then f has a unique *fixed point*, that is a point x such that $f(x) = x$.
- **The Schauder fixed point theorem.** Let E be a Banach space and $K \subset E$ be a nonempty convex and compact subset. Then any continuous function $f : K \to K$ has a fixed point.

1.2.2   The Banach Isomorphism Theorem

- Let E, F be two Banach spaces. If $T : E \to F$ is a linear continuous and bijective map then T^{-1} is continuous.

1.2.3   The Closed Graph Theorem

- Let E, F be two Banach spaces and consider a linear map $T : E \to F$. Assume that the set (called *the graph of T*),

$$G = \{(x, Tx) : x \in E\}$$

  is closed in $E \times F$. Then T is continuous.
- Notice that the converse is also true. If T is continuous then G is closed.
- This result remains true if E and F are Fréchet spaces.

1.2.4   The Banach–Steinhaus Theorem

- Let E be a Banach space, F be a normed space, and Λ be a set. Let $(T_\lambda)_{\lambda \in \Lambda}$ be a family of linear and continuous maps from E to F. Assume that,

$$\forall x \in E, \ \exists C_x > 0: \quad \|T_\lambda x\|_F \leq C_x, \quad \forall \lambda \in \Lambda.$$

Then there exists $C > 0$ such that,

$$\|T_\lambda x\|_F \leq C\|x\|_E, \quad \forall x \in E, \quad \forall \lambda \in \Lambda.$$

This result can be extended to the case where E is a Fréchet space.
- Let $(E, (p_j)_{j \in \mathbf{N}})$ be a Fréchet space and (F, Q) be a locally convex space. Let $(T_\lambda)_{\lambda \in \Lambda}$ be a family of linear and continuous maps from E to F. Assume that,

$$\forall x \in E, \ \forall q \in Q, \ \exists C_{q,x} > 0: \quad q(T_\lambda x) \leq C_{q,x}, \quad \forall \lambda \in \Lambda.$$

Then, for any $q \in Q$, there exists a finite set $J \subset \mathbf{N}$ and $C > 0$ such that,

$$q(T_\lambda x) \leq C \sum_{j \in J} p_j(x), \quad \forall x \in E, \forall \lambda \in \Lambda.$$

1.2.5   The Banach–Alaoglu Theorem

- Let E be a normed space and E' be its dual, that is the space of linear and continuous maps $T: E \to \mathbf{C}$. One can endow E' with the strong topology induced by the norm $\|T\|_{E'} = \sup_{\|x\|_E = 1} |T(x)|$.
- One can also define on E' a weaker topology as follows. For $x \in E$ and $T \in E'$ set $p_x(T) = |T(x)|$. Then p_x is a seminorm on E' and the family $(p_x)_{x \in E}$ defines a topology on E' called the *weak-star topology*. For instance, a sequence $(T_n)_{n \in \mathbf{N}}$ in E' converges to $T \in E'$ in this topology if and only if $T_n(x) \to T(x)$ in $\mathbf{C}$ for every $x \in E$. This is therefore the topology of pointwise convergence.
- **The Banach–Alaoglu theorem.** Let E be a Banach space. Then the set $\{T \in E' : \|T\|_{E'} \leq 1\}$ is compact in the weak-star topology.
- As a Corollary, if E is a separable Banach space then each bounded sequence in E' (for the norm) has a convergent subsequence for the weak-star topology.

1.2.6 The Ascoli Theorem

- Let X be a compact metric space. Let $C^0(X)$ be the Banach space of all continuous functions $f: X \to \mathbf{C}$ endowed with the norm $\|f\| = \sup_{x \in X} |f(x)|$. Let $A \subset C^0(X)$. Assume that,

1. for every $x \in X$, the set $\{f(x) : f \in A\}$ is bounded in $\mathbf{C}$, that is,

$$\forall x \in X, \quad \exists M_x > 0 : |f(x)| \leq M_x, \quad \forall f \in A.$$

2. A is equicontinuous, that is,

$$\forall \varepsilon > 0, \forall x \in X, \quad \exists \delta > 0 : d(x, y) \leq \delta \implies |f(x) - f(y)| \leq \varepsilon, \quad \forall f \in A.$$

Then A has a compact closure in $C^0(X)$.

1.2.7 The Hahn–Banach Theorem

- Let E be a normed vector space and F be a subspace. Let $L : F \to \mathbf{C}$ be a continuous linear form. Then there exists a continuous linear form $\widetilde{L}: E \to \mathbf{C}$ such that $L(x) = \widetilde{L}(x)$ for every $x \in F$ and $\|\widetilde{L}\|_{E'} = \|L\|_{F'}$.

As a consequence we have the following result.

- Let E be a normed vector space and F be a subspace. Then F is dense in E if and only if, for every $L \in E'$,

$$L = 0 \text{ on } F \implies L = 0.$$

1.2.8 Hilbert Spaces

Recall first that a Hilbert space H is a vector space endowed with a scalar product denoted by $(x, y)_H$ and with the corresponding norm $\|x\| = \sqrt{(x, x)_H}$.

- **Weak convergence.** A sequence $(x_n)_{n \in \mathbf{N}}$ in H is said to converge *weakly* to $x \in H$ (and write $x_n \rightharpoonup x$) if $\lim_{n \to +\infty}(x_n, y)_H = (x, y)_H$ for every $y \in H$.

  - The strong convergence (for the norm) implies the weak convergence.
  - If $x_n \rightharpoonup x$ and $\|x_n\| \to \|x\|$ then $(x_n)_{n \in \mathbf{N}}$ converges strongly to x.
  - A sequence $(x_n)_{n \in \mathbf{N}}$ which converges weakly in H is bounded for the norm, that is, there exists $M > 0$ such that $\|x_n\| \leq M$ for every $n \in \mathbf{N}$.
  - Let $(x_n)_{n \in \mathbf{N}}$ be a bounded sequence for the norm. Then there exists a subsequence $(x_{\sigma(n)})_{n \in \mathbf{N}}$ which converges weakly in H.
  - If $(x_n)_{n \in \mathbf{N}}$ converges weakly to x in H then $\|x\|_H \leq \liminf_{n \to +\infty} \|x_n\|_H$.

- **A Riesz theorem**

  - Let H be a complex Hilbert space. An *antilinear form* L on H is a map $L :$ $H \rightarrow \mathbf{C}$ such that for any $x, y \in H, \lambda \in \mathbf{C}$,

  $$L(x + y) = L(x) + L(y), \qquad L(\lambda x) = \bar{\lambda} L(x).$$

  - A Riesz theorem. For every antilinear and continuous map $L \colon H \rightarrow \mathbf{C}$, there exists a unique $x \in H$ such that $L(y) = (x, y)_H$ for every $y \in H$.

- **The Lax–Milgram lemma**

  - Let H be a complex Hilbert space. A *sesquilinear* form on $H \times H$ is a map $a : H \times H \rightarrow \mathbf{C}$ such that for every $y \in H$ the map $x \mapsto a(x, y)$ is linear and for every $x \in H$ the map $y \mapsto a(x, y)$ is antilinear.
  - The Lax–Milgram lemma. Let a be a sesquilinear form on $H \times H$. Assume that,

  $$(i) \ a \text{ is continuous} : \ \exists C > 0 : |a(x, y)| \leq C\|x\|_H \|y\|_H , \quad \forall x, y \in H,$$

  $$(ii) \ a \text{ is coercive} : \ \exists c_0 > 0 : \operatorname{Re} a(x, x) \geq c_0 \|x\|_H^2 , \quad \forall x \in H.$$

  Then for any $L : H \rightarrow \mathbf{C}$ antilinear there exists a unique $x \in H$ such that,

  $$a(x, y) = L(y), \quad \forall y \in H.$$

  - Examples:

  $(i) \ a(x, y) = (x, y)_H.$
  (ii) Take $H = H^1(\mathbf{R}^d) = \left\{ u \in L^2(\mathbf{R}^d) : \partial_j u \in L^2(\mathbf{R}^d), j = 1, \ldots, d \right\}$ and for $u, v \in H$ set,

  $$a(u, v) = \sum_{j,k=1}^{d} \int_{\mathbf{R}^d} a_{jk}(t) \partial_j u(t) \overline{\partial_k v(t)} \, dt + \int_{\mathbf{R}^d} u(t) \overline{v(t)} \, dt,$$

  where $a_{jk} \in L^\infty(\mathbf{R}^d), j, k = 1, \ldots, d$ and,

  $$\operatorname{Re} \sum_{j,k=1}^{d} a_{jk}(t) \zeta_j \overline{\zeta_k} \geq c_0 |\zeta|^2, \quad \forall t \in \mathbf{R}^d, \forall \zeta \in \mathbf{C}^d.$$

1.2.9 Spectral Theory of Self-Adjoint Compact Operators

We first recall several definitions.

- **Compact operators.** Let E, F be two normed vector spaces and $T \in \mathcal{L}(E, F)$ (the space of continuous linear maps). Then T is said to be *compact* if the image of the unit ball in E has a compact closure in F.

  In the particular case where E and F are separable Hilbert spaces, this is equivalent to say that the image by T of a weakly convergent sequence in E is convergent in F for the norm.

- **Adjoint operator.** Let H be a Hilbert space with scalar product $(\cdot, \cdot)_H$. Let $T \in \mathcal{L}(H, H)$. Then there exists a unique operator $T^* \in \mathcal{L}(H, H)$ such that, $(Tx, y)_H = (x, T^*y)_H$ for every $x, y \in H$. It is called the adjoint of T.

  The operator T is said self-adjoint if $T = T^*$.

- **Spectrum.** Let $T \in \mathcal{L}(H, H)$. The *spectrum* of T is the closed subset of $\mathbf{C}$ defined by,

$$\sigma(T) = \{\mu \in \mathbf{C} : T - \mu \,\mathrm{Id} \text{ is not invertible}\},$$

where Id denotes the identity operator.

  If $T - \mu \,\mathrm{Id}$ is non-injective, which means that there exists $x \in H, x \neq 0$ such that $Tx = \mu x$, then μ is called an *eigenvalue* and x an *eigenvector* of T.

- **Spectral theory of compact self-adjoint operators.** Let H be a Hilbert space and $T \in \mathcal{L}(H, H)$ be a compact self-adjoint operator. Then,

  1. the nonzero eigenvalues form a finite subset of $\mathbf{R}$ or a sequence $(\mu_n)_{n \in \mathbf{N}}$ of real numbers converging to zero when $n \to +\infty$,
  2. if μ_n is a nonzero eigenvalue then the space $E_n = \mathrm{Ker}\,(T - \mu_n \mathrm{Id})$ has finite dimension,
  3. $H = \oplus E_n \oplus \mathrm{Ker}\,T$,
  4. if H is finite dimensional then $\sigma(T) = \{\text{eigenvalues}\}$ and if H has infinite dimension then $\sigma(T) = \{0\} \cup \left(\cup_{n=1}^{+\infty} \{\mu_n\}\right)$.

1.2.10 L^p Spaces, $1 \le p \le +\infty$

Let $(X, \mathcal{T}, \mu)$ be a measured space.

- If $1 \le p < +\infty$, $L^p(X)$ is the set of (classes of) measurable functions $f : X \to \mathbf{C}$ such that $\int_X |f(x)|^p \, d\mu < +\infty$. On this space the quantity,

$$\|f\|_{L^p(X)} = \left(\int_X |f(x)|^p \, d\mu\right)^{\frac{1}{p}}$$

is a norm. Endowed with this norm $L^p(X)$ is a Banach space.

The particular space $L^2(X)$ is a Hilbert space when endowed with the scalar product,

$$(f, g)_{L^2(X)} = \int_X f(x)\overline{g(x)}\,d\mu.$$

- $L^\infty(X)$ is the set of (classes of) measurable functions $f : X \to \mathbf{C}$ which are essentially bounded, that means that there exists $M > 0$ such that,

$$\mu\left(\{x \in X : |f(x)| > M\}\right) = 0.$$

By definition $\|f\|_{L^\infty(X)}$ is the infimum of such numbers M. It is characterized as follows:

$$\mu\left(\{x \in X : |f(x)| > \|f\|_{L^\infty(X)}\}\right) = 0 \quad \text{and}$$

$$\forall \alpha < \|f\|_{L^\infty(X)}, \ \mu\left(\{x \in X : |f(x)| > \alpha\}\right) \neq 0.$$

Endowed with this norm $L^\infty(X)$ is a Banach space.

Notice that if f is a continuous and bounded function on an open set $\Omega \subset \mathbf{R}^d$ then $\|f\|_{L^\infty(\Omega)} = \sup_\Omega |f(x)|$.

- If $1 \le p < +\infty$ the dual of $L^p(X)$ is $L^{p'}(X)$ where p' is the conjugate exponent of p defined by,

$$\frac{1}{p} + \frac{1}{p'} = 1.$$

Notice that if $p = 1$ then $p' = +\infty$. Notice also that the dual of $L^\infty(X)$ is NOT $L^1(X)$ but a wider space.
- Let $1 \le p \le +\infty$. Every convergent sequence in $L^p(X)$ has a subsequence which converges almost everywhere.

1.2.11 The Hölder and Young Inequalities

- Let $(X, \mathcal{T}, \mu)$ be a measured space and let $1 \le p, q, r \le +\infty$ be real numbers such that,

$$\frac{1}{p} + \frac{1}{q} = \frac{1}{r}.$$

If $u \in L^p(X)$ and $v \in L^q(X)$ then $uv \in L^r(X)$ and,

$$\|uv\|_{L^r(X)} \le \|u\|_{L^p(X)}\|v\|_{L^q(X)}.$$

- Let $1 \leq p, q \leq +\infty$ and let $r \geq 1$ be such that,

$$\frac{1}{r} = \frac{1}{p} + \frac{1}{q} - 1.$$

If $u \in L^p(\mathbf{R}^d)$ and $v \in L^q(\mathbf{R}^d)$ then for almost all x in $\mathbf{R}^d$ the integral,

$$(u \star v)(x) := \int_{\mathbf{R}^d} u(x - y)v(y)\, dy$$

is convergent, and $u \star v$ belongs to $L^r(\mathbf{R}^d)$. Moreover we have,

$$\|u \star v\|_{L^r(\mathbf{R}^d)} \leq \|u\|_{L^p(\mathbf{R}^d)} \|v\|_{L^q(\mathbf{R}^d)}.$$

1.2.12 Approximation of the Identity

Let $\rho \in C^\infty(\mathbf{R}^d)$ be such that,

$$\operatorname{supp} \rho \subset \left\{ x \in \mathbf{R}^d : |x| \leq 1 \right\}, \quad \rho \geq 0, \quad \int_{\mathbf{R}^d} \rho(x)\, dx = 1.$$

For ε in $(0, 1]$ set,

$$\rho_\varepsilon(x) = \varepsilon^{-d} \rho\left(\frac{x}{\varepsilon}\right).$$

The family $(\rho_\varepsilon)_{\varepsilon \in (0,1]}$ is called an approximation of the identity. Then we have,

- for any $1 \leq p < +\infty$, if u belongs to $\in L^p(\mathbf{R}^d)$, then,

$$\lim_{\varepsilon \to 0} (\rho_\varepsilon \star u) = u \quad \text{in } L^p(\mathbf{R}^d).$$

- If $u \colon \mathbf{R}^d \to \mathbf{C}$ is a bounded uniformly continuous function, then

$$\lim_{\varepsilon \to 0} (\rho_\varepsilon \star u) = u \quad \text{in } L^\infty(\mathbf{R}^d).$$

- For any $1 \leq p \leq +\infty$, if u belongs to $L^p(\mathbf{R}^d)$, then for almost all x in $\mathbf{R}^d$,

$$\lim_{\varepsilon \to 0} (\rho_\varepsilon \star u)(x) = u(x).$$

- For any $1 \leq p \leq +\infty$, if u belongs to $L^p(\mathbf{R}^d)$, then for any $\varepsilon \in (0, 1]$, the function $\rho_\varepsilon \star u$ belongs to $C^\infty(\mathbf{R}^d)$.

1.3 Elements of Distribution Theory

- In what follows Ω will be an open subset of $\mathbf{R}^d$. We shall write $K \subset\subset \Omega$ if K is a compact subset of Ω.
- Recall that the support of a continuous function $f : \Omega \to \mathbf{C}$ is the set,

$$\operatorname{supp} f = \overline{\{x \in \Omega : f(x) \neq 0\}}.$$

- If $x_0 \in \Omega$ we denote by $\mathcal{V}_{x_0}$ the set of open neighborhoods of x_0.

1.3.1 Distributions

- **The space $C_0^\infty(\Omega)$**

  - $C_0^\infty(\Omega)$ is the vector space of functions $f : \Omega \to \mathbf{C}$ which are C^∞ on Ω and whose support is a compact subset of Ω.
  - There is a topology on this space (called a $\mathcal{LF}$-topology) which is useful for applications. We shall not describe it. We just notice that a sequence $(\varphi_j)_{j \in \mathbf{N}} \subset C_0^\infty(\Omega)$ converges to $\varphi \in C_0^\infty(\Omega)$ for this topology if and only if,

    (i) $\exists K \subset\subset \Omega : \operatorname{supp} \varphi_j \subset K, \forall j \in \mathbf{N}$,

    (ii) $\forall \alpha \in \mathbf{N}^d, (\partial^\alpha \varphi_j)_j$ converges uniformly to $\partial^\alpha \varphi$ on K.

  - If $K \subset\subset \Omega$ we set $C_0^\infty(K) = \left\{\varphi \in C_0^\infty(\Omega) : \operatorname{supp} \varphi \subset K\right\}$.

- **The space of distributions $\mathcal{D}'(\Omega)$**

  - **Definition.** $\mathcal{D}'(\Omega)$ is the vector space of linear map $T : C_0^\infty(\Omega) \to \mathbf{C}$ satisfying: for any $K \subset\subset \Omega$, there exist $C_K > 0$ and $k \in \mathbf{N}$ such that,

$$|\langle T, \varphi \rangle| \leq C_K \sum_{|\alpha| \leq k} \sup_K |\partial^\alpha \varphi|, \quad \forall \varphi \in C_0^\infty(K). \tag{1.1}$$

  - If (1.1) is true with the same $k \in \mathbf{N}$ for any compact K we say that T is of order $\leq k$.
  - The set of distributions of order $\leq k$ is denoted by $\mathcal{D}'^{(k)}(\Omega)$.
  - A linear map $T : C_0^\infty(\Omega) \to \mathbf{C}$ is a distribution if and only if for every sequence $(\varphi_j)_{j \in \mathbf{N}}$ converging to zero in $C_0^\infty(\Omega)$, the sequence $\left(\langle T, \varphi_j \rangle\right)_{j \in \mathbf{N}}$ converges to zero in $\mathbf{C}$.
  - **Support.** Let $T \in \mathcal{D}'(\Omega)$. The support of T is defined as follows.

$$x_0 \notin \operatorname{supp} T \iff \exists V \in \mathcal{V}_{x_0} : \langle T, \varphi \rangle = 0 \quad \forall \varphi \in C_0^\infty(V).$$

The support is a closed set.

- **Distributions with compact support.** We denote by $\mathcal{E}'(\Omega)$ the set of distributions T on Ω whose support is a compact set contained in Ω. It can be identified with the set of linear maps $T : C^\infty(\Omega) \to \mathbf{C}$ satisfying: for some $C > 0$, $m \in \mathbf{N}$ and $K \subset\subset \Omega$,

$$|\langle T, \varphi \rangle| \leq C \sum_{|\alpha| \leq m} \sup_K |\partial^\alpha \varphi(x)|, \quad \forall \varphi \in C^\infty(\Omega).$$

- **Product by a C^∞ function.** If $T \in \mathcal{D}'(\Omega)$ and $a \in C^\infty(\Omega)$ we define $aT \in \mathcal{D}'(\Omega)$ by,

$$\langle aT, \varphi \rangle = \langle T, a\varphi \rangle, \quad \forall \varphi \in C_0^\infty(\Omega).$$

- **Differentiation of distributions.** If $T \in \mathcal{D}'(\Omega)$ we define $\partial_{x_j} T \in \mathcal{D}'(\Omega)$ by,

$$\langle \partial_{x_j} T, \varphi \rangle = -\langle T, \partial_{x_j} \varphi \rangle, \quad \forall \varphi \in C_0^\infty(\Omega).$$

Therefore we have $\langle \partial^\alpha T, \varphi \rangle = (-1)^{|\alpha|} \langle T, \partial^\alpha \varphi \rangle$ for $\varphi \in C_0^\infty(\Omega)$.
- For $T \in \mathcal{D}'(\Omega)$ we have $\operatorname{supp} T \subset \{0\}$ if and only if $T = \sum_{|\alpha| \leq m} c_\alpha \partial^\alpha \delta_0$ where δ_0 is the Dirac distribution at zero.
- **Convergence of sequences in $\mathcal{D}'(\Omega)$.** A sequence $(T_j)_{j \in \mathbf{N}} \subset \mathcal{D}'(\Omega)$ converges to $T \in \mathcal{D}'(\Omega)$ if,

$$\lim_{j \to +\infty} \langle T_j, \varphi \rangle = \langle T, \varphi \rangle, \quad \forall \varphi \in C_0^\infty(\Omega).$$

- If $(T_j) \to T$ in $\mathcal{D}'(\Omega)$ then for all $\alpha \in \mathbf{N}^d$, $(\partial^\alpha T_j) \to \partial^\alpha T$ in $\mathcal{D}'(\Omega)$.
- Let $(T_j)_{j \in \mathbf{N}} \subset \mathcal{D}'(\Omega)$. Assume that for every $\varphi \in C_0^\infty(\Omega)$ the sequence $(\langle T_j, \varphi \rangle)$ is convergent in $\mathbf{C}$. Then there exists $T \in \mathcal{D}'(\Omega)$ such that $(T_j) \to T$ in $\mathcal{D}'(\Omega)$.
- **Density.** $C_0^\infty(\Omega)$ is dense is $\mathcal{D}'(\Omega)$.
- **Fundamental solution.** Let $P(D)$ be a differential operator with constant coefficients. A fundamental solution of P is a distribution $E \in \mathcal{D}'(\mathbf{R}^d)$ such that $P(D)E = \delta_0$ (the Dirac distribution at zero).

1.3.2 Tempered Distributions

- **The Schwartz space $S(\mathbf{R}^d)$**

  - **Definition.** $S(\mathbf{R}^d)$ is the vector space of functions $u : \mathbf{R}^d \to \mathbf{C}$ which are C^∞ and satisfy,

$$\forall \alpha, \beta \in \mathbf{N}^d, \ \exists C_{\alpha,\beta} > 0 : |x^\alpha \partial^\beta u(x)| \leq C_{\alpha,\beta}, \ \forall x \in \mathbf{R}^d.$$

Examples: $C_0^\infty(\mathbf{R}^d) \subset S(\mathbf{R}^d)$ and $u(x) = e^{-z|x|^2} \in S(\mathbf{R}^d)$ for $\mathrm{Re}\, z > 0$.

- For $\alpha, \beta \in \mathbf{N}^d$ $p_{\alpha\beta}(u) = \sup_{x \in \mathbf{R}^d} |x^\alpha \partial^\beta u(x)|$ is a seminorm on $S(\mathbf{R}^d)$. Then $\left(S(\mathbf{R}^d), (p_{\alpha\beta})_{\alpha,\beta \in \mathbf{N}^d}\right)$ is a Fréchet space.

- **The space of tempered distributions $S'(\mathbf{R}^d)$**

  - **Definition.** $S'(\mathbf{R}^d)$ is the vector space of linear maps $T: S(\mathbf{R}^d) \to \mathbf{C}$ such that,

  $$\exists k, \ell \in \mathbf{N}, \ \exists C > 0 : |\langle T, \varphi \rangle| \leq C \sum_{\substack{|\alpha| \leq k \\ |\beta| \leq \ell}} p_{\alpha\beta}(\varphi), \quad \forall \varphi \in S(\mathbf{R}^d).$$

  - **Density.** $S(\mathbf{R}^d)$ is dense in $S'(\mathbf{R}^d)$.
  - **Examples.** For $1 \leq p \leq +\infty$, $L^p(\mathbf{R}^d) \subset S'(\mathbf{R}^d)$.
  - **Product.** Let $f \in C^\infty(\mathbf{R}^d)$ be such that,

  $$\forall \alpha \in \mathbf{N}^d, \ \exists C_\alpha > 0, \exists m_\alpha \in \mathbf{R} : |\partial^\alpha f(x)| \leq C_\alpha (1 + |x|)^{m_\alpha}, \quad \forall x \in \mathbf{R}^d$$

  and let $T \in S'(\mathbf{R}^d)$. Then fT defined by,

  $$\langle fT, \varphi \rangle = \langle T, f\varphi \rangle, \quad \forall \varphi \in S(\mathbf{R}^d)$$

  is an element of $S'(\mathbf{R}^d)$.
  - **Derivation.** Let $T \in S'(\mathbf{R}^d)$, then for every $\alpha \in \mathbf{N}^d$, $\partial^\alpha T$ defined by,

  $$\langle \partial^\alpha T, \varphi \rangle = (-1)^{|\alpha|} \langle T, \partial^\alpha \varphi \rangle, \quad \forall \varphi \in S(\mathbf{R}^d)$$

  is an element of $S'(\mathbf{R}^d)$.
  - **Convergence of sequences in $S'(\mathbf{R}^d)$.** A sequence $(T_j)_{j \in \mathbf{N}} \subset S'(\mathbf{R}^d)$ converges to $T \in S'(\mathbf{R}^d)$ if,

  $$\lim_{j \to +\infty} \langle T_j, \varphi \rangle = \langle T, \varphi \rangle, \quad \forall \varphi \in S(\mathbf{R}^d).$$

- If $(T_j) \to T$ in $S'(\mathbf{R}^d)$ then for all $\alpha \in \mathbf{N}^d$, $(\partial^\alpha T_j) \to \partial^\alpha T$ in $S'(\mathbf{R}^d)$.
- Let $(T_j)_{j \in \mathbf{N}} \subset S'(\mathbf{R}^d)$. Assume that for every $\varphi \in S(\mathbf{R}^d)$ the sequence $(\langle T_j, \varphi \rangle)$ is convergent in $\mathbf{C}$. Then there exists $T \in S'(\mathbf{R}^d)$ such that $(T_j) \to T$ in $S'(\mathbf{R}^d)$.

1.3.3 The Fourier Transform

- **The Fourier transform in** $S(\mathbf{R}^d)$. The Fourier transform of $\varphi \in S(\mathbf{R}^d)$ is the function defined by,

$$(\mathcal{F}\varphi)(\xi) = \int e^{-ix\cdot\xi} \varphi(x)\, dx.$$

$\mathcal{F}\varphi$ is often denoted by $\widehat{\varphi}$.

- The map $\mathcal{F}$ is continuous from $S(\mathbf{R}^d)$ to $S(\mathbf{R}^d)$.
- The map $\mathcal{F}$ is invertible. If we introduce for $\psi \in S(\mathbf{R}^d)$,

$$(\mathcal{F}^{-1}\psi)(x) = (2\pi)^{-d} \int e^{ix\cdot\xi} \widehat{\psi}(\xi)\, d\xi,$$

then $\mathcal{F}(\mathcal{F}^{-1}\psi) = \psi$ and $\mathcal{F}^{-1}(\mathcal{F}\varphi) = \varphi$ for all $\psi, \varphi \in S(\mathbf{R}^d)$.

- **The Fourier transform in** $S'(\mathbf{R}^d)$. If $T \in S'(\mathbf{R}^d)$ then $\mathcal{F}T$ defined by,

$$\langle \mathcal{F}T, \varphi \rangle = \langle T, \mathcal{F}\varphi \rangle, \quad \forall \varphi \in S(\mathbf{R}^d)$$

belongs to $S'(\mathbf{R}^d)$. We define $\mathcal{F}^{-1}$ by an analogous formula.

- The maps $\mathcal{F}, \mathcal{F}^{-1}$ are continuous from $S'(\mathbf{R}^d)$ to $S'(\mathbf{R}^d)$.
- $\mathcal{F}(\mathcal{F}^{-1}T) = T$ and $\mathcal{F}^{-1}(\mathcal{F}T) = T$ for all $T \in S'(\mathbf{R}^d)$.
- For $\alpha \in \mathbf{N}^d$ we have,

$$\mathcal{F}(\partial^\alpha T) = (i\xi)^\alpha \mathcal{F}(T) \quad \text{and} \quad \mathcal{F}(x^\alpha T) = (-D_\xi)^\alpha (\mathcal{F}T),$$

where $D_\xi^\alpha = D_{\xi_1}^{\alpha_1} \dots D_{\xi_d}^{\alpha_d}$ with $D_{\xi_j} = \frac{1}{i}\frac{\partial}{\partial \xi_j}$.

- **Fourier transform of distributions with compact support**

  - If $T \in \mathcal{E}'(\mathbf{R}^d)$ then $\mathcal{F}T$ is the C^∞ function on $\mathbf{R}^d$ given by,

$$(\mathcal{F}T)(\xi) = \left\langle T_x, e^{-ix\cdot\xi} \right\rangle.$$

Moreover,

$$\exists k \in \mathbf{N} : \forall \alpha \in \mathbf{N}^d,\ \exists C_\alpha > 0 : |\partial_\xi^\alpha (\mathcal{F}T)(\xi)| \leq C_\alpha (1 + |\xi|)^k \quad \forall \xi \in \mathbf{R}^d.$$

- **Fourier transform in Lebesgue spaces**

  - If $T \in L^1(\mathbf{R}^d)$ then $\mathcal{F}T$ is given by the continuous function $\int e^{-ix\cdot\xi} T(x)\, dx$. Moreover $(\mathcal{F}T)(\xi)$ tends to zero when $|\xi| \to +\infty$.

– The map $T \mapsto (2\pi)^{-\frac{d}{2}} \mathcal{F} T$ is a bijective isometry from $L^2(\mathbf{R}^d)$ to $L^2(\mathbf{R}^d)$.
– More generally, let $1 \leq p \leq 2$ and p' its conjugate, defined by $\frac{1}{p'} = 1 - \frac{1}{p}$. Then $\mathcal{F}$ is continuous from $L^p(\mathbf{R}^d)$ to $L^{p'}(\mathbf{R}^d)$.

• **Fourier transform and convolution**

   – If $T \in \mathcal{S}'(\mathbf{R}^d)$ and $S \in \mathcal{E}'(\mathbf{R}^d)$ then $T \star S \in \mathcal{S}'(\mathbf{R}^d)$ and,

$$\mathcal{F}(T \star S) = (\mathcal{F} T) \cdot (\mathcal{F} S).$$

   – Let p, q be such that,

$$1 \leq p \leq 2, \quad 1 \leq q \leq 2, \quad \frac{1}{p} + \frac{1}{q} \geq \frac{3}{2}.$$

   If $T \in L^p(\mathbf{R}^d)$ and $S \in L^q(\mathbf{R}^d)$ then $T \star S \in L^r(\mathbf{R}^d)$ where $\frac{1}{r} = \frac{1}{p} + \frac{1}{q} - 1$ and,

$$\mathcal{F}(T \star S) = (\mathcal{F} T) \cdot (\mathcal{F} S).$$

   For instance one may take $p = q = 1$ (then $r = 1$) or $p = 1, q = 2$ (then $r = 2$).

• **Bernstein inequalities.** Let p be a real number such that $1 \leq p \leq +\infty$ and let $\theta \in [0, \frac{1}{p}]$. Let $r \geq 1$ be defined by $\frac{1}{r} = \frac{1}{p} - \theta$. Let $a \in L^p(\mathbf{R}^d)$, $R > 0$ and assume that,

$$\operatorname{supp} \widehat{a} \subset \left\{ \xi \in \mathbf{R}^d : |\xi| \leq R \right\}.$$

Then for every $\alpha \in \mathbf{N}^d$ there exists $C_\alpha > 0$ independent of a, R such that,

$$\|\partial^\alpha a\|_{L^r(\mathbf{R}^d)} \leq C_\alpha R^{|\alpha| + d\theta} \|a\|_{L^p(\mathbf{R}^d)}.$$

(See Problem 3.)
   A case which will be used in the problems is $p = 1, \theta = 1, r = +\infty$.

1.3.4   The Stationary Phase Method

It is a method which describes the behavior when $\lambda \to +\infty$ of integrals on the form,

$$I(\lambda) = \int_{\mathbf{R}^d} e^{i\lambda \varphi(x)} a(x) \, dx, \tag{1.2}$$

where φ is a C^∞ function on $\mathbf{R}^d$ with *real* values and $a \in C_0^\infty(\mathbf{R}^d)$ (φ is called the phase, a the amplitude).

The asymptotic of $I(\lambda)$ depends on the behavior of φ on the support of a. A point x such that $\varphi'(x) = 0$ is called a *stationary* (or a critical) point of φ.

- **Nonstationary phase**. Assume that:

$$\varphi'(x) \neq 0, \quad \forall x \in \operatorname{supp} a.$$

Then, for any $N \in \mathbf{N}$, there exists $C_N > 0$ such that,

$$|I(\lambda)| \leq C_N \lambda^{-N}, \quad \forall \lambda \geq 1.$$

- **Stationary phase**. Assume that there exists a unique $x_0 \in \overset{\circ}{\operatorname{supp}} a$ such that,

$$\varphi'(x_0) = 0 \text{ and } \varphi''(x_0) \text{ is nondegenerate},$$

where $\varphi''(x_0)$ denotes the Hessian matrix of φ at x_0 namely, $(\partial_{x_j}\partial_{x_k}\varphi(x_0))_{1 \leq j,k \leq d}$. Then for any $N \in \mathbf{N}$, there exists $b_0, \ldots, b_N$ in $\mathbf{C}$ (depending on a and φ), a function R_N and a constant $C_N > 0$ such that, for all $\lambda \geq 1$ we have,

$$I(\lambda) = e^{i\lambda\varphi(x_0)} \sum_{k=0}^{N} b_k \lambda^{-\frac{d}{2}-k} + R_N(\lambda), \quad |R_N(\lambda)| \leq C_N \lambda^{-\frac{d}{2}-N-1}. \tag{1.3}$$

Moreover

$$b_0 = \frac{(2\pi)^{d/2}}{\sqrt{|\det\varphi''(x_0)|}} \, e^{i\frac{\pi}{4}\operatorname{sgn} \varphi''(x_0)} \, a(x_0),$$

where sgn is the signature.

Comments For a detailed exposition of the classical results in functional analysis we refer to the books by T. Alazard [1], H. Brezis [7], and W. Rudin [24]. Concerning the Distribution theory, including a proof of the stationary phase formula, we refer to the books by L. Hörmander [14] and C. Zuily [32, 33]. □

Chapter 2
Statements of the Problems of Chap. 1

This chapter contains problems about functional analysis.

Problem 1

This problem states the basic results which will be used in the study of spectral theory for the harmonic oscillator in Problem 39.

In all what follows we shall denote $E = E(\mathbf{R}^d)$ if $E = L^2, C_0^\infty$.

Let $V = \{u \in L^2 : \nabla_x u \in L^2 : \langle x \rangle u \in L^2\}$ endowed with the scalar product,

$$(u, v)_V = (\nabla_x u, \nabla_x v)_{L^2} + (\langle x \rangle u, \langle x \rangle v)_{L^2}, \quad \langle x \rangle = (1 + |x|^2)^{\frac{1}{2}},$$

and with the corresponding norm $\|u\|_V = (u, u)_V^{\frac{1}{2}}$. Then V is a Hilbert space.

1. a) Let $V_c = \{u \in V : \text{supp } u \text{ is compact}\}$ endowed with the norm of V. Show that V_c is dense in V.

   b) Show that C_0^∞ is dense in V_c.

   Hint: Let (θ_ε) be an approximation of the identity. Use the fact that $\theta_\varepsilon \star v \to v$ in L^2 for every $v \in L^2$ when $\varepsilon \to 0$.

2. Show that the map $u \mapsto u$ from V to L^2 is compact.

   Hint: Let B be the unit ball in V. For $u \in B$ write $u = \chi_R u + (1 - \chi_R)u$ where $\chi_R(x) = \chi(\frac{x}{R})$, $\chi \in C_0^\infty$ and use the compactness of the map $u \mapsto u$ from H_K^1 to L^2 when K is a compact set in $\mathbf{R}^d$.

© The Editor(s) (if applicable) and The Author(s), under exclusive license
to Springer Nature Switzerland AG 2020
T. Alazard, C. Zuily, *Tools and Problems in Partial Differential Equations*,
Universitext, https://doi.org/10.1007/978-3-030-50284-3_2

3. a) Let e_k be the kth vector of the canonical basis of $\mathbf{R}^d$. For $0 < |h| \leq 1$ and $v \in V$ set

$$D_h v = \frac{1}{h}(v(x - he_k) - v(x)).$$

Prove that $D_h v \in V$.

b) Let $s \in \mathbf{R}$. Prove that for $w \in H^s$ we have, $\lim_{h \to 0} D_h w = -\partial_k w$ in H^{s-1}. (For the definition of the H^s space see next chapter.)

Hint: Use the Fourier transform, the fact that for $\theta \in \mathbf{R}$ we have $|e^{i\theta} - 1| \leq |\theta|$, $|e^{i\theta} - 1 - i\theta| \leq \frac{1}{2}|\theta|^2$ and the dominated convergence theorem.

c) Prove that,

$$\|D_h w\|_{L^2} \leq \|\partial_k w\|_{L^2}, \quad \forall w \in H^1, \quad \|D_h g\|_{H^{-1}} \leq (2\pi)^{\frac{d}{2}}\|g\|_{L^2}, \quad \forall g \in L^2.$$

Hint: Use the same method as in b).

4. Let $L \in V'$. Prove that there exists a unique $u \in V$ such that,

$$\sum_{j=1}^{d} \int_{\mathbf{R}^d} \partial_j v(x) \overline{\partial_j u(x)} \, dx + \int_{\mathbf{R}^d} \langle x \rangle v(x) \overline{\langle x \rangle u(x)} \, dx = L(v), \quad \forall v \in V$$

$$(2.1)$$

and that there exists $C > 0$ such that, $\|u\|_V \leq C\|L\|_{V'}$.

5. In this question we want to give an identification of V'. Let

$$E_1 = \left\{ f_1 \in \mathcal{S}' : \langle \xi \rangle^{-1} \widehat{f_1} \in L^2 \right\}, \quad E_2 = \left\{ f_2 \in \mathcal{S}' : \langle x \rangle^{-1} f_2 \in L^2 \right\},$$

endowed with the norms

$$\|f_1\|_{E_1} = \left\| \langle \xi \rangle^{-1} \widehat{f_1} \right\|_{L^2}, \quad \|f_2\|_{E_2} = \|\langle x \rangle f_2\|_{L^2}.$$

Let $E = E_1 + E_2 := \{ f = f_1 + f_2, f_j \in E_j, j = 1, 2 \}$ endowed with the norm,

$$\|f\|_E = \inf_{\substack{f = f_1 + f_2 \\ f_j \in E_j}} \left(\|f_1\|_{E_1} + \|f_2\|_{E_2} \right).$$

a) Let $f_j \in E_j$, $j = 1, 2$. Show the map,

$$v \mapsto L_{(f_1, f_2)}(v) = (2\pi)^{-\frac{d}{2}} \int_{\mathbf{R}^d} \langle \xi \rangle^{-1} \widehat{f_1}(\xi) \langle \xi \rangle \widehat{v}(-\xi) \, d\xi$$

$$+ \int_{\mathbf{R}^d} \langle x \rangle^{-1} f_1(x) \langle x \rangle v(x) \, dx$$

belongs to V' and,

$$\exists\, C > 0 : |L_{(f_1,f_2)}(v)| \le C(\|f_1\|_{E_1} + \|f_2\|_{E_2})\|v\|_V, \quad \forall v \in V. \qquad (2.2)$$

b) Show that if $v = \varphi \in S$ then $L_{(f_1,f_2)}(\varphi) = \langle f, \varphi\rangle_{S' \times S}$ where $f = f_1 + f_2$.
c) Deduce that if $f_1 + f_2 = g_1 + g_2$ then $L_{(f_1,f_2)} = L_{(g_1,g_2)}$ on V.
d) For $f \in E$ we can therefore set,

$$L_f(v) = (2\pi)^{-\frac{d}{2}} \int_{\mathbf{R}^d} \langle\xi\rangle^{-1}\,\widehat{f_1}(\xi)\,\langle\xi\rangle\,\widehat{v}(-\xi)\,d\xi + \int_{\mathbf{R}^d} \langle x\rangle^{-1}\,f_2(x)\,\langle x\rangle\,v(x)\,dx.$$

Show that $L_f \in V'$ and that there exists $C > 0$ such that $|L_f(v)| \le C\|f\|_E\|v\|_V$.

e) We consider the map $\Phi : E \to V', f \mapsto L_f$. Show that Φ is linear, continuous, and injective.

f) We want to show that Φ is surjective. Let $L \in V'$. Using question 4 show that there exists $u \in V$ such that $\langle -\Delta u + \langle x\rangle^2\, u, \varphi\rangle_{S' \times S} = L(\varphi)$ for all $\varphi \in S$.

   We set $f_1 = -\Delta u$, $f_2 = \langle x\rangle^2\, u$. Show that $f_j \in E_j, j = 1, 2$.
   We set $f = f_1 + f_2$. Show that $L = L_f$ on V.

g) Deduce that $\Phi : E \to V'$ is linear, bijective, continuous, and Φ^{-1} is continuous and that this identifies the dual of V with the space E.

Problem 2

This problem proves a result used in the study of the wave equation. It uses the stationary phase method.

Let $d \ge 2$. For $\lambda > 0$ and $\theta \in S^{d-1}$ (the unit sphere in $\mathbf{R}^d$)) we consider the integral,

$$F(\lambda, \theta) = \int_{S^{d-1}} e^{i\lambda\langle\theta,\omega\rangle}\, d\omega.$$

The goal of this problem is to show that there exists $c_0 \in \mathbf{C}$ such that,

$$F(\lambda, \theta) = \frac{c_0}{\lambda^{\frac{d-1}{2}}}(1 + o(1)), \quad \text{when } \lambda \to +\infty.$$

Part 1. *Preliminaries*
1. Show that for any $d \times d$ orthogonal matrix A we have,

$$F(\lambda, A\theta) = F(\lambda, \theta).$$

Hint: Consider the integral $= \int_{\{y \in \mathbf{R}^d : |y| < r\}} e^{i\lambda \langle A\theta, y \rangle} \, dy$ and use polar coordinates.

2. Deduce that we have $F(\lambda, \theta) = F^+(\lambda, \theta) + F^-(\lambda, \theta)$ where,

$$F^\pm(\lambda, \theta) = \int_{\{\omega' \in \mathbf{R}^{d-1} : |\omega'| < 1\}} e^{\pm i\lambda \sqrt{1 - |\omega'|^2}} \frac{1}{\sqrt{1 - |\omega'|^2}} \, d\omega'.$$

In what follows we shall set $\psi(\omega') = \sqrt{1 - |\omega'|^2}$.

3. Prove that the phase ψ has one and only one nondegenerate critical point at $\omega' = 0$.

Let $\delta > 0$ be a small constant. Let $\chi_1 \in C_0^\infty(\mathbf{R}^{d-1})$, $\chi_1(\omega') = 1$ if $|\omega'| \leq \delta$, $\chi_1(\omega') = 0$ if $|\omega'| \geq 2\delta$. We set $\chi_2 = 1 - \chi_1$ and,

$$G^\pm(\lambda, \theta) = \int_{\{\omega' \in \mathbf{R}^{d-1} : |\omega'| < 1\}} e^{\pm i\lambda \sqrt{1 - |\omega'|^2}} \frac{\chi_1(\omega')}{\sqrt{1 - |\omega'|^2}} \, d\omega',$$

$$H^\pm(\lambda, \theta) = \int_{\{\omega' \in \mathbf{R}^{d-1} : |\omega'| < 1\}} e^{\pm i\lambda \sqrt{1 - |\omega'|^2}} \frac{\chi_2(\omega')}{\sqrt{1 - |\omega'|^2}} \, d\omega',$$

so that,

$$F(\lambda, \theta) = G^+(\lambda, \theta) + G^-(\lambda, \theta) + H^+(\lambda, \theta) + H^-(\lambda, \theta).$$

Part 2. *Study of $G^\pm(\lambda, \theta)$*

1. Using the stationary phase formula prove that there exist constants $c_\pm \in \mathbf{C}$ such that,

$$G^\pm(\lambda, \theta) = \frac{c_\pm}{\lambda^{\frac{d-1}{2}}}(1 + o(1)), \quad \text{when } \lambda \to +\infty.$$

Part 3. *Study of $H^\pm(\lambda, \theta)$*

Our aim is to show that for every $N \in \mathbf{N}$ one can find $C_N > 0$ such that,

$$|H^+(\lambda, \theta) + H^-(\lambda, \theta)| \leq C_N \lambda^{-N}, \quad \text{when } \lambda \to +\infty. \tag{2.3}$$

1. Set $\omega' = (\omega_1, \ldots, \omega_{d-1})$. Consider the vector field $L = \frac{1}{i\lambda} \sum_{j=1}^{d-1} \frac{\partial_{\omega_j} \psi}{|\nabla_{\omega'} \psi|^2} \partial_{\omega_j}$.
Prove that $\pm L e^{\pm i\lambda \psi} = e^{\pm i\lambda \psi}$ and compute L.

2. Using integration by parts with L and an induction, prove that for every $m \in \mathbf{N}$ there exists functions a_m, b_m which are C^∞ for $|\omega'| \leq 1$, with support contained in $\{\omega' : |\omega'| \geq \delta\}$ such that,

$$H^+(\lambda, \theta) + H^-(\lambda, \theta) = \frac{1}{\lambda^m} \int_{|\omega'| < 1} \left(e^{i\lambda \psi(\omega')} + e^{-i\lambda \psi(\omega')} \right) \frac{a_m(\omega')}{\sqrt{1 - |\omega'|^2}} \, d\omega',$$

if m even ,

$$H^+(\lambda, \theta) + H^-(\lambda, \theta) = \frac{1}{\lambda^m} \int_{|\omega'|<1} \left(e^{i\lambda\psi(\omega')} - e^{-i\lambda\psi(\omega')} \right) b_m(\omega')\, d\omega',$$

if m odd .

Hint: Use integration by parts with L and the Gauss–Green formula.
3. Conclude.

Problem 3 *Bernstein Inequalities*

This problem concerns a classical result in the Littlewood–Paley theory.

Let p be a real number such that $1 \leq p \leq +\infty$ and let $\theta \in [0, \frac{1}{p}]$. Let r be defined by $\frac{1}{r} = \frac{1}{p} - \theta$. Let $a \in L^p(\mathbf{R}^d)$, $R > 0$ and assume that supp $\widehat{a} \subset \{\xi \in \mathbf{R}^d : |\xi| \leq R\}$. Prove that for every $\alpha \in \mathbf{N}^d$ there exists $C_\alpha > 0$ independent of a, R such that,

$$\|\partial^\alpha a\|_{L^r(\mathbf{R}^d)} \leq C_\alpha R^{|\alpha|+d\theta} \|a\|_{L^p(\mathbf{R}^d)}.$$

Hint: Let $\chi \in C_0^\infty(\mathbf{R}^d)$ be such that $\chi(\xi) = 1$ if $|\xi| \leq 1$; set $\chi_R(\xi) = \chi\left(\frac{\xi}{R}\right)$. Write $\widehat{\partial^\alpha a} = \widehat{\partial^\alpha a} \cdot \chi_R$, then use the inverse Fourier transform and the Young inequality.

Chapter 3
Functional Spaces

The goal of this chapter is to recall, without proof, some properties of the main functional spaces used in partial differential equations: Sobolev, Hölder, and Zygmund spaces. Other spaces are introduced in the problems.

3.1 Sobolev Spaces

3.1.1 Sobolev Spaces on $\mathbf{R}^d$, $d \geq 1$

3.1.1.1 Definition and First Properties

In what follows we shall set for $\xi \in \mathbf{R}^d$, $\langle \xi \rangle = (1 + |\xi|^2)^{\frac{1}{2}}$.

- For $s \in \mathbf{R}$, $H^s(\mathbf{R}^d)$ is the vector space of all tempered distributions $u \in S'(\mathbf{R}^d)$ such that $\langle \xi \rangle^s \, \widehat{u} \in L^2(\mathbf{R}^d)$. It is endowed with the scalar product and the corresponding norm,

$$(u, v)_{H^s} = \int_{\mathbf{R}^d} \langle \xi \rangle^{2s} \, \widehat{u}(\xi) \, \overline{\widehat{v}(\xi)} \, d\xi \,, \quad \|u\|_{H^s} = \left(\int_{\mathbf{R}^d} \langle \xi \rangle^{2s} \, |\widehat{u}(\xi)|^2 \, d\xi \right)^{\frac{1}{2}}.$$
$$(3.1)$$

- Endowed with this scalar product, $H^s(\mathbf{R}^d)$ is a Hilbert space.
- If $s_1 \geq s_2$, $H^{s_1}(\mathbf{R}^d)$ is continuously embedded in $H^{s_2}(\mathbf{R}^d)$.
- $S(\mathbf{R}^d)$ is contained in $H^s(\mathbf{R}^d)$ for any $s \in \mathbf{R}$.
- $L^1(\mathbf{R}^d) \subset H^s(\mathbf{R}^d)$ if $s < -\frac{d}{2}$. Indeed if $u \in L^1(\mathbf{R}^d)$ we have, $\widehat{u} \in L^\infty(\mathbf{R}^d)$ so $\langle \xi \rangle^s \, \widehat{u} \in L^2(\mathbf{R}^d)$ if $s < -\frac{d}{2}$.
- The Dirac distribution δ_0 belongs to $H^s(\mathbf{R}^d)$ if and only if $s < -\frac{d}{2}$. Indeed $\langle \xi \rangle^s \in L^2(\mathbf{R}^d)$ if and only if $s < -\frac{d}{2}$.

© The Editor(s) (if applicable) and The Author(s), under exclusive license
to Springer Nature Switzerland AG 2020
T. Alazard, C. Zuily, *Tools and Problems in Partial Differential Equations*,
Universitext, https://doi.org/10.1007/978-3-030-50284-3_3

3.1.1.2   Density

- $\mathcal{S}(\mathbf{R}^d)$ is dense in $H^s(\mathbf{R}^d)$ for any $s \in \mathbf{R}$.

3.1.1.3   Operations on $H^s(\mathbf{R}^d)$

- If $\varphi \in \mathcal{S}(\mathbf{R}^d)$ and $u \in H^s(\mathbf{R}^d)$ then $\varphi u \in H^s(\mathbf{R}^d)$ and we have,

$$\|\varphi u\|_{H^s} \leq (2\pi)^{-d} \, 2^{\frac{|s|}{2}} \left(\int \langle \xi \rangle^{|s|} |\widehat{\varphi}(\xi)| \, d\xi \right) \|u\|_{H^s}. \tag{3.2}$$

- If $\alpha \in \mathbf{N}^d$ and $u \in H^s(\mathbf{R}^d)$ then $\partial^\alpha u \in H^{s-|\alpha|}(\mathbf{R}^d)$ and $\|\partial^\alpha u\|_{H^{s-|\alpha|}} \leq \|u\|_{H^s}$.
- For $s > \frac{d}{2}$, $H^s(\mathbf{R}^d)$ is an algebra and there exists $C > 0$ such that,

$$\|uv\|_{H^s} \leq C\|u\|_{H^s}\|v\|_{H^s}, \quad \forall u, v \in H^s(\mathbf{R}^d).$$

- For $s \geq 0$, $H^s(\mathbf{R}^d) \cap L^\infty(\mathbf{R}^d)$ is an algebra and there exists $C > 0$ such that,

$$\|uv\|_{H^s} \leq C(\|u\|_{L^\infty}\|v\|_{H^s} + \|u\|_{H^s}\|v\|_{L^\infty}), \quad \forall u, v \in H^s(\mathbf{R}^d) \cap L^\infty(\mathbf{R}^d).$$

Such an estimate is called a *tame estimate* because the right-hand side is linear with respect to the H^s norm.

- Let $s \geq 0$ and let $f : \mathbf{C} \to \mathbf{C}$ be a C^∞ function such that $f(0) = 0$. Then there exists a nondecreasing function $\mathcal{F} : \mathbf{R}^+ \to \mathbf{R}^+$ such that,

$$\|f(u)\|_{H^s} \leq \mathcal{F}(\|u\|_{L^\infty})\|u\|_{H^s}, \quad \forall u \in H^s(\mathbf{R}^d) \cap L^\infty(\mathbf{R}^d). \tag{3.3}$$

In particular (3.3) holds if $u \in H^s(\mathbf{R}^d)$ for $s > \frac{d}{2}$.

3.1.1.4   Sobolev Embeddings

- We have $\mathcal{E}'(\mathbf{R}^d) \subset \underset{s \in \mathbf{R}}{\cup} H^s(\mathbf{R}^d)$.

Let $C_{\to 0}^k(\mathbf{R}^d)$ be the space of $u \in C^k(\mathbf{R}^d)$ such that $\lim_{|x| \to +\infty} |\partial^\alpha u(x)| = 0$ for all $|\alpha| \leq k$, endowed with the norm $|u|_k = \sum_{|\alpha| \leq k} \sup_{\mathbf{R}^d} |\partial_x^\alpha u|$.

- Let $k \in \mathbf{N}$ and $s \in \mathbf{R}$ be such that $s > \frac{d}{2} + k$. Then,

$$H^s(\mathbf{R}^d) \subset C_{\to 0}^k(\mathbf{R}^d),$$

with continuous embedding. In particular if $s > \frac{d}{2}$ we have $H^s(\mathbf{R}^d) \subset L^\infty(\mathbf{R}^d)$ with continuous embedding.

- Let $s = \frac{d}{2}$. For all $p \in [2, +\infty[$ there exists $C > 0$ such that,

$$\|u\|_{L^p} \leq C\|u\|_{H^s}, \quad \forall u \in H^s(\mathbf{R}^d).$$

- Let $0 \leq s < \frac{d}{2}$. For all p such that $\frac{1}{2} \geq \frac{1}{p} \geq \frac{1}{2} - \frac{s}{d}$ there exists $C > 0$ such that,

$$\|u\|_{L^p} \leq C\|u\|_{H^s}, \quad \forall u \in H^s(\mathbf{R}^d). \tag{3.4}$$

3.1.1.5 Duality

Let $s \in \mathbf{R}$ and $v \in H^{-s}(\mathbf{R}^d)$. If $u \in H^s(\mathbf{R}^d)$ the function $\widehat{u}(\xi)\,\widehat{v}(-\xi)$ belongs to $L^1(\mathbf{R}^d)$. Indeed, $\widehat{u}(\xi)\,\widehat{v}(-\xi) = \langle\xi\rangle^s\,\widehat{u}(\xi)\,\langle\xi\rangle^{-s}\,\widehat{v}(-\xi)$ and the right-hand side is the product of two $L^2(\mathbf{R}^d)$ functions. On the other hand, by the Cauchy–Schwarz inequality we have,

$$\left| \int_{\mathbf{R}^d} \widehat{u}(\xi)\,\widehat{v}(-\xi)\,d\xi \right| \leq \|u\|_{H^s}\,\|v\|_{H^{-s}}. \tag{3.5}$$

Therefore, if $v \in H^{-s}(\mathbf{R}^d)$, the map L_v defined by,

$$u \longmapsto L_v(u) = (2\pi)^{-d} \int_{\mathbf{R}^d} \widehat{u}(\xi)\,\widehat{v}(-\xi)\,d\xi = \int_{\mathbf{R}^d} \widehat{u}(\xi)\,(\mathcal{F}^{-1}v)(\xi)\,d\xi \tag{3.6}$$

is a continuous linear form on $H^s(\mathbf{R}^d)$ (it thus belongs to $(H^s(\mathbf{R}^d))'$) and $\|L_v\|_{(H^s)'} \leq (2\pi)^{-d}\|v\|_{H^{-s}}$. So we have a map,

$$L : H^{-s}(\mathbf{R}^d) \to (H^s(\mathbf{R}^d))', \quad v \longmapsto L_v. \tag{3.7}$$

- The map L, defined by (3.7), is linear, bijective, and bicontinuous. It identifies the dual of $H^s(\mathbf{R}^d)$ with $H^{-s}(\mathbf{R}^d)$.
  We write, for $u \in H^s, v \in H^{-s}$,

$$\langle u, v \rangle_{H^s \times H^{-s}} = L_v(u) = (2\pi)^{-d} \int_{\mathbf{R}^d} \widehat{u}(\xi)\,\widehat{v}(-\xi)\,d\xi.$$

- If $v \in L^2(\mathbf{R}^d)$ and $u \in L^2(\mathbf{R}^d)$ we have,

$$L_v(u) = \int_{\mathbf{R}^d} \widehat{u}(\xi)\,(\mathcal{F}^{-1}v)(\xi)\,d\xi = \int_{\mathbf{R}^d} u(x)\,v(x)\,dx.$$

3.1.1.6 Compactness

- Let K be a compact set in $\mathbf{R}^d$ and let $s, s' \in \mathbf{R}$ be such that $s > s'$. Denote $H_K^s = \{u \in H^s(\mathbf{R}^d) : \operatorname{supp} u \subset K\}$. Then the map $H_K^s \to H^{s'}(\mathbf{R}^d)$, $u \mapsto u$, is compact.

3.1.1.7 Traces

- For all $s > \frac{1}{2}$, the map $u \mapsto u(x', 0)$ from $C_0^\infty(\mathbf{R}^d)$ to $C^\infty(\mathbf{R}^{d-1})$ can be uniquely extended to a linear, continuous, **surjective** map γ, from $H^s(\mathbf{R}^d)$ to $H^{s-\frac{1}{2}}(\mathbf{R}^{d-1})$.

3.1.1.8 Equivalent Norm

- Let $0 < \sigma < 1$. Set,

$$[[u]]_\sigma^2 = \|u\|_{L^2(\mathbf{R}^d)}^2 + \iint_{\mathbf{R}^d \times \mathbf{R}^d} \frac{|u(x) - u(y)|^2}{|x - y|^{d+2\sigma}}\, dx\, dy.$$

Then $\|u\|_{H^\sigma} \sim [[u]]_\sigma$ in the sense that,

$$\exists\, 0 < C_1 \le C_2 : C_1 \|u\|_{H^\sigma} \le [[u]]_\sigma \le C_2 \|u\|_{H^\sigma}, \quad \forall u \in H^\sigma(\mathbf{R}^d).$$

If $s = k + \sigma$ where $k \in \mathbf{N}$ and $0 < \sigma < 1$ we have,

$$\|u\|_{H^s} \sim \sum_{|\alpha| \le k} [[\partial^\alpha u]]_\sigma \quad \forall u \in H^s(\mathbf{R}^d).$$

3.1.2 Local Sobolev Spaces $H_{loc}^s(\mathbf{R}^d)$

- For $s \in \mathbf{R}$ we set,

$$H_{loc}^s(\mathbf{R}^d) = \left\{ u \in S'(\mathbf{R}^d) : \varphi u \in H^s(\mathbf{R}^d), \quad \forall \varphi \in C_0^\infty(\mathbf{R}^d) \right\}.$$

Its topology is given by the semi-norms $p_\varphi(u) = \|\varphi u\|_{H^s}$.

3.1.3 Sobolev Spaces on an Open Subset of $\mathbf{R}^d$

3.1.3.1 Definition and First Properties

- Let Ω be an open subset of $\mathbf{R}^d$ and $k \in \mathbf{N}$. We set,

$$H^k(\Omega) = \left\{ u \in L^2(\Omega) : \partial^\alpha u \in L^2(\Omega), |\alpha| \leq k \right\}.$$

It is endowed with the scalar product and the corresponding norm,

$$(u, v)_{H^k} = \sum_{|\alpha| \leq k} \left(\partial^\alpha u, \partial^\alpha v \right)_{L^2(\Omega)}, \quad \|u\|_{H^k} = \left(\sum_{|\alpha| \leq k} \|\partial^\alpha u\|^2_{L^2(\Omega)} \right)^{\frac{1}{2}}.$$

$$(3.8)$$

- Endowed with this scalar product, $H^k(\Omega)$ is a Hilbert space.
- If $k_2 \geq k_1$, $H^{k_2}(\Omega)$ is continuously embedded in $H^{k_1}(\Omega)$.
- If $s = k \in \mathbf{N}$ and $\Omega = \mathbf{R}^d$, the spaces $H^s(\mathbf{R}^d)$ and $H^k(\Omega)$ coincide and the norms (3.1) and (3.8) are equivalent.

3.1.3.2 Extension

- If Ω has a smooth boundary, then for every $k \in \mathbf{N}$, there exists a linear and continuous map $P_k : H^k(\Omega) \to H^k(\mathbf{R}^d)$ such that $P_k u = u$ in Ω. It is called the extension operator. Consequently, for $k > \frac{d}{2}$, $H^k(\Omega)$ is an algebra.
- Notice that the restriction operator defined by $Ru = u|_\Omega$ is continuous from $H^k(\mathbf{R}^d)$ to $H^k(\Omega)$ for every $k \in \mathbf{N}$.

3.1.3.3 Density

- If $\Omega \neq \mathbf{R}^d$ and $k \geq 1$, $C_0^\infty(\Omega)$ **is not dense** in $H^k(\Omega)$. We denote by $H_0^k(\Omega)$ the closure of $C_0^\infty(\Omega)$ in $H^k(\Omega)$ endowed with the norm (3.8).
- $C_0^\infty(\overline{\Omega})$ is the space of restrictions to Ω of functions in $C_0^\infty(\mathbf{R}^d)$. It is dense in $H^k(\Omega)$ for any $k \in \mathbf{N}$ when Ω has a smooth boundary.

3.1.3.4   Poincaré Inequality

Let Ω be a bounded open subset of $\mathbf{R}^d$ with diameter d. Then we have,

$$\|u\|_{L^2(\Omega)} \le 2d \sum_{j=1}^{d} \left\| \frac{\partial u}{\partial x_j} \right\|_{L^2(\Omega)} \,, \quad \forall u \in H_0^1(\Omega).$$

Actually it is sufficient for Ω to be bounded in one direction for the inequality to hold.

3.1.3.5   Sobolev Embedding

- If $k > p + \frac{d}{2}$ and Ω has a smooth boundary, then $H^k(\Omega)$ is contained in the space $C^p(\overline{\Omega})$ which is the space of restrictions to Ω of functions in $C^p(\mathbf{R}^d)$.

3.1.3.6   Duality

- The space $C_0^\infty(\Omega)$ being, by definition, dense in $H_0^k(\Omega)$ the dual of $H_0^k(\Omega)$ can be embedded in the space of distributions. By definition we set $H^{-k}(\Omega) = \left(H_0^k(\Omega)\right)'$ endowed with the dual norm that is,

$$\|T\|_{-k} = \sup_{\substack{\varphi \in H_0^k(\Omega) \\ \|\varphi\|_{H^k}=1}} |T(\varphi)|.$$

3.1.3.7   Compactness

- Let $\Omega \subset \mathbf{R}^d$ be **bounded** with a smooth boundary. Then for any $k > k'$ the embedding from $H^k(\Omega)$ to $H^{k'}(\Omega)$ is compact.

3.1.3.8   Traces

- The map

$$C_0^\infty(\overline{\Omega}) \to [C^\infty(\partial\Omega)]^k, \quad u \mapsto \left(u|_{\partial\Omega}, \frac{\partial u}{\partial \nu}\Big|_{\partial\Omega}, \dots, \left(\frac{\partial}{\partial \nu}\right)^{k-1} u\Big|_{\partial\Omega} \right)$$

can be extended as a linear, continuous, and surjective map $\gamma = (\gamma_0, \ldots, \gamma_{k-1})$ from $H^k(\Omega)$ to $\prod_{j=0}^{k-1} H^{k-j-\frac{1}{2}}(\partial\Omega)$. Here ν is the exterior normal to $\partial\Omega$ and $\frac{\partial}{\partial\nu}$ is the normal derivative.

- Then $H_0^k(\Omega) = \{u \in H^k(\Omega) : \gamma u = 0\}$.

3.1.4 Sobolev Spaces on the Torus

Consider the d-dimensional torus $\mathbf{T}^d = \mathbf{R}^d/(2\pi\mathbf{Z})^d$. A function $f : \mathbf{T}^d \to \mathbf{C}$ is a function $f : \mathbf{R}^d \to \mathbf{C}$ such that $f(x + 2\pi e_j) = f(x)$ for any $1 \leq j \leq d$, where $(e_1, \ldots, e_d)$ is the canonical basis of $\mathbf{R}^d$. If $f \in L^1(\mathbf{T}^d)$ we define its Fourier coefficient by,

$$\widehat{f}(k) = \int_{(0,2\pi)^d} e^{-ik\cdot x} f(x)\, dx, \quad k \in \mathbf{Z}^d.$$

Then for $s \geq 0$, the Sobolev space $H^s(\mathbf{T}^d)$ is the space of functions f such that,

$$\|f\|_{H^s(\mathbf{T}^d)}^2 := \sum_{k \in \mathbf{Z}^d} (1 + |k|^2)^s |\widehat{f}(k)|^2 < +\infty.$$

For $s > 0$ we define $H^{-s}(\mathbf{T}^d)$ as the dual of $H^s(\mathbf{T}^d)$.

3.2 The Hölder Spaces

3.2.1 Hölder Spaces of Integer Order

3.2.1.1 Definition

- Let $m \in \mathbf{N}$. We set,

$$W^{m,\infty}(\mathbf{R}^d) = \left\{ u \in L^\infty(\mathbf{R}^d) : \partial^\alpha u \in L^\infty(\mathbf{R}^d), \forall |\alpha| \leq m \right\},$$

endowed with the norm,

$$\|u\|_{W^{m,\infty}} = \sum_{|\alpha| \leq m} \|\partial^\alpha u\|_{L^\infty}. \tag{3.9}$$

3.2.1.2   Properties

- Endowed with the norm (3.9), $W^{m,\infty}(\mathbf{R}^d)$ is a Banach space.
- The space $W^{m,\infty}(\mathbf{R}^d)$ is an algebra and there exists $C > 0$ such that,

$$\|uv\|_{W^{m,\infty}} \leq C\|u\|_{W^{m,\infty}}\|v\|_{W^{m,\infty}}, \quad \forall u, v \in W^{m,\infty}(\mathbf{R}^d).$$

- Notice that $W^{1,\infty}(\mathbf{R}^d)$ is the space of bounded Lipschitz functions.

3.2.2   Hölder Spaces of Fractional Order

Let us introduce a notation. For $\sigma \in]0, 1[$ we set,

$$[u]_\sigma = \sup_{\substack{x,y\in\mathbf{R}^d \\ x\neq y}} \frac{|u(x) - u(y)|}{|x - y|^\sigma}.$$

3.2.2.1   Definition

- Let $m \in \mathbf{N}$, $0 < \sigma < 1$ and $\rho = m + \sigma$. We set,

$$W^{\rho,\infty}(\mathbf{R}^d) = \left\{ u : \partial^\alpha u \in L^\infty(\mathbf{R}^d), \ \forall|\alpha| \leq m \text{ and } \sum_{|\alpha|=m} [\partial^\alpha u]_\sigma < +\infty \right\},$$

endowed with the norm,

$$\|u\|_{W^{\rho,\infty}(\mathbf{R}^d)} = \sum_{|\alpha|\leq m} \|\partial^\alpha u\|_{L^\infty} + \sum_{|\alpha|=m} [\partial^\alpha u]_\sigma. \tag{3.10}$$

Notice that these spaces are also denoted in the literature by $C^{m,\sigma}(\mathbf{R}^d)$.

3.2.2.2   Properties

- Endowed with the norm (3.10), $W^{\rho,\infty}(\mathbf{R}^d)$ is a Banach space.

3.3 Characterization of Sobolev and Hölder Spaces in Dyadic Rings

For $j \in \mathbf{N}$ we set,

$$C_j = \left\{ \xi \in \mathbf{R}^d : \frac{1}{2} 2^j \leq |\xi| \leq 2 \cdot 2^j \right\}.$$

Let $\psi \in C_0^\infty(\mathbf{R}^d)$ be such that,

$$\operatorname{supp} \psi \subset \left\{ \xi \in \mathbf{R}^d : |\xi| \leq 1 \right\}, \quad \psi(\xi) = 1 \quad \text{if} \quad |\xi| \leq \frac{1}{2},$$

and set,

$$\varphi(\xi) = \psi(\frac{\xi}{2}) - \psi(\xi). \tag{3.11}$$

Then,

$$\psi(\xi) + \sum_{j=0}^{N-1} \varphi(2^{-j}\xi) = \psi(2^{-N}\xi), \quad \forall \xi \in \mathbf{R}^d, \forall N \geq 1. \tag{3.12}$$

For $u \in \mathcal{S}'(\mathbf{R}^d)$ we set,

$$\Delta_{-1} u = \mathcal{F}^{-1}\left(\psi(\xi)\widehat{u}\right), \quad \Delta_j u = \mathcal{F}^{-1}\left(\varphi(2^{-j}\xi)\widehat{u}\right) \quad \text{for} \quad j \geq 0,$$

and by convention we set $\Delta_j u = 0$ if $j \leq -2$. Then in $\mathcal{S}'(\mathbf{R}^d)$ we have,

$$u = \sum_{j=-1}^{+\infty} \Delta_j u.$$

It is called the *Littlewood–Paley decomposition* of u.

We will be using in the sequel other rings of the form,

$$\widetilde{C}_j = \left\{ \xi \in \mathbf{R}^d : a2^j \leq |\xi| \leq b \, 2^j \right\}, \quad \text{where } 0 < a < b.$$

3.3.1 Characterization of Sobolev Spaces

- Let $s \in \mathbf{R}$. If $u \in H^s(\mathbf{R}^d)$ we have,

$$\|\Delta_j u\|_{L^2(\mathbf{R}^d)} \leq c_j 2^{-js}, \ j \geq -1, \quad \text{where} \quad \left(\sum_{j=-1}^{+\infty} c_j^2 \right)^{\frac{1}{2}} \leq C \|u\|_{H^s(\mathbf{R}^d)}.$$

- For $j \in \mathbf{Z}$, $j \geq -1$ let $u_j \in L^2(\mathbf{R}^d)$ be such that $\operatorname{supp} \widehat{u}_j \subset \widetilde{C}_j$ and,

$$\|u_j\|_{L^2(\mathbf{R}^d)} \leq c_j 2^{-js}, \quad \text{where} \quad \left(\sum_{j=-1}^{+\infty} c_j^2 \right)^{\frac{1}{2}} < +\infty.$$

Set $u = \sum_{j=-1}^{+\infty} u_j$. Then,

$$u \in H^s(\mathbf{R}^d) \quad \text{and} \quad \|u\|_{H^s(\mathbf{R}^d)} \leq C \left(\sum_{j=-1}^{+\infty} 2^{2js} \|u_j\|_{L^2(\mathbf{R}^d)}^2 \right)^{\frac{1}{2}}.$$

- Notice that if $s > 0$ we have the same above statements with the ring C_j replaced by a ball $B(0, C2^j)$.

3.3.2 Characterization of Hölder Spaces

- Let $m \in \mathbf{N}$, $0 < \sigma < 1$ and $\rho = m + \sigma$. If $u \in W^{\rho,\infty}(\mathbf{R}^d)$ we have,

$$\|\Delta_j u\|_{L^\infty(\mathbf{R}^d)} \leq C 2^{-j\rho} \|u\|_{W^{\rho,\infty}(\mathbf{R}^d)}.$$

- For $j \in \mathbf{Z}$, $j \geq -1$ let $u_j \in L^\infty(\mathbf{R}^d)$ be such that $\operatorname{supp} \widehat{u}_j \subset \widetilde{C}_j$ and,

$$\|u_j\|_{L^\infty(\mathbf{R}^d)} \leq C 2^{-j\rho}.$$

Set $u = \sum_{j=-1}^{+\infty} u_j$. Then,

$$u \in W^{\rho,\infty}(\mathbf{R}^d) \quad \text{and} \quad \|u\|_{W^{\rho,\infty}(\mathbf{R}^d)} \leq C \sup_{j \geq -1} \left(2^{j\rho} \|u_j\|_{L^\infty(\mathbf{R}^d)} \right).$$

3.3.3 The Zygmund Spaces

- For $\rho \in \mathbf{R}$ we set,

$$C_*^\rho(\mathbf{R}^d) = \left\{ u \in S'(\mathbf{R}^d) : \sup_{j \geq -1} \left(2^{j\rho} \|\Delta_j u\|_{L^\infty(\mathbf{R}^d)} \right) < +\infty \right\},$$

endowed with the norm $\|u\|_{C_*^\rho} = \sup_{j \geq -1} \left(2^{j\rho} \|\Delta_j u\|_{L^\infty(\mathbf{R}^d)} \right)$.

If $\rho = m + \sigma, m \in \mathbf{N}, 0 < \sigma < 1$ these are the spaces $W^{\rho,\infty}(\mathbf{R}^d)$ defined in the previous paragraph.

If $\rho = m \in \mathbf{N}$ we have $W^{m,\infty}(\mathbf{R}^d) \subset C_*^m(\mathbf{R}^d)$, with strict inclusion.

3.4 Paraproducts

We keep the notations introduced in the beginning of Sect. 3.3 that is,

$$\Delta_{-1} u = \psi(D)u, \quad \Delta_j u = \varphi(2^{-j} D)u, \quad j \geq 0,$$

and we set,

$$S_\ell(u) = \sum_{k=-1}^{\ell-1} \Delta_k u = \psi(2^{-\ell} D)u, \tag{3.13}$$

by (3.12). Notice that $S_\ell(u) = 0$ if $\ell \leq -1$.

If $f \in S'(\mathbf{R}^d)$ we call "spect(f)" (spect. for spectrum) the support of its Fourier transform. According to our choice of the support of φ, ψ for every $a, u \in S'(\mathbf{R}^d)$ we have,

$$\text{spect}(S_\ell(a)) \subset \left\{ \xi \in \mathbf{R}^d : |\xi| \leq 2^\ell \right\},$$
$$\text{spect}(\Delta_j u) \subset \left\{ \xi \in \mathbf{R}^d : 2^{j-1} \leq |\xi| \leq 2^{j+1} \right\}. \tag{3.14}$$

In particular if $\ell \leq j - 2$ we have, $\text{spect}(S_\ell(a)) \cap \text{spect}(\Delta_j(u)) = \emptyset$.

- Let $a, u \in S'(\mathbf{R}^d)$. The paraproduct of a and u is denoted by $T_a u$ and is formally defined by the formula,

$$T_a u = \sum_{j \geq -1} S_{j-2}(a) \Delta_j u. \tag{3.15}$$

We shall give below conditions on a and u under which the above series is convergent.

In (3.15), according to (3.14), the frequencies of a are smaller than those of u. Notice also that the spectrum of $S_{j-2}(a)\Delta_j u$ is contained in a ring $\widetilde{C}_j$.

Now if $a \in L^\infty(\mathbf{R}^d)$ and $u \in H^s(\mathbf{R}^d)$ the product au, when it is defined, does not belong in general to $H^s(\mathbf{R}^d)$. However, we have the following result.

• Let $s \in \mathbf{R}$ and $a \in L^\infty(\mathbf{R}^d)$. There exists $C > 0$ depending on d, s such that,

$$\|T_a u\|_{H^s(\mathbf{R}^d)} \leq C\|a\|_{L^\infty(\mathbf{R}^d)}\|u\|_{H^s(\mathbf{R}^d)}, \tag{3.16}$$

$$\|T_a u\|_{C^s_*(\mathbf{R}^d)} \leq C\|a\|_{L^\infty(\mathbf{R}^d)}\|u\|_{C^s_*(\mathbf{R}^d)}. \tag{3.17}$$

• Let $s \in \mathbf{R}$ and $a \in H^\beta(\mathbf{R}^d)$. Then there exists $C > 0$ such that,

(i) if $\beta > \dfrac{d}{2}$, $\|T_a u\|_{H^s(\mathbf{R}^d)} \leq C\|a\|_{H^\beta(\mathbf{R}^d)}\|u\|_{H^s(\mathbf{R}^d)}$, $\tag{3.18}$

(ii) if $\beta = \dfrac{d}{2}$, $\|T_a u\|_{H^s(\mathbf{R}^d)} \leq C\|a\|_{H^{\frac{d}{2}}(\mathbf{R}^d)}\|u\|_{H^{s+\varepsilon}(\mathbf{R}^d)}, \forall \varepsilon > 0$, $\tag{3.19}$

(iii) if $\beta < \dfrac{d}{2}$, $\|T_a u\|_{H^s(\mathbf{R}^d)} \leq C\|a\|_{H^\beta(\mathbf{R}^d)}\|u\|_{H^{s+\frac{d}{2}-\beta}(\mathbf{R}^d)}$, $\tag{3.20}$

(see Problem 17).

• Let $a, u \in S'(\mathbf{R}^d)$. When the product of a and u is well defined we can write,

$$au = T_a u + T_u a + R(a, u), \quad R(a, u) = \sum_{|j-k|\leq 2} \Delta_k a \Delta_j u.$$

• Let α, β be real numbers such that $\alpha + \beta > 0$ and consider $a \in H^\alpha(\mathbf{R}^d), u \in H^\beta(\mathbf{R}^d)$. Then $R(a, u) \in H^{\alpha+\beta-\frac{d}{2}}(\mathbf{R}^d)$ and there exists $C > 0$ depending on d, α, β such that,

$$\|R(a, u)\|_{H^{\alpha+\beta-\frac{d}{2}}(\mathbf{R}^d)} \leq C\|a\|_{H^\alpha(\mathbf{R}^d)}\|u\|_{H^\beta(\mathbf{R}^d)}.$$

Notice that if $\varepsilon = \alpha - \frac{d}{2} > 0$, then the reminder $R(a, u)$ is ε smoother than u.

We have the same result if $a \in C^\rho_*(\mathbf{R}^d)$ and $u \in H^\beta(\mathbf{R}^d)$ with $\rho + \beta > 0$. The reminder $R(a, u)$ belongs to $H^{\beta+\rho}(\mathbf{R}^d)$ and we have,

$$\|R(a, u)\|_{H^{\beta+\rho}(\mathbf{R}^d)} \leq C\|a\|_{C^\rho_*(\mathbf{R}^d)}\|u\|_{H^\beta(\mathbf{R}^d)}.$$

- Let $\alpha > \frac{d}{2}$ and $F \in C^\infty(\mathbf{C}, \mathbf{C})$. For every $a \in H^\alpha(\mathbf{R}^d)$ we have

$$F(a) - F(0) - T_{F'(a)}a \in H^{2\alpha - \frac{d}{2}}(\mathbf{R}^d)$$

and there exists $C > 0$ depending only on α, d such that,

$$\|F(a) - F(0) - T_{F'(a)}a\|_{H^{2\alpha - \frac{d}{2}}(\mathbf{R}^d)} \leq C \|a\|_{H^\alpha(\mathbf{R}^d)},$$

where F' denotes the differential of F. Notice that $2\alpha - \frac{d}{2} > \alpha$, so again the reminder is smoother than a.
- Let $\alpha > \frac{d}{2}$ and $\rho = \alpha - \frac{d}{2}$. If $a \in H^\alpha(\mathbf{R}^d)$ and $b \in H^\alpha(\mathbf{R}^d)$, then for every $s \in \mathbf{R}$,

$$T_a \circ T_b - T_{ab} \text{ is continuous from } H^s(\mathbf{R}^d) \text{ to } H^{s+\rho}(\mathbf{R}^d),$$

$$(T_a)^* - T_{\bar{a}} \text{ is continuous from } H^s(\mathbf{R}^d) \text{ to } H^{s+\rho}(\mathbf{R}^d).$$

Here $(T_a)^*$ denotes the adjoint of T_a in $L^2(\mathbf{R}^d)$ and $\bar{a}$ the complex conjugate of a.
- We shall introduce later in Chap. 5 a wider class of operators and prove more general results.

3.5 Some Words on Interpolation

We work in the category C of normed paces and we shall denote by $\mathcal{L}(E, F)$ where $E, F \in C$, the set of linear and continuous maps from E to F. Let $A_0, A_1 \in C$.

- Compatible couples. We say that (A_0, A_1) is a compatible couple if there exists a topological vector space $\mathcal{U}$ such that A_0 and A_1 are subspaces of $\mathcal{U}$. Then we can consider the sum $A_0 + A_1 = \{a \in \mathcal{U} : a = a_0 + a_1, a_j \in A_j\}$ and the intersection $A_0 \cap A_1$.
- Intermediate space. It is a space A such that $A_0 \cap A_1 \subset A \subset A_0 + A_1$.
- Interpolation spaces. Let $\mathcal{A} = (A_0, A_1)$, $\mathcal{B} = (B_0, B_1)$ be two compatible couples. We say that a couple (A, B) is *of interpolation* if A, B are intermediate spaces and if,

$$T \in \mathcal{L}(A_0, B_0) \text{ and } T \in \mathcal{L}(A_1, B_1) \implies T \in \mathcal{L}(A, B).$$

Here are some examples.

- Let $(X, \mathcal{T}, \mu)$ be a measured space where μ is a σ-finite positive measure. Let

$$1 \leq p_0, \ p_1 < +\infty, \quad 1 < q_0, \ q_1 \leq +\infty.$$

Set,

$$A_0 = L^{p_0}(X, d\mu), \quad B_0 = L^{q_0}(X, d\mu), \quad A_1 = L^{p_1}(X, d\mu), \quad B_1 = L^{q_1}(X, d\mu),$$

$$\frac{1}{p} = \frac{1-\theta}{p_0} + \frac{\theta}{p_1}, \quad \frac{1}{q} = \frac{1-\theta}{q_0} + \frac{\theta}{q_1}, \quad 0 < \theta < 1.$$

Then

$$(A = L^p(X, d\mu), \ B = L^q(X, d\mu)) \text{ is of interpolation.}$$

Moreover,

$$\|T\|_{\mathcal{L}(A,B)} \le C \|T\|_{\mathcal{L}(A_0,B_0)}^{1-\theta} \|T\|_{\mathcal{L}(A_1,B_1)}^{\theta}.$$

This is the Riesz–Thorin theorem.

A classical application of this theorem is the continuity of the Fourier transform $\mathcal{F}$ on the $L^p(\mathbf{R}^d)$ spaces. Indeed we know that $\mathcal{F}$ is continuous from $L^1(\mathbf{R}^d)$ to $L^\infty(\mathbf{R}^d)$ and from $L^2(\mathbf{R}^d)$ to $L^2(\mathbf{R}^d)$. Taking $p_0 = 1, q_0 = +\infty, p_1 = q_1 = 2$ we see that $\mathcal{F}$ is continuous from $L^p(\mathbf{R}^d)$ to $L^{p'}(\mathbf{R}^d)$ where $1 < p < 2$ and $\frac{1}{p} + \frac{1}{p'} = 1$.

- Let $(X, \mathcal{T}, \mu)$ be a measured space where μ is a σ-finite positive measure. Let $1 \le p \le +\infty$. Let $w_0, w_1, \tilde{w}_0, \tilde{w}_1$ be positive μ-measurable functions and,

$$A_0 = L^p(X, w_0 d\mu), \qquad\qquad B_0 = L^p(X, \tilde{w}_0 d\mu),$$
$$A_1 = L^p(X, w_1 d\mu), \qquad\qquad B_1 = L^p(X, \tilde{w}_1 d\mu).$$

Set for $0 < \theta < 1$,

$$w(x) = w_0(x)^{1-\theta} w_1(x)^\theta, \quad \tilde{w}(x) = \tilde{w}_0(x)^{1-\theta} \tilde{w}_1(x)^\theta.$$

Then the couple

$$(A, B) = (L^p(X, w\, d\mu), L^p(X, \tilde{w}\, d\mu))$$

is of interpolation. Moreover,

$$\|T\|_{\mathcal{L}(A,B)} \le C \|T\|_{\mathcal{L}(A_0,B_0)}^{1-\theta} \|T\|_{\mathcal{L}(A_1,B_1)}^{\theta}.$$

- Let $s_0, \sigma_0, s_1, \sigma_1$ be real numbers. Set

$$A_0 = H^{s_0}(\mathbf{R}^d), \quad B_0 = H^{\sigma_0}(\mathbf{R}^d), \quad A_1 = H^{s_1}(\mathbf{R}^d), \quad B_0 = H^{\sigma_1}(\mathbf{R}^d),$$
$$s = (1-\theta)s_0 + \theta s_1, \quad \sigma = (1-\theta)\sigma_0 + \theta\sigma_1, \quad 0 < \theta < 1.$$

Then the couple,

$$(A = H^s(\mathbf{R}^d), \quad B = H^\sigma(\mathbf{R}^d))$$

is *of interpolation.* Moreover,

$$\|T\|_{\mathcal{L}(A,B)} \le C \|T\|_{\mathcal{L}(A_0,B_0)}^{1-\theta} \|T\|_{\mathcal{L}(A_1,B_1)}^{\theta}.$$

Notice that the result about the Sobolev spaces is a consequence of the second result about the L^p spaces. Indeed $H^s(\mathbf{R}^d)$ can be described as,

$$H^s(\mathbf{R}^d) = \left\{ u \in S'(\mathbf{R}^d) : \hat{u} \in L^2(\mathbf{R}^d, \langle \xi \rangle^{2s} \, d\xi) \right\}, \quad \langle \xi \rangle = (1 + |\xi|^2)^{\frac{1}{2}}.$$

3.6 The Hardy–Littlewood–Sobolev Inequality

• Let $1 < p < +\infty$ and $0 < \alpha < d$. This inequality expresses the fact that the convolution by the function $|x|^{-\alpha}$ maps continuously $L^p(\mathbf{R}^d)$ to $L^r(\mathbf{R}^d)$ for

$$\frac{1}{r} = \frac{1}{p} + \frac{\alpha}{d} - 1.$$

Namely, it states that,

$$\||x|^{-\alpha} \star f\|_{L^r(\mathbf{R}^d)} \le C(\alpha, p, d) \|f\|_{L^p(\mathbf{R}^d)}, \quad \forall f \in L^p(\mathbf{R}^d). \tag{3.21}$$

• The inequality (3.21) is a generalization of the Young inequality which says that for $p, q \ge 1$,

$$\frac{1}{r} = \frac{1}{p} + \frac{1}{q} - 1 \quad \Rightarrow \quad L^p(\mathbf{R}^d) \star L^q(\mathbf{R}^d) \subset L^r(\mathbf{R}^d).$$

The inequality (3.21) is not a consequence of the Young inequality because taking $\frac{1}{q} = \frac{\alpha}{d}$ or $q = \frac{d}{\alpha}$, the function $|x|^{-\alpha}$ does not belong to $L^q(\mathbf{R}^d)$. However, the latter belongs to the *weak L^q* space which is denoted by $L_w^q(\mathbf{R}^d)$.

• The space $L_w^q(\mathbf{R}^d)$: let μ be the Lebesgue measure on $\mathbf{R}^d$. Then $f \in L_w^q(\mathbf{R}^d)$ if f is measurable and,

$$S(f) = \sup_{0 < \lambda < +\infty} \lambda \left(\mu \left\{ x \in \mathbf{R}^d : |f(x)| > \lambda \right\} \right)^{\frac{1}{q}} < +\infty. \tag{3.22}$$

We have $|x|^{-\alpha} \in L_w^{\frac{d}{\alpha}}(\mathbf{R}^d)$ since,

$$\mu\left(\left\{x \in \mathbf{R}^d : |x|^{-\alpha} > \lambda\right\}\right) = \mu\left(\left\{x \in \mathbf{R}^d : |x| < \lambda^{-\frac{1}{\alpha}}\right\}\right) \leq C_d \lambda^{-\frac{d}{\alpha}}.$$

On the other hand, $L^q(\mathbf{R}^d) \subset L_w^q(\mathbf{R}^d)$ since for $0 < \lambda < +\infty$ we have,

$$\int_{\mathbf{R}^d} |f(x)|^q \, dx \geq \int_{\{x:|f(x)|>\lambda\}} |f(x)|^q \, dx \geq \lambda^q \mu\left(\left\{x \in \mathbf{R}^d : |f(x)| > \lambda\right\}\right).$$

- The quantity $S(f)$ defined in (3.22) is not a norm on L_w^q (it does not satisfy the triangle inequality). However, the quantity,

$$\|f\|_{L_w^q(\mathbf{R}^d)} = \sup_A \mu(A)^{\frac{1}{q}-1} \int_A |f(x)| \, dx, \tag{3.23}$$

where A is a measurable set of positive and finite measure, is equivalent to $S(f)$ and it is a norm on $L_w^q(\mathbf{R}^d)$ (see Problem 14).
- The generalization of the inequality (3.21) can be stated as follows.
  Let $1 < p, q < +\infty$ and r be such that,

$$\frac{1}{r} = \frac{1}{p} + \frac{1}{q} - 1.$$

There exists $C = C(p, q, d) > 0$ such that,

$$\|f \star g\|_{L^r(\mathbf{R}^d)} \leq C \|f\|_{L^p(\mathbf{R}^d)} \|g\|_{L_w^q(\mathbf{R}^d)}.$$

Comments For a description of the main functional spaces, the Littlewood–Paley theory, and the paraproducts we refer to H. Bahouri, J.Y. Chemin, and R. Danchin [5] (Chapters 1 and 2) and M. Taylor [31] (Chapter 13). A comprehensive introduction to interpolation theory can be found in J. Bergh and J. Löfström [6]. Different proofs of the Hardy–Littlewood–Sobolev inequality can be found in E. Stein [28] (Chapter VIII, section 4.2) and E. Lieb and M. Loss [19] (Chapter 4). □

Chapter 4
Statements of the Problems of Chap. 3

This chapter contains problems about the main functional spaces.

Problem 4

Let $I = (a, b) \subset \mathbf{R}$, $s \in \mathbf{R}$, and $d \geq 1$. We set $H^s = H^s(\mathbf{R}^d)$ and we consider a family $(u(t, \cdot))_{t \in I} \subset H^s$ such that,

(H1) the map $t \mapsto \|u(t, \cdot)\|_{H^s}$ is continuous from I to $\mathbf{R}$,
(H2) there exists $\delta > 0$ such that $t \to u(t, \cdot)$ is continuous from I to $H^{s-\delta}$.

The goal of this problem is to prove that the map $t \to u(t, \cdot)$ is continuous from I to H^s.

Denote by $\langle \cdot, \cdot \rangle$ the duality bracket between H^s and H^{-s}. Let $t_0 \in I$ and consider a sequence $(t_n) \subset I$ with $t_n \to t_0$.

1. Let $A_n := \langle u(t_n, \cdot) - u(t_0, \cdot), \varphi \rangle$ where $\varphi \in H^{-s}$. Show that $\lim_{n \to +\infty} A_n = 0$.
2. Conclude.
   Hint: Note that if $u, v \in H^s(\mathbf{R}^d)$ then $(u, v)_{H^s} = (2\pi)^d \left(u, \Lambda^{2s} \overline{v} \right)_{H^s \times H^{-s}}$ where $\Lambda = (I - \Delta)^{\frac{1}{2}}$.

Problem 5 *An Interpolation Result*

If $I \subset \mathbf{R}$ is an interval and $s \in \mathbf{R}$ we set,

$$W^s(I) = \left\{ f \in L^2_z(I, H^{s+\frac{1}{2}}(\mathbf{R}^d)) : \partial_z f \in L^2_z(I, H^{s-\frac{1}{2}}(\mathbf{R}^d)) \right\},$$

© The Editor(s) (if applicable) and The Author(s), under exclusive license
to Springer Nature Switzerland AG 2020
T. Alazard, C. Zuily, *Tools and Problems in Partial Differential Equations*,
Universitext, https://doi.org/10.1007/978-3-030-50284-3_4

endowed with its natural norm,

$$\|f\|_{W^s(I)} = \|f\|_{L_z^2(I, H^{s+\frac{1}{2}}(\mathbf{R}^d))} + \|\partial_z f\|_{L_z^2(I, H^{s-\frac{1}{2}}(\mathbf{R}^d))}.$$

Part 1
We assume $I = \mathbf{R}$. We will admit that $C_0^\infty(\mathbf{R} \times \mathbf{R}^d)$ is dense in $W^s(\mathbf{R})$.

1. Let $v \in C_0^\infty(\mathbf{R} \times \mathbf{R}^d)$. Prove that the map $z \mapsto \|v(z, \cdot)\|_{H^s(\mathbf{R}^d)}^2$ is differentiable on $\mathbf{R}$ and that,

$$\frac{d}{dz}\|v(z, \cdot)\|_{H^s(\mathbf{R}^d)}^2 = 2\mathrm{Re}\,(v(z, \cdot), \partial_z v(z, \cdot))_{H^s(\mathbf{R}^d)}.$$

2. Deduce that there exists $C, C' > 0$ such that,

$$\sup_{z \in \mathbf{R}} \|v(z, \cdot)\|_{H^s(\mathbf{R}^d)} \leq C \|v\|_{L_z^2(\mathbf{R}, H^{s+\frac{1}{2}}(\mathbf{R}^d))}^{\frac{1}{2}} \|\partial_z v\|_{L_z^2(\mathbf{R}, H^{s-\frac{1}{2}}(\mathbf{R}^d))}^{\frac{1}{2}}$$
$$\leq C' \|v\|_{W^s(\mathbf{R})}. \tag{4.1}$$

3. Show that if $u \in W^s(\mathbf{R})$ then $u \in C^0(\mathbf{R}, H^s(\mathbf{R}^d)) \cap L^\infty(\mathbf{R}, H^s(\mathbf{R}^d))$ and that an inequality analogue to (4.1) still holds for u.
   Hint: Use question 2 and the density of $C_0^\infty(\mathbf{R} \times \mathbf{R}^d)$ in $W^s(\mathbf{R})$.

Part 2
We assume $I = (0, +\infty)$. Show that there exists $\tilde{u} \in W^s(\mathbf{R})$ such that $u = \tilde{u}$ for $z \in (0, +\infty)$ and $\|\tilde{u}\|_{W^s(\mathbf{R})} \sim \|u\|_{W^s((0,+\infty))}$.

1. Prove that if $u \in W^s((0, +\infty))$ then $u \in (C^0 \cap L^\infty)([0, +\infty), H^s(\mathbf{R}^d))$ and that there exists $C > 0$ (independent of u) such that,

$$\sup_{z \in (0,+\infty)} \|u(z, \cdot)\|_{H^s} \leq C \|u\|_{W^s((0,+\infty))}.$$

2. Deduce that this is still true if $I = (a, +\infty)$ or $(-\infty, b)$, $a, b \in \mathbf{R}$.

Part 3
We assume $I = (-1, 0)$. Let $\chi_j \in C^\infty(\mathbf{R})$, $j = 1, 2$ be such that,

$$\mathrm{supp}\,\chi_1 \subset \left(-\infty, -\frac{1}{4}\right], \quad \mathrm{supp}\,\chi_2 \subset \left[-\frac{3}{4}, +\infty\right), \quad \chi_1(z) + \chi_2(z) = 1 \text{ on } (-1, 0).$$

Let $u \in W(I)$. We write $u = u_1 + u_2$ where $u_j(z, \cdot) = \chi_j(z)u(z, \cdot)$, $j = 1, 2$.

1. Show that $u_1 \in W^s((-1, +\infty))$ and $u_2 \in W^s((-\infty, 0))$.
2. Show that $u \in C^0([-1, 0], H^s(\mathbf{R}^d))$ and there exists $C > 0$ such that,

$$\sup_{z \in [-1,0]} \|u(z, \cdot)\|_{H^s(\mathbf{R}^d)} \leq C \|u\|_{W^s((-1,0))}.$$

3. Deduce the same result if $I = (a, b), a, b \in \mathbf{R}$.

Problem 6 *Spaces of Analytic Functions*

Let $\sigma > 0$ and $s \in \mathbf{R}$. We set $S_\sigma = \{z = x + iy \in \mathbf{C} : |y| < \sigma\}$.
    We consider the space

$$G^{\sigma,s} = \left\{ u \in S'(\mathbf{R}) : e^{\sigma|\xi|} \langle \xi \rangle^s \widehat{u} \in L^2(\mathbf{R}) \right\}, \quad \text{where } \langle \xi \rangle = (1 + |\xi|^2)^{\frac{1}{2}},$$

endowed with the norm:

$$\|u\|_{G^{\sigma,s}} := \|e^{\sigma|\xi|} \langle \xi \rangle^s \widehat{u}\|_{L^2(\mathbf{R})}.$$

Part 1
1. Prove that for $u \in G^{\sigma,s}$ and $z \in S_\sigma$ the function $\xi \mapsto e^{iz\cdot\xi} \widehat{u}(\xi)$ belongs to $L^1(\mathbf{R})$.
    We set for $z \in S_\sigma$

$$U(z) = (2\pi)^{-1} \int_{\mathbf{R}} e^{iz\cdot\xi} \widehat{u}(\xi) \, d\xi.$$

2. Show that U is holomorphic in S_σ and that $U|_{y=0} = u$.
3. Show that there exists $C > 0$ such that,

$$\sup_{|y|<\sigma} \|U(x + iy)\|_{H_x^s(\mathbf{R})} \leq C\|u\|_{G^{\sigma,s}}.$$

Part 2
Let U be a holomorphic function in S_σ such that,

$$M_0 := \sup_{|y|<\sigma} \|U(x + iy)\|_{H_x^s(\mathbf{R})} < +\infty. \tag{4.2}$$

Set $u = U|_{y=0}$. We want to prove that $u \in G^{\sigma,s}$ and there exists $C > 0$ (independent of u, U) such that,

$$\|u\|_{G^{\sigma,s}} \leq C \sup_{|y|<\sigma} \|U(x + iy)\|_{H_x^s(\mathbf{R})}. \tag{4.3}$$

Let $\psi \in C_0^\infty(\mathbf{R}), \psi(\xi) = 1$ if $|\xi| \leq 1$, $\psi(\xi) = 0$ if $|\xi| \geq 2$. For $\lambda > 0$ set $\psi_\lambda(\xi) = \psi\left(\frac{\xi}{\lambda}\right)$, $\varphi_\lambda = \mathcal{F}^{-1}\psi_\lambda$ and,

$$F(z) = (2\pi)^{-1} \int_{\mathbf{R}} e^{iz\cdot\xi} \psi_\lambda(\xi) \widehat{u}(\xi) \, d\xi.$$

1. Show that the function F is well defined and is holomorphic in S_σ.
   For $z = x + iy \in S_\sigma$ and $\lambda > 0$ fixed we set,

$$V(z) = \langle U(z - t), \; \varphi_\lambda(t) \rangle = (U(\cdot + iy) \star \varphi_\lambda)(x).$$

We want to show that V is a holomorphic function in S_σ.

   We have $\varphi_\lambda \in \mathcal{S}(\mathbf{R}) \subset H^{-s}(\mathbf{R})$. Let $(\chi_j) \subset C_0^\infty(\mathbf{R})$ be such that $(\chi_j) \to \varphi_\lambda$ in $H^{-s}(\mathbf{R})$.

2. a) Set $V_j(z) = \langle U(z - t), \; \chi_j(t) \rangle$. Show that V_j is holomorphic in S_σ.
   Hint: Let $K \subset S_\sigma$ be compact. Use the fact that $V_j(z) = \int_{\mathbf{R}} U(z - t)\chi_j(t)\,dt$ for $z \in K$.
   b) Show that for $z \in S_\sigma$, $|V_j(z) - V(z)| \leq M_0 \|\chi_j - \varphi_\lambda\|_{H^{-s}(\mathbf{R})}$ and conclude.
3. Prove that $V|_{y=0} = F|_{y=0}$ and deduce that $V(z) = F(z)$ for all z in S_σ.
4. Deduce that $\mathcal{F}_x(U(\cdot + iy))(\xi) = e^{-y \cdot \xi} \widehat{u}(\xi)$ and that,

$$\sup_{|y| < \sigma} \int_{\mathbf{R}} e^{-2y \cdot \xi} \langle \xi \rangle^{2s} |\widehat{u}(\xi)|^2 \, d\xi \leq M_0^2,$$

where M_0 is defined by (4.2).
5. Choosing appropriately y and using the Fatou lemma deduce (4.3).

Part 3
We fix $s_0 > \frac{1}{2}$.

1. Prove that there exists $C > 0$ such that,

$$\|au\|_{G^{\sigma,0}} \leq C \|a\|_{G^{\sigma,s_0}} \|u\|_{G^{\sigma,0}}, \quad \forall a \in G^{\sigma,s_0}, \quad \forall u \in G^{\sigma,0}.$$

2. a) Show that for all $s \geq 0$ there exists $C > 0$ such that,

$$\|au\|_{G^{\sigma,s}} \leq C \|a\|_{G^{\sigma,s_0}} \|u\|_{G^{\sigma,s}} + \|a\|_{G^{\sigma,s}} \|u\|_{G^{\sigma,s_0}}, \quad \forall a, u \in G^{\sigma,s_0} \cap G^{\sigma,s}.$$

   b) Deduce that for $s > \frac{1}{2}$ the space $G^{\sigma,s}$ is an algebra.

Comments These spaces appear to be useful if one wants to deal with analytic solutions in a microlocal setting.

Problem 7 *Uniformly Local Sobolev Spaces*

The goal of this problem is to study a family of spaces introduced by Tosio Kato.

Part 1

1. Show that there exists a positive function $\chi \in C^\infty(\mathbf{R}^d)$ such that,

$$(i) \quad \mathrm{supp}\, \chi \subset [-1, 1]^d, \quad (ii) \quad \sum_{q \in \mathbf{Z}^d} \chi_q(x) = 1, \quad \forall x \in \mathbf{R}^d, \qquad (4.4)$$

where $\chi_q(x) = \chi(x - q)$.

For $s \in \mathbf{R}$, $H_{ul}^s(\mathbf{R}^d)$ is the space of elements $u \in H_{loc}^s(\mathbf{R}^d)$ such that,

$$\|u\|_{H_{ul}^s(\mathbf{R}^d)} := \sup_{q \in \mathbf{Z}^d} \|\chi_q u\|_{H^s(\mathbf{R}^d)} < +\infty. \qquad (4.5)$$

2. Show that endowed with the norm (4.5) $H_{ul}^s(\mathbf{R}^d)$ is a Banach space.
3. Give simple examples of functions belonging to $L_{ul}^2(\mathbf{R}^d)$ but not to $L^2(\mathbf{R}^d)$.

Part 2

The goal of this part is to prove that the above definition H_{ul}^s does not depend on the function χ introduced in (4.4).

1. a) Let $\rho > 0$. Show that for any $\alpha \in \mathbf{N}^d$ and all $x \in \mathbf{R}^d$ we have $\partial_x^\alpha (1 + |x|^2)^\rho = (1 + |x|^2)^{\rho - |\alpha|} P_\alpha(x)$ where P_α is a polynomial of degree $|\alpha|$.
   b) Deduce that for any $\alpha \in \mathbf{N}^d$ there exists $C_\alpha > 0$ such that,

$$|\partial_x^\alpha (1 + |x|^2)^\rho| \le C_\alpha (1 + |x|^2)^{\rho - \frac{1}{2}|\alpha|}, \quad \forall x \in \mathbf{R}^d.$$

   c) Show, by induction on $|\alpha|$, that for all $\alpha \in \mathbf{N}^d$ there exists $C_\alpha > 0$ such that,

$$|\partial_x^\alpha (1 + |x|^2)^{-\rho}| \le C_\alpha (1 + |x|^2)^{-\rho - \frac{1}{2}|\alpha|}, \quad \forall x \in \mathbf{R}^d.$$

   Hint: Use the fact that $(1 + |x|^2)^\rho (1 + |x|^2)^{-\rho} = 1$.
   d) Deduce that for any $\alpha \in \mathbf{N}^d$ there exists $C_\alpha > 0$ such that for all $a \in \mathbf{R}^d$ and all $x \in \mathbf{R}^d$ we have,

$$|\partial_x^\alpha (1 + |x - a|^2)^{-\rho}| \le C_\alpha (1 + |x - a|^2)^{-\rho - \frac{1}{2}|\alpha|}.$$

2. Let $\tilde{\chi} \in C_0^\infty(\mathbf{R}^d)$ and $\tilde{\chi}_k(x) = \tilde{\chi}(x - k)$. For $k, q \in \mathbf{Z}^d$ and $N \in \mathbf{N}$ we consider the function,

$$\Phi_{k,q}(x) = \frac{\langle k - q \rangle^N}{\langle x - q \rangle^N} \tilde{\chi}_k, \quad \text{where } \langle y \rangle = (1 + |y|^2)^{\frac{1}{2}}.$$

   Show that for any $M \in \mathbf{N}$ there exists $C = C_{N,M,\tilde{\chi}} > 0$ independent on k, q such that,

$$\|\Phi_{k,q}\|_{H^M(\mathbf{R}^d)} \le C.$$

3. Let $s \in \mathbf{R}$ and $M \in \mathbf{N}$, $M \geq |s|$. Using the fact that,

$$\|\theta u\|_{H^s(\mathbf{R}^d)} \leq C \left(\int \langle \xi \rangle^{|s|} |\theta(\xi)| \, d\xi \right) \|u\|_{H^s(\mathbf{R}^d)}$$

and writing,

$$\widetilde{\chi}_k \chi_q u(x) = \langle k - q \rangle^{-N} \, \Phi_{k,q}(x) \left\{ \langle x - q \rangle^N \, \psi_q(x) \right\} \chi_q u(x),$$

where $\psi \in C_0^\infty(\mathbf{R}^d)$ is equal to one on the support of χ, $\psi_q(x) = \psi(x - q)$ and $N > d + 1$ show that,

$$\|\widetilde{\chi}_k u\|_{H^s(\mathbf{R}^d)} \leq \sum_{q \in \mathbf{Z}^d} \|\widetilde{\chi}_k \chi_q u\|_{H^s(\mathbf{R}^d)} \leq C \sum_{q \in \mathbf{Z}^d} \langle k - q \rangle^{-N} \sup_{q \in \mathbf{Z}^d} \|\chi_q u\|_{H^s(\mathbf{R}^d)}.$$

4. Conclude.

Comments The uniformly local Sobolev spaces have been introduced by T. Kato in [17]. They are useful if one wants to work with solutions having a local Sobolev regularity but do not decrease at infinity. For instance, it contains in a unified setting the Sobolev spaces $H^s(\mathbf{R}^d)$ and the periodic Sobolev spaces $H^s(\mathbf{T}^d)$. It also contains Hölder spaces, as it will be shown in the next problem.

Problem 8 *Link Between H_{ul}^s and C_*^s*

The goal of this problem is to explore the link between the uniformly local Sobolev spaces and the Zygmund spaces.

1. Prove that for $m \in \mathbf{N}$ we have, $W^{m,\infty}(\mathbf{R}^d) \subset H_{ul}^m(\mathbf{R}^d)$.
2. Let $m \in \mathbf{N}$. Prove that for every $\varepsilon > 0$ we have $C_*^{m+\varepsilon}(\mathbf{R}^d) \subset H_{ul}^m(\mathbf{R}^d)$.

   Let $s = m + \sigma$, $m \in \mathbf{N}$, $0 < \sigma < 1$. The goal of the following questions is to show that for $\delta > 0$, $C_*^{s+\delta}(\mathbf{R}^d) \subset H_{ul}^s(\mathbf{R}^d)$.
3. a) Let $s = m + \sigma$, $m \in \mathbf{N}$, $0 < \sigma < 1$ and let $\varepsilon > 0$ be so small that $0 < \sigma + \varepsilon < 1$. Then $s + \varepsilon \notin \mathbf{N}$ and $C_*^{s+\varepsilon}(\mathbf{R}^d) = W^{s+\varepsilon,\infty}(\mathbf{R}^d)$. Prove that,

$$\left(W^{\sigma+\varepsilon,\infty}(\mathbf{R}^d) \subset H_{ul}^\sigma(\mathbf{R}^d) \right) \Longrightarrow \left(W^{s+\varepsilon,\infty}(\mathbf{R}^d) \subset H_{ul}^s(\mathbf{R}^d) \right).$$

We recall that,

$$\begin{cases} \|\chi_q u\|_{H^\sigma(\mathbf{R}^d)}^2 \approx \|\chi_q u\|_{L^2(\mathbf{R}^d)}^2 + \iint f_q(x,y) \, dx \, dy = (1) + (2), \\[2mm] \text{where,} \quad f_q(x,y) = \dfrac{|\chi_q(x)u(x) - \chi_q(y)u(y)|^2}{|x - y|^{d+2\sigma}}. \end{cases}$$

b) Prove that $(1) \le C_1 \|u\|^2_{L^\infty(\mathbf{R}^d)}$.

We write

$$(2) = \iint_{|x-y|>1} f_q(x, y)\, dx\, dy + \iint_{|x-y|<1} f_q(x, y)\, dx\, dy = (3) + (4).$$

c) Estimating in f_q the difference by the sum prove that $(3) \le C_2 \|u\|^2_{L^\infty(\mathbf{R}^d)}$.

d) Prove that,

$$(4) = \iint_\Omega f_q(x, y)\, dx\, dy,$$

$$\Omega = \left\{ (x, y) \in \mathbf{R}^d \times \mathbf{R}^d : |x - y| < 1, |y - q| < 2 \right\}.$$

e) Prove that, $f_q(x, y) \le C \left(g_q(x, y) + h_q(x, y) \right)$ where,

$$g_q(x, y) = |\chi_q(x)|^2 |u(x) - u(y)|^2, \quad h_q(x, y) = |u(y)|^2 |\chi_q(x) - \chi_q(y)|^2.$$

f) Prove that,

$$\iint_\Omega g_q(x, y)\, dx\, dy \le C_3 \|u\|^2_{W^{\sigma+\varepsilon,\infty}(\mathbf{R}^d)},$$

and

$$\iint_\Omega h_q(x, y)\, dx\, dy \le C_4 \|u\|^2_{L^\infty(\mathbf{R}^d)}.$$

g) Deduce that $C^{s+\varepsilon}_*(\mathbf{R}^d) \subset H^s_{ul}(\mathbf{R}^d)$.

Problem 9

Let $d \ge 2$ and let E be the closure of $C^\infty_0(\mathbf{R}^d \setminus \{0\})$ in $H^1(\mathbf{R}^d)$. Our goal is to show that $E = H^1(\mathbf{R}^d)$, in other words that $C^\infty_0(\mathbf{R}^d \setminus \{0\})$ is dense in $H^1(\mathbf{R}^d)$.

Part 1
We assume here $d \ge 3$.

1. Let $\chi \in C^\infty(\mathbf{R}^d)$ be such that $\chi(x) = 0$ for $|x| \le 1$, $\chi(x) = 1$ for $|x| \ge 2$ and $0 \le \chi \le 1$. Set $\chi_\varepsilon(x) = \chi(\frac{x}{\varepsilon})$ for $\varepsilon > 0$. Let $u \in C^\infty_0(\mathbf{R}^d)$ and set $u_\varepsilon = \chi_\varepsilon u$. Prove that (u_ε) converges to u in $H^1(\mathbf{R}^d)$ when $\varepsilon \to 0$.
2. Prove that $C^\infty_0(\mathbf{R}^d \setminus \{0\})$ is dense in $H^1(\mathbf{R}^d)$.

Part 2

We assume in what follows that $d \geq 2$.

Let $k \in \mathbf{N}$ and $P(\xi) = \sum_{|\alpha| \leq k} c_\alpha \xi^\alpha$, $\xi \in \mathbf{R}^d, c_\alpha \in \mathbf{C}$. We set $P_j(\xi) = \sum_{|\alpha|=j} c_\alpha \xi^\alpha$ and we suppose that P is exactly of order k which, P_k being homogeneous, reads,

$$(*) \quad \exists \omega_0 \in \mathbf{S}^{d-1} : |P_k(\omega_0)| := c_0 > 0.$$

1. Prove that there exists V_{ω_0} a neighborhood of ω_0 in $\mathbf{S}^{d-1}$ and $r_0 > 0$ such that,

$$|P(r\omega)| \geq \frac{c_0}{4} r^k, \quad \forall \omega \in V_{\omega_0}, \quad \forall r \geq r_0.$$

2. Let $U \in H^{-1}(\mathbf{R}^d)$ be such that supp $U \subset \{0\}$. Show that $U = 0$.
   Let $E^\perp$ be the orthogonal of E in $H^1(\mathbf{R}^d)$ and Δ be the Laplacian in $\mathbf{R}^d$.
3. Let $u \in E^\perp$. Prove that supp $(-\Delta u + u) \subset \{0\}$.
4. Deduce that $-\Delta u + u = 0$ in $\mathcal{S}'(\mathbf{R}^d)$ and then that $u = 0$.
5. Prove that $C_0^\infty(\mathbf{R}^d \setminus \{0\})$ is dense in $H^1(\mathbf{R}^d)$.
6. Show that, for $d \geq 2$, the space $H^1(\mathbf{R}^d)$ cannot be continuously embedded in the space $C^0(\overline{B}(0, \varepsilon))$ where $\overline{B}(0, \varepsilon)$ is the closed ball in $\mathbf{R}^d$ with center 0 and radius $\varepsilon > 0$.

Part 3

Let $d \geq 2$ and let Ω be an open subset of $\mathbf{R}^d$ with smooth boundary such that $0 \in \Omega$. We shall denote by $C_0^\infty(\overline{\Omega} \setminus \{0\})$ the set of restrictions to Ω of functions in $C_0^\infty(\mathbf{R}^d \setminus \{0\})$.

1. Prove that $C_0^\infty(\overline{\Omega} \setminus \{0\})$ is dense in $H^1(\Omega)$.
   Hint: Use the extension and the restriction operators P_1 and R (see Sect. 3.1.3 in Chap. 2).

Remark In all this problem we can of course replace the point 0 by any point $x_0 \in \mathbf{R}^d$ (or $x_0 \in \Omega$).

Problem 10 *A Poincaré Inequality*

The second part of this problem uses the result proved in Part 3 of Problem 9.

In this problem we shall denote by $|A|$ the Lebesgue measure of a measurable set $A \subset \mathbf{R}^d$.

Part 1

1. Let Ω be a bounded open subset of $\mathbf{R}^d$.

   Show that for every $\alpha > 0$ there exists $C_\alpha > 0$ such that for every $u \in H^1(\Omega)$ such that,

$$(\star) \quad |\{x \in \Omega : u(x) = 0\}| \geq \alpha$$

we have,

$$\int_\Omega |u(x)|^2\, dx \leq C_\alpha \int_\Omega |\nabla u(x)|^2\, dx. \tag{4.6}$$

Hint: Argue by contradiction and use the compactness of the map $u \mapsto u$ from $H^1(\Omega)$ to $L^2(\Omega)$.

Part 2

The purpose of this part is to see what happens if we relax the condition $(\star)$.

1. Prove by a simple example that the condition

$$(\star\star) \quad |\{x \in \Omega : u(x) = 0\}| = 0$$

is not sufficient to have (4.6).

2. We assume $d = 1, \Omega = (a, b)$. Show that one can find $C > 0$ depending only on a, b such that for every $u \in H^1((a, b))$ satisfying,

$$(\star\star\star) \quad \{x \in (a, b) : u(x) = 0\} \neq \emptyset,$$

we have,

$$\int_a^b |u(x)|^2\, dx \leq C \int_a^b |u'(x)|^2\, dx.$$

3. Let $d \geq 2$ and let Ω be a bounded open subset of $\mathbf{R}^d$ with smooth boundary. We assume for simplicity that $0 \in \Omega$. Show that the condition

$$(\star\star\star\star) \quad |\{x \in \Omega : u(x) = 0\}| > 0$$

is not sufficient to have (4.6) with a constant $C > 0$ independent of u.

   Hint: Use Part 3 of Problem 9.

Problem 11 *A Composition Rule in Sobolev Space*

Part 1

Let $p \geq 1, m \geq 1$ be two integers and let $G \colon \mathbf{R}^p \to \mathbf{C}$ be a C^∞ function such that,

$$|\partial^\alpha G(y)| \leq M|y|^{m-|\alpha|}, \quad \forall y \in \mathbf{R}^p, \quad \forall |\alpha| \leq m. \tag{4.7}$$

For $u = (u_1, \ldots, u_p) \colon \mathbf{R}^d \to \mathbf{R}^p$ and any norm $\| \cdot \|$ we shall set $\|u\| = \sum_{j=1}^p \|u_j\|$.

1. a) Prove that if $m = 1$ and $d \geq 1$ we have,

$$\|G(u)\|_{H^1} \leq K \|u\|_{H^1}, \quad \forall u \in (H^1(\mathbf{R}^d))^p.$$

b) Prove that if $m = 2$ and $1 \leq d \leq 3$ we have,

$$\|G(u)\|_{H^2} \leq K \|u\|_{H^2}^2 \quad \forall u \in (H^2(\mathbf{R}^d))^p.$$

Hint: Use the Sobolev embeddings.

c) Prove, by induction on $m \geq 2$, that if $1 \leq d \leq 3$ we have,

$$\|G(u)\|_{H^m} \leq K \|u\|_{H^m}^m \quad \forall u \in (H^m(\mathbf{R}^d))^p$$

Hint: Use the fact that $\partial_{x_k}(G(u)) = (\nabla_y G)(u) \cdot \partial_{x_k} u$ and the induction.

Part 2

Let $p \geq 1, m \geq 1$ be two integers and let $F \colon \mathbf{R}^p \to \mathbf{C}$ be a function in $C^\infty(\mathbf{R}^p \setminus \{0\})$ homogeneous of degree m, that is:

$$F(\lambda y) = \lambda^m F(y), \quad \forall \lambda \geq 0, \quad \forall y \in \mathbf{R}^p. \tag{4.8}$$

The goal of this part is to prove that there exists $K > 0$ such that, for all $u = (u_1, \ldots, u_p) \in H^m(\mathbf{R}^d)^p$, $F(u) \in H^m(\mathbf{R}^d)$ and,

$$\|F(u)\|_{H^m} \leq K \|u\|_{H^m}^m,$$

where we assume $d \geq 1$ if $m = 1$ and $1 \leq d \leq 3$ if $m \geq 2$.

1. Prove that for all $m \geq 1$ we have,

$$|\partial^\alpha F(y)| \leq \|\partial^\alpha F\|_{L^\infty(\mathbf{S}^{p-1})} |y|^{m-|\alpha|}, \quad \forall y \in \mathbf{R}^p \setminus \{0\}, \quad |\alpha| \leq m,$$

where $\mathbf{S}^{p-1}$ is the unit ball in $\mathbf{R}^p$. We set $M = \sum_{|\alpha| \leq m} \|\partial^\alpha F\|_{L^\infty(\mathbf{S}^{p-1})}$.

2. Let $\chi \in C_0^\infty(\mathbf{R}^p)$ be such that $\chi(y) = 1$ for $|y| \leq 1$, $\chi(y) = 0$ for $|y| \geq 2$ and $0 \leq \chi \leq 1$. For $\ell \in \mathbf{N}$, $\ell \geq 1$, define $F_\ell \in C^\infty(\mathbf{R}^p)$ by,

$$F_\ell(y) = (1 - \chi(\ell y)) F(y).$$

Prove that for $\alpha \in \mathbf{N}^p$, $|\alpha| \leq m$ there exists $C = C(\chi, \alpha) > 0$ such that,

$$|\partial^\alpha F_\ell(y)| \leq C M |y|^{m - |\alpha|}, \quad \forall y \in \mathbf{R}^p, \forall \ell \in \mathbf{N}, \ell \geq 1.$$

3. Prove that there exists $K > 0$ such that if $m = 1, d \geq 1$ or $m \geq 2, 1 \leq d \leq 3$,

$$\|F_\ell(u)\|_{H^m} \leq K \|u\|_{H^m}^m \quad \forall \ell \geq 1, \forall u \in (H^m(\mathbf{R}^d))^p.$$

4. a) Let $m \geq 1$, $\varphi \in C_0^\infty(\mathbf{R}^d)$ and $u \in (H^m(\mathbf{R}^d))^p$. Prove that $F(u)\varphi \in L^1(\mathbf{R}^d)$.
   b) Let $I = \left| \int_{\mathbf{R}^d} F(u(x))\varphi(x)\, dx \right|$. Prove that there exists $K > 0$ such that, for all $u \in H^m(\mathbf{R}^d)^p$ we have $I \leq K \|u\|_{H^m}^m \|\varphi\|_{H^{-m}}$.
   Deduce that $F(u) \in H^m(\mathbf{R}^d)$ and $\|F(u)\|_{H^m} \leq K \|u\|_{H^m}^m$.

Problem 12

Let $H = 1_{(0,+\infty)}$ be the Heaviside function. Show that for any $0 \leq s \leq 1$ there exists $C > 0$ such that for all $u \in H^1(\mathbf{R})$ satisfying $u(0) = 0$ we have,

$$\cdot \; \|Hu\|_{H^s(\mathbf{R})} \leq C \|u\|_{H^s(\mathbf{R})}.$$

Problem 13 *Hardy Inequality*

The goal of this part is to show that in dimension $d \geq 3$ we have,

$$\left(\int_{\mathbf{R}^d} \frac{|u(x)|^2}{|x|^2}\, dx \right)^{\frac{1}{2}} \leq \frac{2}{d-2} \left(\sum_{i=1}^d \int_{\mathbf{R}^d} |\frac{\partial u}{\partial x_i}(x)|^2\, dx \right)^{\frac{1}{2}}, \tag{4.9}$$

for all $u \in C_0^\infty(\mathbf{R}^d)$.

1. Let $\omega \in \mathbf{S}^{d-1}$. Prove that,

$$\int_0^{+\infty} \frac{|u(r\omega)|^2}{r^2} r^{d-1}\, dr = -\frac{2}{d-2} \mathrm{Re} \int_0^{+\infty} u(r\omega) \, (\omega \cdot \nabla_x u) \, (r\omega) r^{d-2}\, dr, \tag{4.10}$$

where $\nabla_x u = (\partial_{x_i} u)_{i=1,\dots,d}$ denotes the gradient of u.
Hint: Use the fact that $r^{d-3} = \frac{1}{d-2} \frac{d}{dr} r^{d-2}$.

2. Using the Cauchy–Schwarz inequality and the polar coordinates $(r, \omega) \in (0, +\infty) \times \mathbf{S}^{d-1}$ in $\mathbf{R}^d$ deduce the inequality (4.9).

Problem 14 *Norm in the Weak Lebesgue Space*

Let $q > 1$ and μ be the Lebesgue measure on $\mathbf{R}^d$. We denote by $L_w^q(\mathbf{R}^d)$ the space of Lebesgue measurable functions $f : \mathbf{R}^d \to \mathbf{C}$ such that,

$$S(f) := \sup_{0<\lambda<+\infty} \lambda \left(\mu \left\{ x \in \mathbf{R}^d : |f(x)| > \lambda \right\} \right)^{\frac{1}{q}} < +\infty.$$

On the other hand, we introduce,

$$\|f\|_{L_w^q(\mathbf{R}^d)} = \sup_A \left(\mu(A)^{\frac{1}{q}-1} \int_A |f(x)|\, dx \right),$$

where the sup is taken on all measurable sets with positive and finite measure.

Our goal is to prove that the quantity $S(f)$ is equivalent to $\|f\|_{L_w^q(\mathbf{R}^d)}$ and that the latter is a norm.

1. Show that $S(f) \le \|f\|_{L_w^q(\mathbf{R}^d)}$.
2. a) Let A be a measurable set. Prove that,

$$\int_A |f(x)|\, dx = \int_0^{+\infty} \mu\{x \in A : |f(x)| > \lambda\}\, d\lambda,$$

$$\le \int_0^{+\infty} \min \left(\mu \left\{ x \in \mathbf{R}^d : |f(x)| > \lambda \right\}, \mu(A) \right) d\lambda.$$

b) Deduce that there exists $C(q) > 0$ such that,

$$\|f\|_{L_w^q(\mathbf{R}^d)} \le C(q) S(f).$$

c) Show that $\|f\|_{L_w^q(\mathbf{R}^d)}$ is a norm on $L_w^q(\mathbf{R}^d)$.

Problem 15 *Functions of Bounded Mean Oscillations*

We recall that $L_{loc}^1(\mathbf{R}^d)$ denotes the space of functions $f : \mathbf{R}^d \to \mathbf{C}$ which are integrable on each compact subset of $\mathbf{R}^d$.

In what follows we shall denote by B a ball in $\mathbf{R}^d$, $B = \{x \in \mathbf{R}^d : |x - x_0| < r\}$ where $r > 0$, $x_0 \in \mathbf{R}^d$ and by $|B|$ its Lebesgue measure. Moreover, for $f \in L^1_{loc}(\mathbf{R}^d)$ we set,

$$f_B = \frac{1}{|B|} \int_B f(x)dx.$$

We shall denote by BMO the space of functions f such that,

$$\begin{cases} (i) \quad f \in L^1_{loc}(\mathbf{R}^d), \\[2mm] (ii) \quad \exists A > 0 : \dfrac{1}{|B|} \displaystyle\int_B |f(x) - f_B|dx \le A, \quad \text{for every ball } B \subset \mathbf{R}^d. \end{cases}$$

1. Show that $L^\infty(\mathbf{R}^d) \subset BMO$.
2. Let $f \in BMO$ and $\delta > 0$. We denote by f_δ the function $x \mapsto f(\delta x)$. Prove that $f_\delta \in BMO$ for all $\delta > 0$.
3. We consider the space $\widetilde{E}$ of functions $f : \mathbf{R}^d \to \mathbf{C}$ such that,

$$\begin{cases} (iii) \quad f \in L^1_{loc}(\mathbf{R}^d), \\[2mm] (iv) \quad \exists A' > 0 : \forall B \quad \exists c_B > 0 : \dfrac{1}{|B|} \displaystyle\int_B |f(x) - c_B|dx \le A'. \end{cases}$$

We have obviously $BMO \subset \widetilde{E}$ (take $c_B = f_B$).

a) Let $f \in \widetilde{E}$, B be a ball, A' and c_B satisfying (iv). Prove that $|c_B - f_B| \le A'$.
b) Deduce that $\widetilde{E} = BMO$.

4. Our purpose is to prove that if $f \in BMO$ then for all $\varepsilon > 0$ the function g_ε defined by $g_\varepsilon(x) \doteq \dfrac{f(x)}{(1+|x|)^{d+\varepsilon}}$ belongs to $L^1(\mathbf{R}^d)$. For $k \in \mathbf{N}$ we set, $B_k = \{x \in \mathbf{R}^d : |x| \le 2^k\}$.

a) Show that it is enough to prove that,

$$(1) \quad \int_{|x| \ge 1} \frac{|f(x) - f_{B_0}|}{|x|^{d+\varepsilon}} dx < +\infty.$$

b) Let $f \in BMO$. Prove that there exists $C_1 > 0$ such that for all $k \in \mathbf{N}$,

$$(2) \quad \int_{B_k} |f(x) - f_{B_k}|dx \le C_1 2^{nk} A$$

where A is linked to f by (ii).

c) Prove that there exists $C_2 > 0$ such that for all $k \in \mathbf{N}$,

$$(3) \quad |f_{B_{k+1}} - f_{B_k}| \leq C_2 A.$$

Deduce that there exists $C_3 > 0$ such that for all $k \in \mathbf{N}$,

$$(4) \quad |f_{B_k} - f_{B_0}| \leq kC_3 A.$$

d) Prove that there exists $C_4 > 0$ such that for all $k \in \mathbf{N}$,

$$(5) \quad \int_{B_k} |f(x) - f_{B_0}| dx \leq kC_4 A 2^{nk}.$$

e) Writing,

$$\left\{ x \in \mathbf{R}^d : |x| \geq 1 \right\} = \cup_{k=1}^{+\infty} \left\{ x \in \mathbf{R}^d : 2^{k-1} \leq |x| < 2^k \right\}$$

and using question 4d) prove our claim.

5. For $a \in \mathbf{R}^d$ we denote by $1_{B(a,1)}$ the indicator function of the ball $B(a, 1) = \left\{ x \in \mathbf{R}^d : |x - a| \leq 1 \right\}$.

a) Let $(a_k) \subset \mathbf{R}^d$ be a sequence such that $\lim_{k \to +\infty} a_k = a \in \mathbf{R}^d$. Prove that $1_{B(a_k,1)} \to 1_{B(a,1)}$ almost everywhere.

b) Let $f \in L^1_{loc}(\mathbf{R}^d)$. Prove that the map $\Phi : \mathbf{R}^d \to \mathbf{C}$ defined by $\Phi(a) = \int_{B(a,1)} f(x) dx$ is continuous.

c) Let f be the function defined on $\mathbf{R} \setminus \{0\}$ by $f(x) = \ln |x|$. Show that $f \in L^1_{loc}(\mathbf{R})$.

d) Prove that there exists $A > 0$ such that for all $|y_0| \leq 2$ we have,

$$\int_{|y-y_0|<1} |\ln |y|| dy \leq A.$$

e) Prove that there exists $A' > 0$ such that for all $|y_0| > 2$ we have,

$$\int_{|y-y_0|<1} |\ln |y| - \ln |y_0|| dy \leq A'.$$

f) Deduce that $f \in BMO$.
   Hint: Use question 3b).

Problem 16 *A Product Estimate*

Let $m : \mathbf{R} \to \mathbf{R}$ be a nondecreasing C^∞ function such that $\lim_{x \to -\infty} m = 0$ and $\lim_{x \to +\infty} m = 1$. We shall denote by m' the derivative of m.

The goal of this problem is to prove that for all $s \in [0, \frac{1}{2})$ there exists $C > 0$ such that,

$$\|mu\|_{H^s} \leq C \|m'\|_{L^1} \|u\|_{H^s}. \tag{4.11}$$

We shall use the Littlewood–Paley decomposition and denote by C various universal constants depending only on s. The notation $A \lesssim B$ means that, $A \leq CB$ for such a constant C.

1. Prove that m' belongs to $L^1(\mathbf{R})$ and that the L^∞-norm of m is controlled by the L^1-norm of m'. Deduce that,

$$\|mu\|_{L^2} \lesssim \|m'\|_{L^1} \|u\|_{L^2}.$$

This proves the inequality (4.11) when $s = 0$. In the following we assume that $s \in (0, \frac{1}{2})$.

2. Prove that,

$$\|T_m u\|_{\dot{H}^s} \lesssim \|m\|_{L^\infty} \|u\|_{\dot{H}^s}.$$

3. Prove that,

$$\|\Delta_j m\|_{L^2} \lesssim 2^{-\frac{1}{2}j} \|m'\|_{L^1} \quad \text{and} \quad \|\Delta_p u\|_{L^\infty} \lesssim 2^{\frac{1}{2}p} \|\Delta_p u\|_{L^2}.$$

4. Prove that there exists a constant C such that, for all $3 \leq k$,

$$\left\| \sum_{j=k-1}^{\infty} (\Delta_{j-k} u) \Delta_j m \right\|_{\dot{H}^s}^2 \leq C 2^{(2s-1)k} \|m'\|_{L^1}^2 \|u\|_{\dot{H}^s}^2.$$

5. In this section we use the convention that $\Delta_\ell u$ for any $\ell \leq -2$. Verify that $T_u m$ can be written under the form

$$T_u m = \sum_{k=3}^{\infty} \sum_{j=k-1}^{\infty} (\Delta_{j-k} u) \Delta_j m.$$

and conclude that $T_u m$ belongs to $\dot{H}^s(\mathbf{R})$ together with the estimate,

$$\|T_u m\|_{\dot{H}^s} \lesssim \|m'\|_{L^1} \|u\|_{\dot{H}^s}.$$

6. Estimate similarly the reminder $um - T_u m - T_m u$ and conclude the proof of (4.11).
7. The goal of this question is to prove that, in general, there does not exist a constant $C > 0$ such that, for any u in $H^1(\mathbf{R})$,

$$\|mu\|_{H^1} \leq C\|m'\|_{L^1}\|u\|_{H^1}. \qquad (4.12)$$

a) Consider the function $f : [0, +\infty) \to \mathbf{R}$ defined by

$$f(x) = \left(\frac{2}{\pi}\arctan(x)\right)^{\frac{1}{2}}.$$

Prove that the function f is bounded on $[0, +\infty)$, is C^∞ on $(0, +\infty)$, that f' is nonincreasing, and that $f' \in L^1((0, +\infty))$.

b) Let $\rho : \mathbf{R} \to (0, +\infty)$ be a C^∞ function supported in $[0, 1]$ such that $\int \rho(z)\, dz = 1$. Set for $\varepsilon \in (0, 1]$, $\rho_\varepsilon(x) = \frac{1}{\varepsilon}\rho(\frac{x}{\varepsilon})$ and for $x \in \mathbf{R}$,

$$m_\varepsilon(x) = \rho_\varepsilon \star \tilde{f},$$

where $\tilde{f}(x) = 0$ for $x < 0$ and $\tilde{f}(x) = f(x)$ for $x \geq 0$.
Prove that m_ε is a C^∞ function on $\mathbf{R}$ and that,

$$m_\varepsilon(x) = \int_{-\infty}^{\frac{x}{\varepsilon}} \rho(z) f(x - \varepsilon z)\, dz.$$

Deduce that, for any $\varepsilon > 0$, m_ε is a nondecreasing C^∞ function such that $\lim_{x \to -\infty} m_\varepsilon = 0$ and $\lim_{x \to +\infty} m_\varepsilon = 1$.
Prove that,

$$\sup_{\varepsilon \in (0,1]} \|m_\varepsilon\|_{L^\infty(\mathbf{R})} < +\infty.$$

c) Compute the function m'_ε and show that it belongs to $L^p(\mathbf{R})$ for all $\varepsilon \in (0, 1]$ and all $p \in [1, +\infty)$. Prove that,

$$\sup_{\varepsilon \in (0,1]} \|m'_\varepsilon\|_{L^1(\mathbf{R})} < +\infty.$$

d) Prove that,

$$\sup_{\varepsilon\in(0,1]} \int_0^1 m_\varepsilon'(x)^2\,dx = +\infty.$$

Hint: Show that for all x in $[\varepsilon, 1]$, one has $m_\varepsilon'(x) \geq \frac{C}{\sqrt{x}}$ for some $C > 0$.
e) Use this family of functions to contradict (4.12).

Problem 17 *Continuity of Paraproducts*

The purpose of this problem is to investigate the continuity of a paraproduct T_a when a belongs to a Sobolev space $H^\beta(\mathbf{R}^d)$. More precisely our goal is to prove that if α, β, γ are real numbers such that,

$$(i)\quad \alpha \leq \gamma, \quad (ii)\quad \alpha \leq \beta+\gamma-\frac{d}{2}, \quad (iii)\quad \alpha < \gamma \text{ if } \beta=\frac{d}{2}, \qquad (4.13)$$

then there exists a constant $C > 0$ depending only on α, β, γ, d such that,

$$\|T_a u\|_{H^\alpha(\mathbf{R}^d)} \leq C\|a\|_{H^\beta(\mathbf{R}^d)}\|u\|_{H^\gamma(\mathbf{R}^d)}. \qquad (4.14)$$

We recall that $T_a u = \sum_{j\geq-1} S_{j-2}(a)\Delta_j u$ and we set $v_j = S_{j-2}(a)\Delta_j u$.

1. Prove (4.14) when $\beta > \frac{d}{2}$.
2. We assume $\beta = \frac{d}{2}$, so $\alpha < \gamma$.

   a) Prove that

   $$\|S_{j-2}(a)\|_{H^{\frac{d}{2}+\gamma-\alpha}(\mathbf{R}^d)} \leq C2^{j(\gamma-\alpha)}\|a\|_{H^{\frac{d}{2}}(\mathbf{R}^d)}.$$

   Hint: Use (3.13).
   b) Using the fact that $\frac{d}{2}+\gamma-\alpha > \frac{d}{2}$ prove (4.14).
3. We assume $\beta < \frac{d}{2}$.

   a) Prove that

   $$\|v_j\|_{L^2(\mathbf{R}^d)} \leq \sum_{k=-1}^{j-2} \|\Delta_k a\|_{L^\infty(\mathbf{R}^d)}\|\Delta_j u\|_{L^2(\mathbf{R}^d)}.$$

b) Prove that

$$\|\Delta_k a\|_{L^\infty(\mathbf{R}^d)} \leq C \|2^{kd}\chi(2^k\cdot)\|_{L^2_x(\mathbf{R}^d)} \|\Delta_k a\|_{L^2(\mathbf{R}^d)},$$

where $\chi \in S(\mathbf{R}^d)$.

c) Deduce that,

$$\|v_j\|_{L^2(\mathbf{R}^d)} \leq C \left(\sum_{k=-1}^{j-2} 2^{-2k\beta} \|2^{kd}\chi(2^k\cdot)\|^2_{L^2(\mathbf{R}^d)} \right)^{\frac{1}{2}}$$

$$\times \left(\sum_{k=-1}^{j-2} 2^{2k\beta} \|\Delta_k a\|^2_{L^2(\mathbf{R}^d)} \right)^{\frac{1}{2}} \|\Delta_j u\|_{L^2(\mathbf{R}^d)}.$$

d) Prove that,

$$\|v_j\|_{L^2(\mathbf{R}^d)} \leq 2^{j(\alpha-\gamma+\frac{d}{2}-\beta)} 2^{j\gamma} \|\Delta_k u\|_{L^2(\mathbf{R}^d)} \|a\|_{H^\beta(\mathbf{R}^d)},$$

and deduce (4.14).

e) Show that (4.14) can be summarized as follows: if $a \in H^\beta(\mathbf{R}^d)$ then T_a is of order

$$\text{zero if}\quad \beta > \frac{d}{2}, \qquad \frac{d}{2}-\beta \quad \text{if } \beta < \frac{d}{2}, \qquad \varepsilon > 0 \quad \text{if}\quad \beta = \frac{d}{2}.$$

Problem 18 *Products in Sobolev Spaces*

This problem uses the result (4.14) in Problem 17.

Let $d \in \mathbf{N}, d \geq 1$ and $s_j \in \mathbf{R}, j = 1, 2$ be such that $s_1 + s_2 > 0$. Let s_0 be such that,

$$(i)\quad s_0 \leq s_j, \quad j = 1, 2, \quad (ii)\quad s_0 \leq s_1 + s_2 - \frac{d}{2},$$

where the inequality in (ii) is strict if $s_1 = \frac{d}{2}$ or $s_2 = \frac{d}{2}$ or $s_0 = -\frac{d}{2}$.

The purpose of this problem is to prove that if $u_j \in H^{s_j}(\mathbf{R}^d), j = 1, 2$ then $u_1 u_2 \in H^{s_0}(\mathbf{R}^d)$ and there exists $C > 0$ depending only on s_1, s_2, d such that,

$$\|u_1 u_2\|_{H^{s_0}(\mathbf{R}^d)} \leq C \|u_1\|_{H^{s_1}(\mathbf{R}^d)} \|u_2\|_{H^{s_2}(\mathbf{R}^d)}. \qquad (4.15)$$

We write,

$$u_1 u_2 = T_{u_1} u_2 + T_{u_2} u_1 + R,$$

where T_a denotes the paraproduct by a and $R = \sum_{|j-k|\leq 2} \Delta_j u_1 \Delta_k u_2$.

Part 1

1. Using (4.14) prove that there exists $C > 0$ such that,

$$\|T_{u_1} u_2\|_{H^{s_0}(\mathbf{R}^d)} + \|T_{u_2} u_1\|_{H^{s_0}(\mathbf{R}^d)} \leq C\|u_1\|_{H^{s_1}(\mathbf{R}^d)}\|u_2\|_{H^{s_2}(\mathbf{R}^d)}. \qquad (4.16)$$

Part 2

The goal of this part is to prove that there exists $C > 0$ such that,

$$\|R\|_{H^{s_1+s_2-\frac{d}{2}}(\mathbf{R}^d)} \leq C\|u_1\|_{H^{s_1}(\mathbf{R}^d)}\|u_2\|_{H^{s_2}(\mathbf{R}^d)}. \qquad (4.17)$$

We have

$$R = \sum_j R_j, \quad R_j = \left(\sum_{k=j-2}^{j+2} \Delta_k u_1\right)\Delta_j u_2.$$

Notice that the spectrum of R_j is contained in a ball $B(0, C2^j)$ (not in a ring). So we write, $R = \sum_{p\geq-1} \Delta_p R$.

1. a) Show that there exists $N_0 \in \mathbf{N}$ such that,

$$\Delta_p R = \sum_{j\geq p-N_0} \Delta_p R_j.$$

b) Prove that $\Delta_p R_j = 2^{pd}(\mathcal{F}^{-1}\varphi)(2^p\cdot) \star R_j$ and that there exists $C_1 > 0$ such that,

$$\|\Delta_p R_j\|_{L^2(\mathbf{R}^d)} \leq C_1 2^{\frac{pd}{2}}\|R_j\|_{L^1(\mathbf{R}^d)}.$$

c) Deduce that there exists $C_2 > 0$ such that,

$$\|\Delta_p R_j\|_{L^2(\mathbf{R}^d)} \leq C_2 2^{\frac{pd}{2}} 2^{-j(s_1+s_2)}\alpha_j \quad \text{with}$$

$$\alpha_j = \sum_{k=j-2}^{j+2} 2^{js_1}\|\Delta_k u_1\|_{L^2(\mathbf{R}^d)} 2^{js_2}\|\Delta_j u_2\|_{L^2(\mathbf{R}^d)},$$

and prove that there exists $C_3 > 0$, $(d_j) \in \ell^2$ such that,

$$\alpha_j \leq C_3 \, d_j \, \|u_1\|_{H^{s_1}(\mathbf{R}^d)} \|u_2\|_{H^{s_2}(\mathbf{R}^d)}.$$

2. a) We set $\beta_p = 2^{p(s_1+s_2-\frac{d}{2})} \|\Delta_p R\|_{L^2(\mathbf{R}^d)}$. Prove that,

$$\beta_p \leq C_2 \sum_{j \geq p - N_0} 2^{(p-j)(s_1+s_2)} \alpha_j.$$

b) Deduce that there exists $C_4 > 0$ such that,

$$\sum_{p \geq -1} \beta_p^2 \leq C_4 \sum_{j \geq -1} \alpha_j^2.$$

c) Deduce (4.17).
d) Conclude.

Chapter 5
Microlocal Analysis

In this chapter we recall, without proof, the definitions and the basic results in microlocal analysis.

5.1 Symbol Classes

5.1.1 Definition and First Properties

- For $m \in \mathbf{R}$, $S^m(\mathbf{R}^d)$ is the set of all complex-valued functions $a \in C^\infty(\mathbf{R}^d \times \mathbf{R}^d)$ such that,

$$\forall \alpha, \beta \in \mathbf{N}^d, \exists C_{\alpha\beta} > 0 : \left| \partial_x^\alpha \partial_\xi^\beta a(x, \xi) \right| \leq C_{\alpha\beta}(1+|\xi|)^{m-|\beta|}, \forall (x, \xi) \in \mathbf{R}^d \times \mathbf{R}^d.$$

- This set is a vector space endowed with the semi-norms,

$$p_{\alpha,\beta}(a) = \sup_{(x,\xi) \in \mathbf{R}^d \times \mathbf{R}^d} (1 + |\xi|)^{-m+|\beta|} \left| \partial_x^\alpha \partial_\xi^\beta a(x, \xi) \right|, \quad \alpha, \beta \in \mathbf{N}^d. \quad (5.1)$$

We will denote this space by S^m if there is no confusion. Then S^m is a Fréchet space. Its elements are called "symbols."
- If $m \geq m'$ we have $S^{m'} \subset S^m$. We set $S^{+\infty} = \cup_{m \in \mathbf{R}} S^m$ and $S^{-\infty} = \cap_{m \in \mathbf{R}} S^m$.
- If $a \in S^m$ then $\partial_{x_j} a \in S^m$ and $\partial_{\xi_j} a \in S^{m-1}$ for all $j = 1, \ldots, d$.
- If $a \in S^m$ and $b \in S^{m'}$ then $ab \in S^{m+m'}$.
- If $F : \mathbf{C}^k \to \mathbf{C}$ is a C^∞ function and if $a_1, \ldots, a_k \in S^0$ then $F(a_1, \ldots, a_k) \in S^0$.

© The Editor(s) (if applicable) and The Author(s), under exclusive license
to Springer Nature Switzerland AG 2020
T. Alazard, C. Zuily, *Tools and Problems in Partial Differential Equations*,
Universitext, https://doi.org/10.1007/978-3-030-50284-3_5

5.1.2 Examples

- If $a = a(\xi)$ belongs to the Schwartz class $\mathcal{S}(\mathbf{R}^d)$, then $a \in S^{-\infty}(\mathbf{R}^d)$.
- Let $m \in \mathbf{N}$ and for $\alpha \in \mathbf{N}^d$, $|\alpha| \leq m$ let c_α be $C^\infty(\mathbf{R}^d)$ functions all of whose derivatives are bounded. Then $a(x, \xi) = \sum_{|\alpha| \leq m} c_\alpha(x)\xi^\alpha \in S^m$.
- For $m \in \mathbf{R}$, $a(\xi) = (1 + |\xi|^2)^{\frac{m}{2}} \in S^m$.

5.1.3 Classical Symbols

- A symbol $a \in S^m$ is called "classical" if there exists a sequence $(a_j)_{j \in \mathbf{N}}$, where $a_j \in S^{m-j}$ is a homogeneous function in ξ of order $m - j$ for $|\xi| \geq 1$ (that is, $a_j(x, \lambda\xi) = \lambda^{m-j}a_j(x, \xi), |\xi| \geq 1, \lambda \geq 1$) such that for all $N \in \mathbf{N}$ we have,

$$a - \sum_{j=0}^{N-1} a_j \in S^{m-N}.$$

We write $a \sim \sum_j a_j$.
- Let $a_j \in S^{m_j}$ where $(m_j)_{j \in \mathbf{N}}$ is a strictly decreasing sequence with $\lim_{j \to +\infty} m_j = -\infty$. Then there exists $a \in S^{m_0}$ such that $a \sim \sum_j a_j$.

5.2 Pseudo-Differential Operators

5.2.1 Definition and First Properties

- Let $a \in S^m$ and A be the operator defined on $\mathcal{S}(\mathbf{R}^d)$ by the formula,

$$Au(x) = (2\pi)^{-d} \int e^{ix\cdot\xi} a(x, \xi)\widehat{u}(\xi)\,d\xi.$$

A is called a pseudo-differential operator (ΨDO) with symbol a. We denote sometimes $A = \mathrm{Op}(a)$.
- Such an operator is continuous from $\mathcal{S}(\mathbf{R}^d)$ to $\mathcal{S}(\mathbf{R}^d)$. It can be extended as a continuous operator from $\mathcal{S}'(\mathbf{R}^d)$ to $\mathcal{S}'(\mathbf{R}^d)$.
- For $u \in \mathcal{S}'(\mathbf{R}^d)$ we have,

$$\mathrm{sing\,supp}(Au) \subset \mathrm{sing\,supp}(u), \tag{5.2}$$

where sing supp, the singular support, is defined as follows:

$$x \notin \text{sing supp(u)} \iff u \text{ is a } C^\infty \text{ function in a neighborhood of } x.$$

We say then that the ΨDO are *pseudo-local*.

Notice that the analogue of the inclusion (5.2) for the support is false unless A is a differential operator (see Problem 23).

5.2.2 Kernel of a ΨDO

- Consider a symbol $a \in S^m$ and set $A = \text{Op}(a)$. There exists a distribution $K = K(x, y) \in \mathcal{D}'(\mathbf{R}^d \times \mathbf{R}^d)$ such that, for all $u, v \in C_0^\infty(\mathbf{R}^d)$ we have,

$$\langle Au, \varphi \rangle = \langle K(x, y), u(y) \otimes v(x) \rangle.$$

- If $m < -d$ we have,

$$K(x, y) = (2\pi)^{-d} \int e^{i(x-y)\cdot\xi} a(x, \xi) \, d\xi,$$

where the integral converges absolutely.
- The kernel K is C^∞ outside the diagonal $D = \{(x, y) \in \mathbf{R}^d \times \mathbf{R}^d : x = y\}$.

5.2.3 Image of a ΨDO by a Diffeomorphism

- Let χ be a C^∞ diffeomorphism from $\mathbf{R}_x^d$ to $\mathbf{R}_y^d$ all of whose derivatives are bounded. Let $A = \text{Op}(a)$ where $a \in S^m(\mathbf{R}_x^d)$. Let B be defined for $u \in S(\mathbf{R}_y^d)$ by the formula,

$$Bu = A(u \circ \chi) \circ \chi^{-1}.$$

Then $B = \text{Op}(b)$ where $b \in S^m(\mathbf{R}_y^d)$ and,

$$b(\chi(x), \eta) \sim \sum_{\alpha \in \mathbf{N}^d} \frac{1}{\alpha!} \phi_\alpha(x, \eta)(\partial_\xi^\alpha a)(x, {}^t\chi'(x)\eta),$$

where χ' is the differential of χ, ${}^t\chi'$ is its transpose, and

$$\phi_\alpha(x, \eta) = D_z^\alpha \left[\exp\left(i(\langle \chi(z) - \chi(x) - \chi'(x)(z - x), \eta \rangle) \right) \right]\big|_{z=x}.$$

- In particular if A is a classical ΨDO with principal symbol a_m then B is classical with principal symbol,

$$b_m(\chi(x), \eta) = a_m\left(x, {}^t\chi'(x)\eta\right).$$

5.2.4   Symbolic Calculus

5.2.4.1   Composition

- If $a_1 \in S^{m_1}, a_2 \in S^{m_2}$ then $\text{Op}(a_1) \circ \text{Op}(a_2) = \text{Op}(b)$ where $b := a_1 \# a_2 \in S^{m_1+m_2}$ is such that,

$$\left(b - \sum_{|\alpha| \leq N-1} \frac{1}{i^{|\alpha|}\alpha!} \partial_\xi^\alpha a_1 \partial_x^\alpha a_2\right) \in S^{m_1+m_2-N}, \quad \forall N \geq 1.$$

- In particular if $a_1 \in S^{m_1}, a_2 \in S^{m_2}$ the commutator,

$$[\text{Op}(a_1), \text{Op}(a_2)] = \text{Op}(a_1) \circ \text{Op}(a_2) - \text{Op}(a_2) \circ \text{Op}(a_1)$$

is a ΨDO of order $m_1 + m_2 - 1$ and its symbol b satisfies,

$$b - \frac{1}{i}\{a_1, a_2\} \in S^{m_1+m_2-2}$$

where $\{\cdot, \cdot\}$ denotes the *Poisson bracket*, defined by,

$$\{f, g\} = \sum_{j=1}^{d}\left(\frac{\partial f}{\partial \xi_j}\frac{\partial g}{\partial x_j} - \frac{\partial f}{\partial x_j}\frac{\partial g}{\partial \xi_j}\right).$$

5.2.4.2   Adjoint

Let A be a ΨDO. Its adjoint in the $L^2(\mathbf{R}^d)$ sense is an operator denoted by A^* and defined by,

$$(Au, v)_{L^2(\mathbf{R}^d)} = \left(u, A^*v\right)_{L^2(\mathbf{R}^d)}, \quad \forall u \in \mathcal{S}'(\mathbf{R}^d), \forall v \in \mathcal{S}(\mathbf{R}^d). \tag{5.3}$$

- Let $a \in S^m$. There exists $a^* \in S^m$ such that if set we $A^* = \mathrm{Op}(a^*)$ then A^* satisfies (5.3). Moreover,

$$\left(a^* - \sum_{|\alpha| \leq N-1} \frac{1}{i^{|\alpha|}\alpha!} \partial_\xi^\alpha \partial_x^\alpha \overline{a} \right) \in S^{m-N}, \quad \forall N \geq 1.$$

5.2.5   Action of the ΨDO on Sobolev Spaces

- Let $m \in \mathbf{R}$ and $s \in \mathbf{R}$. If $a \in S^m$ then $\mathrm{Op}(a)$ is continuous from $H^s(\mathbf{R}^d)$ to $H^{s-m}(\mathbf{R}^d)$. Moreover, its operator norm can be bounded by a finite number of semi-norms of the symbol defined by (5.1).

5.2.6   Garding Inequalities

The Garding inequalities explore the link between the positivity of a symbol and the positivity of the corresponding operator.

5.2.6.1   The Weak Inequality

- Let $a \in S^{2m}$ satisfying,

$$\exists c_0 > 0, \exists R > 0 : \quad \mathrm{Re}\, a(x, \xi) \geq c_0 |\xi|^{2m}, \quad \forall (x, \xi) \in \mathbf{R}^d \times \mathbf{R}^d \text{ with } |\xi| \geq R.$$

Then there exists $C > 0$ such that for all $N \in \mathbf{N}$ there exists $C_N > 0$ such that,

$$\mathrm{Re}\ (\mathrm{Op}(a)u, u)_{L^2(\mathbf{R}^d)} + C_N \|u\|^2_{H^{-N}(\mathbf{R}^d)} \geq C\|u\|^2_{H^m(\mathbf{R}^d)}.$$

5.2.6.2   The "Sharp Garding" Inequality

- Let $a \in S^{2m}$ satisfying:

$$\exists R > 0 : \quad \mathrm{Re}\, a(x, \xi) \geq 0, \quad \forall (x, \xi) \in \mathbf{R}^d \times \mathbf{R}^d, |\xi| \geq R.$$

Then

$$\exists C > 0 : \quad \mathrm{Re}\ (\mathrm{Op}(a)u, u)_{L^2(\mathbf{R}^d)} + C\|u\|^2_{H^{m-\frac{1}{2}}(\mathbf{R}^d)} \geq 0.$$

5.2.6.3 The Fefferman–Phong Inequality

- Let $a \in S^{2m}$ be a real valued symbol satisfying:

$$a(x, \xi) \geq 0, \quad \forall (x, \xi) \in \mathbf{R}^d \times \mathbf{R}^d.$$

Then

$$\exists C > 0: \quad \text{Re} \ (\text{Op}(a)u, u)_{L^2(\mathbf{R}^d)} + C\|u\|^2_{H^{m-1}(\mathbf{R}^d)} \geq 0.$$

5.3   Invertibility of Elliptic Symbols

Let $a \in S^m$. We say that a is an elliptic symbol if

$$\exists c_0 > 0, \exists R > 0: \quad \text{Re} \ a(x, \xi) \geq c_0 |\xi|^{2m}, \quad \forall (x, \xi) \in \mathbf{R}^d \times \mathbf{R}^d, \quad |\xi| \geq R.$$

- If a is an elliptic symbol there exists $b \in S^{-m}$ such that,

$$\text{Op}(a) \circ \text{Op}(b) - \text{Id} \in \text{Op}(S^{-\infty}), \quad \text{Op}(b) \circ \text{Op}(a) - \text{Id} \in \text{Op}(S^{-\infty}).$$

5.4   Wave Front Set of a Distribution

5.4.1   Definition and First Properties

- If $\xi_0 \in \mathbf{R}^d \setminus \{0\}$, a conical neighborhood of ξ_0 is a set of the form,

$$\Gamma_{\xi_0} = \left\{ \xi \in \mathbf{R}^d \setminus \{0\} : \left| \frac{\xi}{|\xi|} - \frac{\xi_0}{|\xi_0|} \right| < \varepsilon \right\}, \quad \varepsilon > 0.$$

- Let $u \in \mathcal{D}'(\mathbf{R}^d)$ and $(x_0, \xi_0) \in \mathbf{R}^d \times (\mathbf{R}^d \setminus \{0\})$. We say that $(x_0, \xi_0) \notin \text{WF}(u)$ if there exists a neighborhood V_{x_0} of x_0, a conical neighborhood of Γ_{ξ_0} of ξ_0, a function $\varphi \in C_0^\infty(V_{x_0})$ equal to 1 in a neighborhood of x_0 such that,

$$\exists R \geq 1 : \forall N \in \mathbf{N}, \exists C_N > 0: \quad |\widehat{\varphi u}(\xi)| \leq C_N |\xi|^{-N}, \quad \forall \xi \in \Gamma_{\xi_0}, \quad |\xi| \geq R.$$

- WF(u) is called the "wave front set" of u. It is a closed subset of $\mathbf{R}^d \times (\mathbf{R}^d \setminus \{0\})$.
- WF(u) is a conical set, that is $(x_0, \xi_0) \in \text{WF}(u) \implies (x_0, t\xi_0) \in \text{WF}(u), \forall t > 0$.

- Let $\pi_1 : \mathbf{R}^d \times \mathbf{R}^d \setminus \{0\} \to \mathbf{R}^d$ be the map $(x, \xi) \mapsto x$. Then,

$$\pi_1(\mathrm{WF}(u)) = \mathrm{sing\,supp}(u).$$

5.4.2 Wave Front Set and ΨDO

- Let $u \in \mathcal{D}'(\mathbf{R}^d)$ and $(x_0, \xi_0) \in \mathbf{R}^d \times (\mathbf{R}^d \setminus \{0\})$. If $(x_0, \xi_0) \notin \mathrm{WF}(u)$ then,

$\exists \varphi \in C_0^\infty(\mathbf{R}^d), \varphi = 1$ near x_0, $\exists a \in S^0 : a(x_0, \xi_0) \neq 0$ such that ,

$$\mathrm{Op}(a)(\varphi u) \in C^\infty(\mathbf{R}^d).$$

- Let $u \in \mathcal{D}'(\mathbf{R}^d)$ and $(x_0, \xi_0) \in \mathbf{R}^d \times (\mathbf{R}^d \setminus \{0\})$. Assume that,

$\exists \varphi \in C_0^\infty(\mathbf{R}^d), \varphi = 1$ near x_0, $\exists a \in S^0 : a(x_0, \xi_0) \neq 0$ such that

$$\mathrm{Op}(a)(\varphi u) \in C^\infty(\mathbf{R}^d).$$

Then $(x_0, \xi_0) \notin \mathrm{WF}(u)$.

- Let $u \in \mathcal{S}'(\mathbf{R}^d)$ and $a \in S^m$ be a classical symbol , $a \sim \sum_{j \geq 0} a_{m-j}$. The set

$$\Sigma = \left\{ (x, \xi) \in \mathbf{R}^d \times (\mathbf{R}^d \setminus \{0\}) : a_m(x, \xi) = 0 \right\}$$

is called the *characteristic set* of a. We have then,

$$\mathrm{WF}(u) \subset \mathrm{WF}(\mathrm{Op}(a)u) \cup \Sigma.$$

5.4.3 The Propagation of Singularities Theorem

5.4.3.1 Bicharacteristics

- Let $a \in S^m$ be a real symbol. Let $\rho_0 = (x_0, \xi_0) \in \mathbf{R}^d \times (\mathbf{R}^d \setminus \{0\})$. The bicharacteristic of a issued from the point ρ_0 is the parametric curve $(x(t), \xi(t))$ such that for $1 \leq j \leq d$,

$$\dot{x}_j(t) = \frac{\partial a}{\partial \xi_j}(x(t), \xi(t)), \quad x_j(0) = x_{0,j},$$

$$\dot{\xi}_j(t) = -\frac{\partial a}{\partial x_j}(x(t), \xi(t)) \quad \xi_j(0) = \xi_{0,j}.$$

The Cauchy–Lipschitz theorem implies the existence of a maximal solution of this system.

• If we define the *Hamiltonian* of a as the vector field on $\mathbf{R}^d \times (\mathbf{R}^d \setminus \{0\})$,

$$H_a = \sum_{j=1}^{d} \left(\frac{\partial a}{\partial \xi_j} \frac{\partial}{\partial x_j} - \frac{\partial a}{\partial x_i} \frac{\partial}{\partial \xi_j} \right),$$

we see that the bicharacteristics are the integral curves of this field, that is the curves along which the field is tangent at every point.

• The symbol a is constant along the bicharacteristic curves. Those where $a = 0$ are called the *null bicharacteristics*.

5.4.3.2 The Propagation Theorem

• Let $a \in S^m$ be a real symbol. Let $u \in S'(\mathbf{R}^d)$. Let $\rho_0 = (x_0, \xi_0) \in \mathbf{R}^d \times (\mathbf{R}^d \setminus \{0\})$ and let γ_{ρ_0} be the maximal null bicharacteristic issued from this point. Assume that $\gamma_{\rho_0} \cap \mathrm{WF}(Pu) = \emptyset$. Then,

$$\rho_0 \in \mathrm{WF}(u) \iff \gamma_{\rho_0} \subset \mathrm{WF}(u),$$

$$\rho_0 \notin \mathrm{WF}(u) \iff \gamma_{\rho_0} \cap \mathrm{WF}(u) = \emptyset.$$

5.5 Paradifferential Calculus

Paradifferential operators are the generalization of pseudo-differential operators to the case where the symbol has a limited regularity with respect to x.

Recall (see Sect. 3.2) that for $k \in \mathbf{N}$, $W^{k,\infty}(\mathbf{R}^d)$ denotes the space of functions whose derivatives up to the order k belong to $L^\infty(\mathbf{R}^d)$ and for $\rho = k + \sigma, \sigma \in (0, 1)$, $W^{\rho,\infty}(\mathbf{R}^d)$ is the space of functions whose derivatives up to the order k belong to $L^\infty(\mathbf{R}^d)$ and are uniformly Hölder continuous with exponent σ.

5.5.1 Symbols Classes

• Given $\rho \geq 0$ and $m \in \mathbf{R}$, $\Gamma_\rho^m(\mathbf{R}^d)$ denotes the space of locally bounded functions $a(x, \xi)$ on $\mathbf{R}^d \times (\mathbf{R}^d \setminus \{0\})$, which are C^∞ with respect to ξ for $\xi \neq 0$ and such that, for all $\alpha \in \mathbf{N}^d$ and all $\xi \neq 0$, the function $x \mapsto \partial_\xi^\alpha a(x, \xi)$ belongs to $W^{\rho,\infty}(\mathbf{R}^d)$ and there exists a constant C_α such that for every $|\xi| \geq \frac{1}{2}$,

$$\left\| \partial_\xi^\alpha a(\cdot, \xi) \right\|_{W^{\rho,\infty}(\mathbf{R}^d)} \leq C_\alpha (1 + |\xi|)^{m-|\alpha|}.$$

The elements of $\Gamma_\rho^m(\mathbf{R}^d)$ are called symbols.

- For $a \in \Gamma_\rho^m(\mathbf{R}^d)$ we introduce the following semi-norms:

$$M_\rho^m(a) = \sup_{|\alpha| \leq 2(d+2)+\rho} \sup_{|\xi| \geq 1/2} \left\| (1 + |\xi|)^{|\alpha|-m} \partial_\xi^\alpha a(\cdot, \xi) \right\|_{W^{\rho,\infty}(\mathbf{R}^d)}. \qquad (5.4)$$

5.5.2 Paradifferential Operators

- Given a symbol a we define the paradifferential operator T_a by,

$$\widehat{T_a u}(\xi) = (2\pi)^{-d} \int \chi(\xi - \eta, \eta) \widehat{a}(\xi - \eta, \eta) \psi(\eta) \widehat{u}(\eta) \, d\eta, \qquad (5.5)$$

where $\widehat{a}(\theta, \xi) = \int e^{-ix\cdot\theta} a(x, \xi) \, dx$ is the Fourier transform of a with respect to the first variable and χ, ψ are two fixed C^∞ functions such that,

$$\psi(\eta) = 0 \quad \text{for } |\eta| \leq \frac{1}{5}, \qquad \psi(\eta) = 1 \quad \text{for } |\eta| \geq \frac{1}{4}, \qquad (5.6)$$

and $\chi(\theta, \eta)$ satisfies, for $0 < \varepsilon_1 < \varepsilon_2$ small enough,

$$\chi(\theta, \eta) = 1 \quad \text{if} \quad |\theta| \leq \varepsilon_1 |\eta|, \qquad \chi(\theta, \eta) = 0 \quad \text{if} \quad |\theta| \geq \varepsilon_2 |\eta|,$$

and,

$$\forall(\theta, \eta): \qquad \left| \partial_\theta^\alpha \partial_\eta^\beta \chi(\theta, \eta) \right| \leq C_{\alpha,\beta} (1 + |\eta|)^{-|\alpha|-|\beta|}.$$

Notice that in the above integral we have taken the frequencies of a (with respect to its first variable) smaller than those of u since on the support of χ we have $|\xi - \eta| \leq \varepsilon_2 |\eta|$.

5.5.3 The Symbolic Calculus

Here are the main features of the symbolic calculus for paradifferential operators.

Let $m \in \mathbf{R}$. An operator T is said to be of order m if, for all $\mu \in \mathbf{R}$, it is bounded from $H^\mu(\mathbf{R}^d)$ to $H^{\mu-m}(\mathbf{R}^d)$.

- If $a \in \Gamma_0^m(\mathbf{R}^d)$, then T_a is of order m. Moreover, for all $\mu \in \mathbf{R}$ there exists a constant K such that,

$$\|T_a\|_{H^\mu \to H^{\mu-m}} \leq K M_0^m(a). \tag{5.7}$$

- Let $m \in \mathbf{R}$ and $\rho > 0$. If $a \in \Gamma_\rho^m(\mathbf{R}^d), b \in \Gamma_\rho^{m'}(\mathbf{R}^d)$ then $T_a T_b - T_{a \# b}$ is of order $m + m' - \rho$ where,

$$a \# b = \sum_{|\alpha| < \rho} \frac{1}{i^{|\alpha|}\alpha!} \partial_\xi^\alpha a \, \partial_x^\alpha b.$$

Moreover, for all $\mu \in \mathbf{R}$ there exists a constant K such that,

$$\|T_a T_b - T_{a \# b}\|_{H^\mu \to H^{\mu-m-m'+\rho}} \leq K \left(M_\rho^m(a) M_0^{m'}(b) + M_0^m(a) M_\rho^{m'}(b) \right). \tag{5.8}$$

- Let $m \in \mathbf{R}$ and $\rho > 0$. If $a \in \Gamma_\rho^m(\mathbf{R}^d)$ we denote by $(T_a)^*$ the adjoint operator of T_a. Then $(T_a)^* - T_{a^*}$ is of order $m - \rho$ where,

$$a^* = \sum_{|\alpha| < \rho} \frac{1}{i^{|\alpha|}\alpha!} \partial_\xi^\alpha \partial_x^\alpha \overline{a}$$

and $\overline{a}$ is the complex conjugate of a.

Moreover, for all μ there exists a constant K such that,

$$\left\| (T_a)^* - T_{a^*} \right\|_{H^\mu \to H^{\mu-m+\rho}} \leq K M_\rho^m(a). \tag{5.9}$$

We can also consider symbols with negative regularity.

- For $m \in \mathbf{R}$ and $\rho < 0$, $\Gamma_\rho^m(\mathbf{R}^d)$ denotes the space of distributions $a(x, \xi)$ on $\mathbf{R}^d \times (\mathbf{R}^d \setminus 0)$, which are C^∞ with respect to ξ and such that, for all $\alpha \in \mathbf{N}^d$ and all $\xi \neq 0$, the function $x \mapsto \partial_\xi^\alpha a(x, \xi)$ belongs to $C_*^\rho(\mathbf{R}^d)$ and there exists a constant C_α such that,

$$\forall |\xi| \geq \frac{1}{2}, \quad \left\| \partial_\xi^\alpha a(\cdot, \xi) \right\|_{C_*^\rho} \leq C_\alpha (1 + |\xi|)^{m-|\alpha|}. \tag{5.10}$$

For $a \in \Gamma_\rho^m$, we define

$$M_\rho^m(a) = \sup_{|\alpha| \le 2(d+2)+|\rho|} \sup_{|\xi| \ge 1/2} \left\| (1 + |\xi|)^{|\alpha|-m} \partial_\xi^\alpha a(\cdot, \xi) \right\|_{C_*^\rho(\mathbf{R}^d)}. \tag{5.11}$$

Then we have the following continuity result.

• Let $\rho < 0$, $m \in \mathbf{R}$, and $a \in \Gamma_\rho^m$. Then the operator T_a is of order $m - \rho$ and we have,

$$\|T_a\|_{H^s \to H^{s-(m-\rho)}} \le C M_\rho^m(a), \qquad \|T_a\|_{C_*^s \to C_*^{s-(m-\rho)}} \le C M_\rho^m(a). \tag{5.12}$$

5.5.4   Link with the Paraproducts

When $a = a(x)$ is an L^∞ function we have defined in Chap. 3, Sect. 3.4, the paraproduct T_a by the formula (3.15). We have also defined T_a by the formula (5.5), so the question of the link between the two formulas arises. The answer is the following. If $a \in W^{\rho,\infty}(\mathbf{R}^d)$ where $\rho \in (0, 1]$ then the difference of the two operators is of order $-\rho$, which means that it sends $H^\mu(\mathbf{R}^d)$ to $H^{\mu+\rho}(\mathbf{R}^d)$ for every $\mu \in \mathbf{R}$. So the difference is ρ regularizing.

5.6   Microlocal Defect Measures

Notations We denote by $L_c^2(\mathbf{R}^d)$ the set of $L^2(\mathbf{R}^d)$ functions with compact support. We denote by $S_c^0(\mathbf{R}^d)$ the set of symbols $a = a(x, \xi) \in S^0$ positively homogeneous of degree zero in ξ for $|\xi| \ge 1$ and compactly supported in x, that is,

    i)   there is a compact $K \subset \mathbf{R}^d$ such that $a(x, \xi) = 0$ if $x \notin K$,

    ii)   for all $\lambda \ge 1$, $a(x, \lambda\xi) = a(x, \xi)$ for all $x \in \mathbf{R}^d$ and $|\xi| \ge 1$.

Let $(u_k)_{k \in \mathbf{N}}$ be a bounded sequence in $L_{\text{loc}}^2(\mathbf{R}^d)$, that is

$$\sup_{k \in \mathbf{N}} \int_K |u_k(x)|^2 \, dx < +\infty, \qquad \text{for all compact subset } K \subset \mathbf{R}^d.$$

We assume that $(u_k)_{k \in \mathbf{N}}$ converges weakly to zero that is,

$$\lim_{k \to +\infty} \int_{\mathbf{R}^d} u_k(x)\varphi(x) \, dx = 0, \qquad \forall \varphi \in L_c^2(\mathbf{R}^d). \tag{5.13}$$

The goal of this section is to describe the lack of strong convergence in $L^2_{loc}(\mathbf{R}^d)$ of the sequence $(u_k)_{k \in \mathbf{N}}$ by means of a positive measure.

Here are the main results.

- There exists a subsequence $(u_{\sigma(k)})_{k \in \mathbf{N}}$ and a positive measure μ on $\mathbf{R}^d \times S^{d-1}$ such that, for every pseudo-differential operator $A = \mathrm{Op}(a)$ with symbol a in $S^0_c(\mathbf{R}^d)$, we have

$$\lim_{k \to +\infty} \left(A(\chi u_{\sigma(k)}), \chi u_{\sigma(k)}\right)_{L^2} = \int\int_{\mathbf{R}^d \times S^{d-1}} a(x, \xi) \, d\mu(x, \xi), \qquad (5.14)$$

where $\chi \in C^\infty_0(\mathbf{R}^d)$ is any function equal to one on the support (in x) of the symbol of A.

The measure μ is called **the** *microlocal defect measure* of the sequence $(u_{\sigma(k)})$ or **a** *microlocal defect measure* of the sequence (u_k).

- Notice that, taking for A the multiplication by a $C^\infty_0(\mathbf{R}^d)$ function in (5.14) we find that,

$$\text{for every compact } K, \quad \lim_{k \to +\infty} \int_K |u_{\sigma(k)}|^2 \, dx = 0 \Longleftrightarrow \mu = 0.$$

- Here are two classical examples.

  (i) Let $u_k(x) = b(x) e^{ikx \cdot \xi_0}$, $\xi_0 \in \mathbf{R}^d \setminus \{0\}$, $b \in L^2_{loc}(\mathbf{R}^d)$. Then this sequence is bounded in L^2_{loc}, converges to zero weakly in the sense of (5.13), and the sequence itself has the defect measure defined by,

$$\langle \mu, a_0 \rangle = \int_{\mathbf{R}^d} |b(x)|^2 a_0\left(x, \frac{\xi_0}{|\xi_0|}\right) dx,$$

  which can be written as, $\mu = |b(x)|^2 dx \otimes \delta_{\frac{\xi_0}{|\xi_0|}}$ (see Problem 36).

  (ii) Let $u_k(x) = k^{\frac{d}{2}} f(k(x - x_0))$ where $f \in L^2(\mathbf{R}^d)$, $x_0 \in \mathbf{R}^d$. Then this sequence is bounded in L^2, converges weakly to zero in the sense of (5.13) and the sequence itself has the defect measure defined by,

$$\langle \mu, a_0 \rangle = \int_{S^{d-1}} h(\omega) a(x_0, \omega) \, d\omega,$$

  where $h(\omega) = (2\pi)^{-d} \int_0^{+\infty} |\widehat{f}(r\omega)|^2 r^{d-1} \, dr$, which can be written as $\mu = \delta_{x_0} \otimes h(\omega) d\omega$.

- **Elliptic regularity.** Let P be a differential operator of order m on $\mathbf{R}^d$ and let $(u_k)_{k \in \mathbf{N}}$ be a bounded sequence in $L^2_{loc}(\mathbf{R}^d)$ converging weakly to zero in the sense of (5.13). Let μ be a microlocal defect measure of (u_k).

Assume that $\lim_{k \to +\infty} P u_k = 0$ in $H_{\text{loc}}^{-m}(\mathbf{R}^d)$. Then we have,

$$\text{supp}\,\mu \subset \left\{ (x, \xi) \in \mathbf{R}^d \times \mathbf{S}^{d-1} : p_m(x, \xi) = 0 \right\}. \qquad (5.15)$$

- **Propagation of the support.** Let P be a differential operator of order m with real principal symbol p_m. Let $(u_k)_{k \in \mathbf{N}}$ be a bounded sequence in $L_{\text{loc}}^2(\mathbf{R}^d)$ converging to zero weakly in the sense of (5.13) and let μ be a defect measure of (u_k).

  Assume that $\lim_{k \to +\infty} P u_k = 0$ in $H_{\text{loc}}^{-(m-1)}(\mathbf{R}^d)$.

  Then the support of μ is a union of curves of type $s \to \left(x(s), \frac{\xi(s)}{|\xi(s)|} \right)$ where $s \to (x(s), \xi(s))$ is a bicharacteristic of p_m.

Comments A detailed exposition of the microlocal analysis theory can be found for instance in the books by T. Alazard [1], S. Alinhac, and P. Gérard [3] and M.Taylor [30]. A comprehensive treatment of the paradifferential calculus can be found in the books by G. Métivier [21] and M. Taylor [31]. There are different notions of defect measures. The one we are using here have been developed by P. Gérard in [10]. They have many applications, in particular to control theory (see the book by M. Zworski [34]).   □

Chapter 6
Statements of the Problems of Chap. 5

This chapter contains problems on microlocal analysis.

Problem 19

For $m \in \mathbf{R}$ and $\rho > 0$ we set,

$$S_\rho^m = \left\{ a \in C^\infty(\mathbf{R}^{2d}) : \forall \alpha, \beta \in \mathbf{N}^d \; \exists C_{\alpha\beta} > 0 : \right.$$

$$\left. |\partial_x^\alpha \partial_\xi^\beta a(x,\xi)| \le C_{\alpha\beta}(1 + |\xi|)^{m - \rho|\beta|}, \; \forall(x,\xi) \in \mathbf{R}^{2d} \right\}.$$

1. Assume $m < 0$, $\rho > 1$. Let $a \in S_\rho^m$.

   a) Prove, by induction on k, that for any $\alpha, \beta \in \mathbf{N}^d$ there exists $C_{\alpha\beta k} > 0$ such that,

   $$(H_k) \quad |\partial_x^\alpha \partial_\xi^\beta a(x,\xi)| \le C_{\alpha\beta k}(1 + |\xi|)^{m - \rho|\beta| - k(\rho - 1)}, \quad \forall(x,\xi) \in \mathbf{R}^{2d}.$$

   Hint: Use the fact that a function is the integral of its derivative.

   b) Deduce that $S_\rho^m = S^{-\infty}$.

2. a) Let $m \in \mathbf{N}$, $m \ge 1$ and let $p(\xi)$ be a polynomial of degree $m - 1$. Set $\rho = 1 + \varepsilon$, $\varepsilon > 0$. Prove that if $\varepsilon(m - 1) \le 1$ we have $p \in S_\rho^m$ but $p \notin S^{-\infty}$.

© The Editor(s) (if applicable) and The Author(s), under exclusive license
to Springer Nature Switzerland AG 2020
T. Alazard, C. Zuily, *Tools and Problems in Partial Differential Equations*,
Universitext, https://doi.org/10.1007/978-3-030-50284-3_6

b) Give a simple example with $m = 0$ showing that the result of question 1b) is false in this case.

Problem 20

Let $a \in S^1$ be a real valued symbol satisfying,

$$\exists c > 0 : a(x, \xi) \geq c \langle \xi \rangle, \quad \forall (x, \xi) \in \mathbf{R}^d \times \mathbf{R}^d, \quad \langle \xi \rangle = (1 + |\xi|^2)^{\frac{1}{2}}.$$

Let $\sigma > 0$. Prove that for any $\alpha, \beta \in \mathbf{N}^d$ we have,

$$(\partial_x^\alpha \partial_\xi^\beta e^{-\sigma a})(x, \xi) = \sum_{j=0}^{|\alpha|+|\beta|} \sigma^j d_{j,\alpha\beta}(x, \xi) e^{-\sigma a}, \quad d_{j,\alpha\beta} \in S^{j-|\beta|}.$$

Problem 21 *A Commutator Estimate*

If P is a first order differential operator with constant coefficients, then

$$\|P(fu) - f P u\|_{L^2} \leq \|\nabla f\|_{L^\infty} \|u\|_{L^2}.$$

The goal of this problem is to prove a similar result when P is replaced by a pseudo differential operator with constant coefficients and L^2 is replaced by a Sobolev space H^σ.

More precisely, consider real numbers σ_0 and m such that $\sigma_0 > 1 + \frac{d}{2}$ and $m \in [1, \sigma_0]$. Let $p = p(\xi) \in S^m$ and $P = \mathrm{Op}(p)$.

We want to prove that, for all $\sigma \in [-\sigma_0 + m, 0] \cup [m - 1, \sigma_0 - 1]$ there exists $K > 0$ such that,

$$\|P(fu) - f P u\|_{H^{\sigma-(m-1)}} \leq K \|f\|_{H^{\sigma_0}} \|u\|_{H^\sigma}. \tag{6.1}$$

for all $f, u \in \mathcal{S}(\mathbf{R}^d)$.

Part 1
Assume $\sigma \in [m - 1, \sigma_0 - 1]$.

1. Set $v := P(fu) - f P u$ with f, u in $\mathcal{S}(\mathbf{R}^d)$. Verify that $v = v_L + v_H$ with,

$$\widehat{v_L}(\xi) = \int_{|\xi-\eta| \leq \frac{1}{2}|\eta|} \big(p(\xi) - p(\eta)\big) \widehat{f}(\xi - \eta) \widehat{u}(\eta)\, d\eta,$$

$$\widehat{v_H}(\xi) = \int_{|\xi-\eta| > \frac{1}{2}|\eta|} \big(p(\xi) - p(\eta)\big) \widehat{f}(\xi - \eta) \widehat{u}(\eta)\, d\eta.$$

2. Verify that there exists C such that for all (ξ, η) satisfying $|\xi - \eta| \leq \frac{1}{2} |\eta|$ we have,

$$|p(\xi) - p(\eta)| \leq C \, |\xi - \eta| \, \langle \eta \rangle^{m-1}.$$

3. a) Prove that,

$$\langle \xi \rangle^{\sigma-m+1} |\widehat{v_L}(\xi)| \leq K \int |\xi - \eta| \, |\widehat{f}(\xi - \eta)| \, \langle \eta \rangle^{\sigma} \, |\widehat{u}(\eta)| \, d\eta.$$

b) Deduce that,

$$\|v_L\|_{H^{\sigma-m+1}} \leq C \, \|f\|_{H^{\sigma_0}} \, \|u\|_{H^{\sigma}}.$$

4. Prove that

$$\langle \xi \rangle^{\sigma-m+1} |\widehat{v_H}(\xi)| \leq K \int \langle \xi - \eta \rangle^{\sigma_0} \, |\widehat{f}(\xi - \eta)| \frac{\langle \eta \rangle^{\sigma}}{\langle \eta \rangle^{\sigma_0-1}} \, |\widehat{u}(\eta)| \, d\eta,$$

and deduce that

$$\|v_H\|_{H^{\sigma-m+1}} \leq C \, \|f\|_{H^{\sigma_0}} \, \|u\|_{H^{\sigma}},$$

and conclude that the estimate (6.1) holds for $\sigma \in [m-1, \sigma_0 - 1]$.

Part 2

Assume $\sigma \in [-\sigma_0 + m, 0]$.

1. Define the operator C by $Cu = P(fu) - f Pu$. By using the previous result for $\sigma \in [-1, \sigma_0]$ and a duality argument, show that $C^* \in \mathcal{L}(H^{-\sigma+m-1}, H^{-\sigma})$ with operator norm bounded by $K \, \|f\|_{H^{\sigma_0}}$.

Deduce that the estimate (6.1) holds for $\sigma \in [-\sigma_0 + m, 0)$.

Problem 22

The goal of this problem is to study the commutator between a function in the Hölder space $W^{\nu,\infty}(\mathbf{R})$ where $0 < \nu < 1$ and the Hilbert transform $\mathcal{H}$. Recall that $\mathcal{H}$ is the operator acting on tempered distribution u by,

$$\widehat{\mathcal{H}u}(\xi) := \frac{\xi}{i \, |\xi|} \widehat{u}(\xi).$$

We will use the paradifferential calculus. In particular, given a function $f = f(x)$, we denote by T_f the paraproduct by f.

1. Given two functions f and g verify that,

$$fg - T_f g = \sum_{j \geq -1} (S_{j+2}g)\Delta_j f.$$

Deduce that, for all real numbers $0 < \theta < \nu < 1$, there is a constant C such that, for all $g \in H^{-\theta}(\mathbf{R})$, we have,

$$\|fg - T_f g\|_{L^2} \leq C \|f\|_{W^{\nu,\infty}} \|g\|_{H^{-\theta}}.$$

2. Let $q = q(\xi)$ be a symbol in S^0 and $Q = \mathrm{Op}(q)$.
   Prove that, for $0 < \theta < \nu < 1$, there exists a constant K such that for all $f \in W^{\nu,\infty}(\mathbf{R})$, and all $u \in H^{-\theta}(\mathbf{R})$ we have,

$$\|Q(fu) - fQu\|_{L^2} \leq K \|f\|_{W^{\nu,\infty}} \|u\|_{H^{-\theta}}.$$

3. Deduce that, for $0 < \theta < \nu < 1$, there exists a constant K such that for all $f \in W^{\nu,\infty}(\mathbf{R})$, and all u in $H^{-\theta}(\mathbf{R})$,

$$\|\mathcal{H}(fu) - f\mathcal{H}u\|_{L^2} \leq K \|f\|_{W^{\nu,\infty}} \|u\|_{H^{-\theta}}.$$

Problem 23 *Local ΨDO Are Differential*

The purpose of this problem is to show that the only local pseudodifferential operators are the differential operators.

Part 1
We consider in this part pseudo-differential operators whose symbols belong to the S^m classes. We say that an operator is *local* if it satisfies,

$$\mathrm{supp}\, Pu \subset \mathrm{supp}\, u, \quad \forall u \in S(\mathbf{R}^d).$$

1. Prove that a differential operator with C^∞-bounded coefficients is local.
2. Prove that if P and Q are local operators the same is true for PQ and $[P, Q] = PQ - QP$.
3. Prove that if P is local the same is true for its adjoint P^*.
4. a) Prove that if $P^*P \equiv 0$ then $P \equiv 0$.
   b) Let $\ell \geq 2$. Show that if $(P^*P)^\ell \equiv 0$ then $(P^*P)^{\ell-1} \equiv 0$. Deduce then that $P \equiv 0$.

Part 2

Let $a \in S^m$ and $A = \text{Op}(a)$.

The goal of this part is to prove that if A is local then A is a differential operator.

1. Let $u \in S(\mathbf{R}^d)$ and $x_0 \in \mathbf{R}^d$. Show that there exists a sequence $(u_k)_{k \in \mathbf{N}} \subset S(\mathbf{R}^d)$, whose elements vanish identically in a neighborhood of x_0, which converges to u in $L^2(\mathbf{R}^d)$.

2. Assume $m < -\frac{d}{2}$. Show that there exists $C > 0$ such that,

$$\sup_{x \in \mathbf{R}^d} |Av(x)| \leq C\|v\|_{L^2(\mathbf{R}^d)}, \quad \forall v \in S(\mathbf{R}^d).$$

3. Assume that A is local and $m < -\frac{d}{2}$. Let $x_0 \in \mathbf{R}^d$. Show that $Au(x_0) = 0$ for every $u \in S(\mathbf{R}^d)$. Deduce that $A \equiv 0$.

4. Assume that A is local and $m < 0$. Show that $A \equiv 0$.

    Hint: We may assume that $-\frac{d}{2} \leq m < 0$. Let $\ell \in \mathbf{N}$ such that $2\ell m < -\frac{d}{2}$. Consider the operator $(A^*A)^\ell$.

    Notation: If P and Q are two operators we set $(ad\,P)(Q) = [P, Q] = PQ - QP$.

5. Let $k \in \mathbf{N}^*$. Let $a \in S^m$ where $m < k$. Let $g_1, \ldots, g_k$ be $C^\infty(\mathbf{R}^d)$-bounded. Prove that $(ad\,g_1)(ad\,g_2)\cdots(ad\,g_k)(A) = 0$.

6. Let $a \in S^m$, $w \in C_0^\infty(\mathbf{R}^d)$ and $g \in C_0^\infty(\mathbf{R}^d)$ be such that $g(x_0) = 0$. Show, by induction on $\ell \geq 1$ that for x near x_0 we have,

$$A(g^\ell w)(x) = (-1)^\ell (ad\,g)^\ell (A)w(x) + O(|x - x_0|).$$

7. Let $k \in \mathbf{N}^*$, $x_0 \in \mathbf{R}^d$ and $u \in C_0^\infty(\mathbf{R}^d)$. Assume that $\partial_x^\alpha u(x_0) = 0$, $|\alpha| \leq k-1$.

    a) Show that $u(x) = \sum_{|\beta|=k}(x - x_0)^\beta v_\beta(x)$ where $v_\beta \in C^\infty(\mathbf{R}^d)$.

    b) Let $\theta \in C_0^\infty(\mathbf{R}^d)$, $\theta = 1$ on supp u. Noticing that $\theta^{k+1}u = u$ show that,

$$Au(x) = \sum_{|\beta|=k}(-1)^{|\beta|}(ad\,f_1)^{\beta_1}(ad\,f_2)^{\beta_2}\cdots(ad\,f_d)^{\beta_d}(A)(\theta v_\beta)+O(|x-x_0|),$$

    where $f_j = (x_j - x_{0,j})\theta$.
    Hint: Use question 6.

    c) Let $a \in S^m$ be such that $A = \text{Op}(A)$ is local. Let $k > m$. Show that $Au(x_0) = 0$.

8. Let $u \in C_0^\infty(\mathbf{R}^d)$. Setting

$$w(x) = u(x) - \sum_{|\alpha| \leq k-1} \frac{1}{\alpha!}(x - x_0)^\alpha \chi(x)\partial_x^\alpha u(x_0),$$

where $\chi \in C_0^\infty(\mathbf{R}^d)$ is equal to 1 near x_0, show that

$$Au(x_0) = \sum_{|\alpha| \leq k-1} c_\alpha \partial_x^\alpha u(x_0),$$

and conclude.

Problem 24 *Continuity of the ΨDO in the $H_{ul}^s(\mathbf{R}^d)$ Spaces*

This problem follows Problem 7.

Let P be pseudo-differential operator with symbol $p \in S^m(\mathbf{R}^d)$ and $s \in \mathbf{R}$. The goal of this problem is to show that P is continuous from $H_{ul}^{s+m}(\mathbf{R}^d)$ to $H_{ul}^s(\mathbf{R}^d)$.

The continuity of the pseudo-differentials operators on the classical Sobolev spaces $H^s(\mathbf{R}^d)$ will be assumed. We keep the notations of Problem 7.

We write for $k \in \mathbf{Z}^d$ (with $|x| = \sup_{1 \leq j \leq d}|x_j|$),

$$\chi_k P u = \sum_{|k-q|\leq 2} \chi_k P \chi_q u + \sum_{|k-q|\geq 3} \chi_k P \chi_q u := A_k + \sum_{|k-q|\geq 3} B_{k,q}. \qquad (6.2)$$

1. Show that there exists a positive constant C independent of k, q such that,

$$\|A_k\|_{H^s(\mathbf{R}^d)} \leq C\|u\|_{H_{ul}^{s+m}(\mathbf{R}^d)}, \quad \forall u \in H_{ul}^{s+m}(\mathbf{R}^d).$$

The following questions will be devoted to prove the estimate,

$$\|B_{k,q}\|_{H^s(\mathbf{R}^d)} \leq \frac{C}{|k-q|^{d+1}}\|u\|_{H_{ul}^{s+m}(\mathbf{R}^d)}.$$

2. Let $n_0 \in \mathbf{N}, n_0 \geq s$. Then $\|B_{k,q}\|_{H^s(\mathbf{R}^d)} \leq \sum_{|\alpha|\leq n_0} \|\partial_x^\alpha B_{k,q}\|_{L^2(\mathbf{R}^d)}$.
   Show that there exists $C > 0$ such that for all $k, q \in \mathbf{Z}^d$ and all $\alpha \in \mathbf{N}^d$ we have,

$$\|\partial_x^\alpha B_{k,q}\|_{L^2(\mathbf{R}^d)} \leq C\|\partial_x^\alpha B_{k,q}\|_{L^\infty(\mathbf{R}^d)}.$$

3. Prove that for $|\alpha| \leq n_0$, $\partial_x^\alpha B_{kq}(x)$ is a finite linear combination of terms of the form,

$$C_{kq\alpha}(x) =: \partial_x^{\alpha_1} \chi_k(x) \int e^{ix\cdot\xi} a_{\alpha_2}(x,\xi)\widehat{\chi_q u}(\xi)\,d\xi, \quad |\alpha_1| + |\alpha_2| \leq |\alpha|,$$

where, $a_{\alpha_2} \in S^{m+|\alpha_2|}$.

4. Let $\theta \in C_0^\infty(\mathbf{R}^d)$ be such that $\operatorname{supp} \theta \subset \{\xi : |\xi| \leq 1\}$. Show that,

$$C_{kq\alpha}(x) = \lim_{\varepsilon \to 0} \int K_{kq\alpha}^\varepsilon(x, y) \chi_q u(y)\, dy := \lim_{\varepsilon \to 0} C_{kq\alpha}^\varepsilon(x), \text{ where,}$$

$$K_{kq\alpha}^\varepsilon(x, y) = \partial_x^{\alpha_1} \chi_k(x) \widetilde{\chi}_q(y) \int e^{i(x-y)\cdot\xi} a_{\alpha_2}(x, \xi)\theta(\varepsilon\xi)\, d\xi,$$

where $\widetilde{\chi} \in C_0^\infty(\mathbf{R}^d)$, $\operatorname{supp} \widetilde{\chi} \subset \left\{ \xi : |\xi| \leq \frac{3}{2} \right\}$, $\widetilde{\chi} = 1$ on the support of χ.

5. Let $n_1 \in \mathbf{N}$, $n_1 \geq -(m+s)$. Show that,

$$|C_{kq\alpha}^\varepsilon(x)| \leq \|K_{kq\alpha}^\varepsilon(x, \cdot)\|_{H^{n_1}(\mathbf{R}^d)} \|u\|_{H_{ul}^{s+m}(\mathbf{R}^d)}.$$

6. Show that there exists $C > 0$ independent of $k, q, \alpha, \varepsilon$ such that,

$$\|K_{kq\alpha}^\varepsilon(x, \cdot)\|_{H^{n_1}(\mathbf{R}^d)} \leq C \sum_{|\beta|\leq n_1} \|\partial_y^\beta K_{kq\alpha}^\varepsilon(x, \cdot)\|_{L^\infty(\mathbf{R}^d)}.$$

7. Show that $\partial_y^\beta K_{kq\alpha}^\varepsilon(x, y)$ is a finite linear combination of terms of the form,

$$K_{kq\alpha\beta}^\varepsilon(x, y) = \partial_x^{\alpha_1} \chi_k(x) \partial_y^{\beta_1} \widetilde{\chi}_q(y) \int e^{i(x-y)\cdot\xi} b_{\alpha_2,\beta_2}(x, \xi)\theta(\varepsilon\xi)\, d\xi,$$

where $b_{\alpha_2,\beta_2} \in S^{m+|\alpha_2|+|\beta_2|}$ and $|\alpha_1| + |\alpha_2| \leq |\alpha|$, $\beta_1 + \beta_2 = \beta$.

8. Show that if $|k - q| \geq 3$ we have $|x - y| \geq \frac{1}{6}|k - q|$ on the support of $K_{kq\alpha\beta}^\varepsilon(x, y)$.

9. Let $M(x) = \frac{x-y}{|x-y|^2}$ and $L = M(x) \cdot \nabla_\xi = \sum_{j=1}^d M_j(x)\partial_{\xi_j}$.

   a) Show that for $N \in \mathbf{N}$, $L^N = \sum_{|\gamma|=N} c_\gamma M^\gamma \partial_\xi^\gamma$ and that $L^N e^{i(x-y)\cdot\xi} = e^{i(x-y)\cdot\xi}$.

   b) Integrating by parts with L^N with N large, in the integral giving $K_{kq\alpha\beta}^\varepsilon(x, y)$, show that there exists $C > 0$ independent of $x, y, k, q, \alpha, \beta, \varepsilon$ such that,

$$|K_{kq\alpha\beta}^\varepsilon(x, y)| \leq \frac{C}{|k - q|^{d+1}} \|u\|_{H_{ul}^{s+m}(\mathbf{R}^d)}, \quad \forall x, y \in \mathbf{R}^d.$$

10. Conclude.

Problem 25

This problem uses Problem 20.

Let $I = (0, 1)$, $d \geq 1$, and $\mu < \frac{1}{2}$. Let $a = a(\xi) \in S^1$ be a real symbol such that

$$\exists c_0 > 0 : a(\xi) \geq c_0 \langle\xi\rangle , \quad \forall \xi \in \mathbf{R}^d, \quad \langle\xi\rangle = (1 + |\xi|^2)^{\frac{1}{2}}.$$

The goal of this problem is to show that for every $\rho \in \mathbf{R}$ the map $u \mapsto e^{-za(D_x)}u$ is continuous from $C_*^{\rho}(\mathbf{R}^d)$ to $L_z^2(I, C_*^{\rho+\mu}(\mathbf{R}^d))$.

We recall (see Problem 20) that, $\partial_\xi^\alpha \left(e^{-za(\xi)}\right) = \sum_{j=0}^{|\alpha|} z^j q_{j\alpha}(\xi)e^{-za(\xi)}$, where $q_{j\alpha} \in S^{j-|\alpha|}$.

Let $\mu < \frac{1}{2}$ and $p(\xi) = z^\mu \langle\xi\rangle^\mu e^{-za(\xi)}$.

1. Show that for every $\alpha \in \mathbf{N}^d$ there exists $C_\alpha > 0$ such that,

$$|\partial_\xi^\alpha p(\xi)| \leq C_\alpha z^\mu \langle\xi\rangle^{\mu-|\alpha|} e^{-\frac{1}{2}za(\xi)}, \quad \forall \xi \in \mathbf{R}^d.$$

2. Deduce that $|x|^{d+1}|\widehat{p}(x)| \leq C_d, \quad \forall|x| \geq 1$.

3. a) We set $\langle z, \eta\rangle = (z^2 + |\eta|^2)^{\frac{1}{2}}$. Prove that,

$$|x^\alpha \widehat{p}(x)| \leq C_\alpha z^{|\alpha|-d} \int \langle z, \eta\rangle^{\mu-|\alpha|} e^{-\frac{c_0}{2}\langle z,\eta\rangle} d\eta.$$

b) Prove that for $|\alpha| = d$ and $|\alpha| = d - 1$ there exists $M > 0$ such that,

$$\int \langle z, \eta\rangle^{\mu-|\alpha|} e^{-\frac{c_0}{2}\langle z,\eta\rangle} d\eta \leq M.$$

c) Prove that for every $\varepsilon \in]0, 1]$ we have,

$$|x|^{d-\varepsilon}|\widehat{p}(x)| \leq C_\varepsilon z^{-\varepsilon}, \quad \forall|x| \leq 1.$$

d) Deduce that for every $\varepsilon \in]0, 1]$ there exists $C_\varepsilon > 0$ such that,

$$\|\widehat{p}\|_{L^1(\mathbf{R}^d)} \leq C_\varepsilon z^{-\varepsilon}.$$

4. Prove that for every $\varepsilon \in]0, 1]$ there exists $C_\varepsilon > 0$ such that,

$$\|\Delta_j e^{-za(D_x)}u\|_{L^\infty(\mathbf{R}^d)} \leq C_\varepsilon z^{-\mu-\varepsilon} 2^{-j(\mu+\rho)} \|u\|_{C_*^\rho}.$$

5. Conclude.

Problem 26

Let $\alpha \in]0, +\infty[$ with $\alpha \neq 1$ and let $S(t) = e^{-it|D_x|^\alpha}$ be the operator defined on $S'(\mathbf{R}^d)$ by $S(t)u = \mathcal{F}^{-1}(e^{-it|\xi|^\alpha}\widehat{u})$.

The goal of the first two parts problem is to prove the following result.

Let $s, \sigma \in \mathbf{R}$. Assume that there exists $t_0 \neq 0$ such that $S(t_0)$ is continuous from $C_*^\sigma(\mathbf{R}^d)$ to $C_*^s(\mathbf{R}^d)$. Then

$$s \leq \sigma - \frac{d\alpha}{2}. \qquad (6.3)$$

In a third part we shall prove that this result does not hold for $\alpha = 1$.

Part 1. *Preliminaries*

Let $K \in C^0(\mathbf{R}^d) \cap L^1(\mathbf{R}^d)$. Let T be the operator defined by,

$$Tu(x) = \int K(x - y)u(y)dy.$$

1. Show that T is continuous from $L^\infty(\mathbf{R}^d)$ to $L^\infty(\mathbf{R}^d)$ and that,

$$\|T\|_{L^\infty \to L^\infty} = \int_{\mathbf{R}^d} |K(z)| \, dz.$$

2. For $\lambda \in \mathbf{R}$ let $M(\lambda) = (m_{jk}(\lambda))_{1 \leq j,k \leq d}$, where $m_{jk}(\lambda) = \delta_{jk} + \lambda \omega_j \omega_k$, $\omega \in \mathbf{S}^{d-1}$, be a $d \times d$ matrix. Set $F(\lambda) = \det M(\lambda)$. We want to show that $F(\lambda) = 1 + \lambda$. We have trivially $F(0) = 1$.

   a) Denote by $\ell_i(\lambda)$ the ith line of the matrix $M(\lambda)$. Noticing that,

$$F'(\lambda) = \sum_{k=1}^{d} \det(\ell_1(\lambda), \ldots, \ell_k'(\lambda), \ldots, \ell_d(\lambda)),$$

   show that $F'(0) = 1$.

   b) Noticing that $\ell_i(\lambda)$ is linear with respect to λ show that $F''(\lambda) \equiv 0$ and conclude.

Part 2

We assume first that $t_0 = -1$.

1. Let $u \in L^\infty(\mathbf{R}^d)$. Consider the canonical decomposition of u (see Sect. 3.3),

$$u = \Delta_{-1}u + \sum_{j=0}^{+\infty} \Delta_j u.$$

   Show that there exist $C, C' > 0$ such that for all $j \in \mathbf{N}$ we have, $\Delta_j u \in C_*^\sigma(\mathbf{R}^d)$ and,

$$\|\Delta_j u\|_{C_*^\sigma(\mathbf{R}^d)} \leq C 2^{j\sigma} \|\Delta_j u\|_{L^\infty(\mathbf{R}^d)} \leq C' 2^{j\sigma} \|u\|_{L^\infty(\mathbf{R}^d)}.$$

2. Show that there exists $C > 0$ such that,

$$2^{js}\|S(-1)\Delta_j\Delta_j u\|_{L^\infty(\mathbf{R}^d)} \le C2^{j\sigma}\|u\|_{L^\infty(\mathbf{R}^d)} \quad \forall u \in L^\infty(\mathbf{R}^d), \quad \forall j \in \mathbf{N}.$$

3. We fix j and we set, $T_j = S(-1)\Delta_j\Delta_j$, $h = 2^{-j}$.

   a) Check that,

$$T_j u(x) = \int_{\mathbf{R}^d} K_h(x - y)u(y)dy,$$

   where,

$$K_h(z) = (2\pi h)^{-d} \int_{\mathbf{R}^d} e^{\frac{i}{h}(z\cdot\eta + h^{1-\alpha}|\eta|^\alpha)}\varphi^2(\eta)d\eta.$$

   b) Show that $K_h \in C^0(\mathbf{R}^d) \cap L^1(\mathbf{R}^d)$.
   c) Deduce that,

$$\|T_j\|_{L^\infty \to L^\infty} = h^{d(1-\alpha)} \int_{\mathbf{R}^d} |\tilde{K}_h(s)|\, ds,$$

   with,

$$\tilde{K}_h(s) = (2\pi h)^{-d} \int_{\mathbf{R}^d} e^{ih^{-\alpha}\phi(s,\eta)}\varphi^2(\eta)d\eta, \quad \phi(s,\eta) = s\cdot\eta + |\eta|^\alpha.$$

4. Assume that,

$$|s| \le \frac{1}{2}\frac{\alpha}{2^{|1-\alpha|}}.$$

   Using the vector field

$$L = \frac{h^\alpha}{i}\frac{1}{|\partial_\eta\phi|^2}\sum_{k=1}^{d}\frac{\partial\phi}{\partial\eta_k}\frac{\partial}{\partial\eta_k},$$

   show that,

$$|\tilde{K}_h(s)| \le C_N h^N, \quad \forall N \in \mathbf{N}.$$

5. Assume now that $|s| \ge 2^{1+|\alpha-1|}\alpha$. Check that on the support of φ we have,

$$\left|\frac{\partial\phi}{\partial\eta}(s,\eta)\right| \ge |s| - \alpha|\eta|^{\alpha-1} \ge \frac{1}{2}|s|.$$

Using the same vector field show that,

$$|\tilde{K}_h(s)| \le C_N |s|^{-N} h^N, \quad \forall N \in \mathbf{N}.$$

6. Assume now that,

$$\frac{1}{2}\frac{\alpha}{2^{|1-\alpha|}} \le |s| \le 2^{1+|\alpha-1|}\alpha.$$

a) Check that the phase ϕ has a critical point given by $\eta_c = c_\alpha s |s|^{\frac{2-\alpha}{\alpha-1}}$, and that

$$\frac{\partial^2 \phi}{\partial \eta_j \partial \eta_k}(s, \eta) = \alpha |\eta|^{\alpha-2} m_{jk} \quad m_{jk} = \delta_{jk} - (\alpha - 2)\omega_j \omega_k, \quad \omega = \frac{\eta}{|\eta|}.$$

b) Prove that the determinant of the matrix $\left(\frac{\partial^2 \phi}{\partial \eta_j \partial \eta_k}\right)$ is different from zero and

that its value at the critical point η_c is equal to $C_{\alpha,d} |s|^{\frac{(\alpha-2)d}{(\alpha-1)}}$.

c) Check that the stationary phase formula implies that there exists $C_{\alpha,d} > 0$ such that for h small enough we have,

$$\tilde{K}_h(s) = C_{\alpha,d}\, h^{-d} h^{\frac{\alpha d}{2}} \left\{ \frac{e^{i h^{-\alpha} \phi(s,\eta_c)}}{|s|^{\frac{(\alpha-2)d}{2(\alpha-1)}}} \varphi^2(\eta_c) + O(h^\alpha) \right\}.$$

7. Deduce from the previous questions that for $h = 2^{-j}$ small enough we have,

$$\|T_j\|_{L^\infty \to L^\infty} \ge C h^{d(1-\alpha)} h^{-d} h^{\frac{d\alpha}{2}} - C_N h^N \ge C' h^{-\frac{d\alpha}{2}} = C' 2^{j\frac{d\alpha}{2}}.$$

8. Conclude.
9. We assume that there exists $t_0 \ne 0$ such that $S(t_0)$ is continuous from $C_*^\sigma(\mathbf{R}^d)$ to $C_*^s(\mathbf{R}^d)$.

   a) If $t_0 < 0$ show that $\Delta_j S(-1)u(x) = \Delta_j^\lambda S(t_0)u_{\frac{1}{\lambda}}(\lambda x)$ where $\lambda^\alpha = -t_0$, $\Delta_j^\lambda v = \mathcal{F}^{-1}(\varphi(2^{-j}\lambda \xi)\hat{v})$, $u_{\frac{1}{\lambda}}(x) = u(\frac{x}{\lambda})$.
   b) If $t_0 > 0$ show that $\Delta_j S(-t_0)u = \overline{\Delta_j S(t_0)\bar{u}}$.
   c) Deduce that $s \le \sigma - \frac{d\alpha}{2}$.

Part 3
The purpose of this part is to show that the inequality (6.3) is false when $\alpha = 1$. We assume for simplicity that $d = 3$. For $u_0 \in S'(\mathbf{R}^3)$, $t \ge 0$ and $x \in \mathbf{R}^3$, set $u(t, x) = (e^{-it|D_x|}u_0)(x)$.

1. Show that u is a solution of the problem,

$$\Box u = 0, \quad \text{in } (0, +\infty) \times \mathbf{R}^3, \quad u|_{t=0} = u_0 \qquad \partial_t u|_{t=0} = -i|D_x|u_0,$$

where $\Box = \partial_t^2 - \Delta_x$.

2. Using the formula (7.10) in Chap. 4 prove that if $u_0 \in C_*^\sigma(\mathbf{R}^3)$ then the function $x \mapsto u(t_0, x)$ belongs to $C_*^{\sigma-1}(\mathbf{R}^3)$ and conclude.

Problem 27 *A Carleman Inequality*

Notations If $f \in C^\infty(\mathbf{R}_t \times \mathbf{R}_x^d)$ we denote by $f(t)$ the function $x \mapsto f(t, x)$. On the other hand, for $T > 0$ we set,

$$E_T = \left\{ u \in C^\infty(\mathbf{R}, \mathcal{S}(\mathbf{R}^d)) : \operatorname{supp} u \subset (0, T) \right\}.$$

Notice that for $u \in E_T$ we have $u(0, \cdot) = u(T, \cdot) = 0$.

Part 1

Let $a \in S^1(\mathbf{R}^d)$ be a **real** symbol. We set $A = \operatorname{Op}(a)$ and we consider the operator $P = D_t + A$, $\quad D_t = \frac{1}{i}\frac{\partial}{\partial t}$.

1. Let $k \geq 1$, $T > 0$ and $u \in C^\infty(\mathbf{R} \times \mathbf{R}^d)$. Set $v = e^{\frac{k}{2}(t-T)^2}u$. Show that,

$$e^{\frac{k}{2}(t-T)^2} Pu = \tilde{P}v, \quad \tilde{P} = X + Y, \quad X = D_t + A, \quad Y = ik(t - T).$$

2. a) Justify the fact that $A^* = A + B_0$, where $B_0 \in \operatorname{Op}(S^0)$.

   b) Prove that there exists $T_0 > 0$ small enough such that for all $T \leq T_0$,

$$2\operatorname{Re} \int_0^T (Xv(t), Yv(t))_{L^2(\mathbf{R}^d)}\, dt \geq \frac{k}{2} \int_0^T \|v(t)\|_{L^2(\mathbf{R}^d)}^2\, dt, \quad \forall v \in E_T.$$

3. Deduce that there exists $C > 0$ and $T_0 > 0$ such that for all $T \leq T_0$,

$$k \int_0^T e^{k(t-T)^2}\|u(t)\|_{L^2(\mathbf{R}^d)}^2\, dt \leq C \int_0^T e^{k(t-T)^2}\|Pu(t)\|_{L^2(\mathbf{R}^d)}^2\, dt, \quad \forall v \in E_T.$$

4. a) Let $a(\xi) = (1 + |\xi|^2)^{\frac{1}{2}}$ and $A = \operatorname{Op}(a)$. Show that $A^2 = -\Delta_x + \operatorname{Id}$.

   b) What is the operator $Q = (D_t - A)(D_t + A)$?

   c) Noticing that $(D_t \pm A)u \in E_T$ if $u \in E_T$, show that there exist positive constants C, T_0, k_0 such that for $T \leq T_0$, $k \geq k_0$ and for all $u \in E_T$, we have,

$$k^2 \int_0^T e^{k(t-T)^2}\|u(t)\|_{L^2(\mathbf{R}^d)}^2\, dt \leq C \int_0^T e^{k(t-T)^2}\|\Box u(t)\|_{L^2(\mathbf{R}^d)}^2\, dt,$$

where $\Box = \partial_t^2 - \sum_{j=1}^d \partial_{x_j}^2$ is the wave operator.

Part 2

Let $a, b \in S^1(\mathbf{R}^d)$ be two **real** symbols. We assume that b is elliptic in the following sense:

$$\exists c_0 > 0 : b(\xi) \geq c_0(1 + |\xi|), \quad \forall \xi \in \mathbf{R}^d.$$

Set $B = \mathrm{Op}(b)$ and consider the operator $P = D_t + A + iB$, $D_t = \frac{1}{i}\frac{\partial}{\partial t}$.

1. For $u \in C^\infty(\mathbf{R} \times \mathbf{R}^d)$ set $v = e^{\frac{k}{2}(t-T)^2} u$. Show that

$$e^{\frac{k}{2}(t-T)^2} Pu = \tilde{P}v, \quad \tilde{P} = X + iY, \quad X = D_t + A, \quad Y = iB + ik(t-T).$$

2. Let $r > 0$. Prove that there exists $T_0 > 0$ such that for $T \leq T_0$ and all $v \in C^\infty(\mathbf{R} \times \mathbf{R}^d)$ with supp $v \subset \{(t, x) \in \mathbf{R} \times \mathbf{R}^d : 0 \leq t \leq T, |x| \leq r\}$ we have,

$$2\mathrm{Re} \int_0^T (Xv(t), Yv(t))_{L^2(\mathbf{R}^d)} \, dt = (1) + (2) + (3),$$

$$(1) \geq \frac{k}{2} \int_0^T \|v(t)\|^2_{L^2(\mathbf{R}^d)} \, dt, \quad (2) = 2\mathrm{Re} \int_0^T (Av(t), iBv(t))_{L^2(\mathbf{R}^d)} \, dt,$$

$$(3) = 2\mathrm{Re} \int_0^T (D_t v(t), iBv(t))_{L^2(\mathbf{R}^d)} \, dt.$$

3. a) Set $C = A^*B$. Show that $C \in \mathrm{Op}(S^2)$ and that its symbol c satisfies $c = c_2 + r_1$ where c_2 is real and $r_1 \in S^1$. Deduce that $C - C^* \in \mathrm{Op}(S^1)$.

   b) Show that there exists $C > 0$ such that,

$$|(2)| \leq C \int_0^T \|v(t)\|_{H^1(\mathbf{R}^d)} \|v(t)\|_{L^2(\mathbf{R}^d)} \, dt.$$

4. a) Show that there exists $C > 0$ such that,

$$|(3)| \leq C \int_0^T \|v(t)\|_{L^2(\mathbf{R}^d)} \|D_t v(t)\|_{L^2(\mathbf{R}^d)} \, dt.$$

   b) Writing $D_t = D_t + A - A$, deduce that there exists $C > 0$ and for any $\varepsilon > 0$ there exists $C_\varepsilon > 0$ such that,

$$|(3)| \leq C\varepsilon \int_0^T \|Xv(t)\|^2_{L^2(\mathbf{R}^d)} \, dt + C_\varepsilon \int_0^T \|v(t)\|^2_{L^2(\mathbf{R}^d)} \, dt$$

$$+ C \int_0^T \|v(t)\|_{L^2(\mathbf{R}^d)} \|v(t)\|_{H^1(\mathbf{R}^d)} \, dt.$$

5. Show that,

$$\int_0^T \|\tilde{P}v(t)\|_{L^2(\mathbf{R}^d)}^2 \, dt \geq \frac{1}{2} \int_0^T \|Xv(t)\|_{L^2(\mathbf{R}^d)}^2 \, dt + \int_0^T \|Yv(t)\|_{L^2(\mathbf{R}^d)}^2 \, dt$$

$$+ \frac{k}{2} \int_0^T \|v(t)\|_{L^2(\mathbf{R}^d)}^2 \, dt - C \int_0^T \|v(t)\|_{L^2(\mathbf{R}^d)} \|v(t)\|_{H^1(\mathbf{R}^d)} \, dt.$$

6. a) Using the ellipticity of B show that there exists $C > 0$ such that,

$$\|v(t)\|_{H^1(\mathbf{R}^d)} \leq C \left(\|Yv(t)\|_{L^2(\mathbf{R}^d)} \, dt + (1+kT) \|v(t)\|_{L^2(\mathbf{R}^d)} \right). \tag{6.4}$$

b) Deduce that for any $\varepsilon > 0$ there exists $C_\varepsilon > 0$ such that,

$$\int_0^T \|v(t)\|_{L^2(\mathbf{R}^d)} \|v(t)\|_{H^1(\mathbf{R}^d)} \, dt \leq \varepsilon \int_0^T \|Yv(t)\|_{L^2(\mathbf{R}^d)}^2 \, dt$$

$$+ C_\varepsilon (1+kT) \int_0^T \|v(t)\|_{L^2(\mathbf{R}^d)}^2.$$

7. Prove that there exists $C > 0$ and $T_0 > 0$ such that for all $T \leq T_0$,

$$\int_0^T \|Xv(t)\|_{L^2(\mathbf{R}^d)}^2 \, dt + \int_0^T \|Yv(t)\|_{L^2(\mathbf{R}^d)}^2 \, dt + k \int_0^T \|v(t)\|_{L^2(\mathbf{R}^d)}^2$$

$$\leq C \int_0^T \|\tilde{P}v(t)\|_{L^2(\mathbf{R}^d)}^2 \, dt.$$

8. a) Using (6.4) show that there exists $C > 0$ such that,

$$\int_0^T \|v(t)\|_{H^1(\mathbf{R}^d)}^2 \, dt \leq C(1+kT^2) \int_0^T \|\tilde{P}v(t)\|_{L^2(\mathbf{R}^d)}^2 \, dt.$$

b) Deduce that,

$$\int_0^T \|D_t v(t)\|_{L^2(\mathbf{R}^d)}^2 \, dt \leq C(1+kT^2) \int_0^T \|\tilde{P}v(t)\|_{L^2(\mathbf{R}^d)}^2 \, dt.$$

c) Set $I = \int_0^T e^{k(t-T)^2} \|Pu(t)\|_{L^2(\mathbf{R}^d)}^2 \, dt$. Show that there exists $C > 0$ such that for all $u \in E_T$,

$$k \int_0^T e^{k(t-T)^2} \|u(t)\|_{L^2(\mathbf{R}^d)}^2 \, dt \leq CI,$$

$$\int_0^T e^{k(t-T)^2} \left(\|u(t)\|_{H^1(\mathbf{R}^d)}^2 + \|D_t u(t)\|_{L^2(\mathbf{R}^d)}^2 \right) dt \leq C(1+kT^2)I.$$

9. Let $a = 0$, $b(\xi) = (1 + |\xi|^2)^{\frac{1}{2}}$, $B = \mathrm{Op}(b)$.

  a) What is the operator $Q = (D_t - iB)(D_t + iB)$?

  b) Using the estimates proved in question 8b) show that there exist positive constants C, T_0, k_0 such that for all $T \leq T_0, k \geq k_0$ and all $u \in E_T$ we have,

$$\int_0^T e^{k(t-T)^2} \left(\|u(t)\|_{H^1(\mathbf{R}^d)}^2 + \|D_t u(t)\|_{L^2(\mathbf{R}^d)}^2 \right) dt$$

$$\leq C(\frac{1}{k} + T^2) \int_0^T e^{k(t-T)^2} \|Qu(t)\|_{L^2(\mathbf{R}^d)}^2 dt,$$

where $Q = \partial_t^2 + \sum_{j=1}^d \partial_{x_j}^2$ is the Laplacian.

10. Deduce that we have the same estimate as in question 9b) for the operator

$$L = Q + c_0(t, x)\partial_t + \sum_{j=1}^d c_j(t, x)\partial_{x_j} + d(t, x),$$

where c_0, c_j, d belong to $L^\infty((0, T) \times \mathbf{R}^d)$.

Comments Carleman inequalities are L^2 estimates with weights of the form $e^{k\varphi}$. They have been introduced by T. Carleman in 1939 (see his article [8]). These inequalities are a basic tool for many subjects in PDE, like uniqueness in the Cauchy problem, control theory, inverse problems. A comprehensive treatment of this tool can be found in the book [18] by N. Lerner.   □

Problem 28 *Smoothing for the Heat Equation*

This problem uses Problem 20.

Let $T > 0$ and $a \in S^1$ be a real symbol satisfying,

$$\exists c > 0 : a(x, \xi) \geq c \langle \xi \rangle \quad \forall (x, \xi) \in \mathbf{R}^d \times \mathbf{R}^d, \quad \langle \xi \rangle = (1 + |\xi|^2)^{\frac{1}{2}}.$$

The goal of this problem is to prove that if $u \in C^0([0, T], L^2(\mathbf{R}^d))$ is a solution of the problem,

$$\partial_t u + \mathrm{Op}(a)u = f \in C^0([0, T], L^2(\mathbf{R}^d)), \quad u(0) = 0, \tag{6.5}$$

then for all $\varepsilon > 0$ the function $u(T) = u|_{t=T}$ belongs to $H^{1-\varepsilon}(\mathbf{R}^d)$.

Let $\sigma > 0$. We recall (see Problem 20) that for all $\alpha, \beta \in \mathbf{N}^d$ we have,

$$(\partial_x^\alpha \partial_\xi^\beta e^{-\sigma a})(x, \xi) = \sum_{j=0}^{|\alpha|+|\beta|} \sigma^j d_{j\alpha\beta}(x, \xi) e^{-\sigma a}, \quad d_{j\alpha\beta} \in S^{j-|\beta|}.$$

1. For $t \in [0, T)$ set $e_t(x, \xi) = e^{(t-T)a(x,\xi)}$. Show that for all $m \in [0, +\infty[$ and all $\alpha, \beta \in \mathbf{N}^d$ there exists $C_{m,\alpha\beta} > 0$ such that,

$$|\partial_x^\alpha \partial_\xi^\beta e_t(x, \xi)| \leq \frac{C_{m,\alpha\beta}}{(T - t)^m}(1 + |\xi|)^{-m-|\beta|}, \quad \forall t \in [0, T), \ \forall (x, \xi) \in \mathbf{R}^d \times \mathbf{R}^d.$$

2. Deduce that for all $\varepsilon > 0$ there exists $K > 0$ such that for all $v \in L^2(\mathbf{R}^d)$ and all $t \in [0, T)$ we have,

$$\| \operatorname{Op}(e_t)v \|_{H^{1-\varepsilon}(\mathbf{R}^d)} \leq \frac{K}{(T - t)^{1-\varepsilon}} \|v\|_{L^2(\mathbf{R}^d)},$$

$$\| (\operatorname{Op}(\partial_t e_t) - \operatorname{Op}(e_t) \operatorname{Op}(a)) v \|_{H^{1-\varepsilon}(\mathbf{R}^d)} \leq \frac{K}{(T - t)^{1-\varepsilon}} \|v\|_{L^2(\mathbf{R}^d)}.$$

3. Let u be the solution of (6.5). Show that

$$u(T) = \int_0^T (\operatorname{Op}(\partial_t e_t) - \operatorname{Op}(e_t) \operatorname{Op}(a)) u(t) \, dt + \int_0^T \operatorname{Op}(e_t) f(t) \, dt.$$

4. Conclude.

Problem 29 *A Characterization of Ellipticity*

Part 1
We introduce the space

$$E = \left\{ \varphi \in S(\mathbf{R}^d) : \widehat{\varphi} \in C_0^\infty(\mathbf{R}^d), \widehat{\varphi}(\eta) = 1 \text{ if } |\eta| \leq \frac{1}{2}, \widehat{\varphi}(\eta) = 0 \text{ if } |\eta| \geq 1 \right\}.$$

Let $m > 0$. We consider a family $(r_\lambda)_{\lambda \geq 1}$ of C^∞ functions on $\mathbf{R}^d \times \mathbf{R}^d$ such that: there exists $\lambda_0 \geq 1$ such that for all $\alpha \in \mathbf{N}^d$ there exists $C_\alpha > 0$ such that,

$$|D_\eta^\alpha r_\lambda(y, \eta)| \leq C_\alpha \lambda^{m-\frac{1}{2}}(1 + |y|), \quad \forall y \in \mathbf{R}^d, \ \forall |\eta| \leq 1, \ \forall \lambda \geq \lambda_0. \tag{6.6}$$

We consider $R_\lambda = \text{Op}(r_\lambda)$ the ΨDO with symbol r_λ defined on E. Show that for any $\varphi \in E$ there exists $\lambda_0 > 0$, $C(\varphi) > 0$ such that,

$$\|R_\lambda \varphi\|_{L^2(\mathbf{R}^d)} \leq C(\varphi)\lambda^{m-\frac{1}{2}}, \quad \forall \lambda \geq \lambda_0.$$

Hint: Prove a pointwise estimate of $y_j^N(R_\lambda \varphi)(y)$ for all $N \in \mathbf{N}$.

Part 2

Let $m > 0$ and $p \in S^m$ be a classical symbol, which means that $p \sim \sum_{j \geq 0} p_{m-j}$ where p_{m-j} is homogeneous of degree $m - j$. Assume in addition that $P = \text{Op}(p)$ satisfies the following condition: there exist two constants $0 \leq \delta < \frac{1}{2}$ and $C > 0$ such that

$$\|u\|_{H^{m-\delta}(\mathbf{R}^d)} \leq C\left(\|Pu\|_{L^2(\mathbf{R}^d)} + \|u\|_{L^2(\mathbf{R}^d)}\right), \quad \forall u \in C_0^\infty(\mathbf{R}^d). \tag{6.7}$$

The goal of this part is to prove that P is elliptic, which means that

$$\exists c > 0: \quad |p_m(x, \xi)| \geq c|\xi|^m, \quad \forall x \in \mathbf{R}^d, \forall \xi \neq 0.$$

Let $x^0 \in \mathbf{R}^d$, $\xi^0 \in \mathbf{S}^{d-1}$ and let $\varphi \in E$ be such that $\|\varphi\|_{L^2(\mathbf{R}^d)} = 1$. For $\lambda \geq 1$ we set,

$$u_\lambda(x) = \lambda^{\frac{d}{4}} e^{i\lambda x \cdot \xi^0} \varphi(\lambda^{\frac{1}{2}}(x - x^0)).$$

1. Compute the Fourier transform of u_λ in terms of that of φ.
2. Deduce that there exists $c_0 > 0$ such that $\|u_\lambda\|_{H^{m-\delta}(\mathbf{R}^d)} \geq c_0 \lambda^{m-\delta}$.
3. Show that,

$$e^{-i\lambda x \cdot \xi^0} P u_\lambda(x) = \lambda^{\frac{d}{4}}\left(Q_\lambda(y, D_y)\varphi\right)\left(\lambda^{\frac{1}{2}}(x - x^0)\right),$$

where Q_λ has symbol

$$q_\lambda(y, \eta) = p\left(x^0 + \lambda^{-\frac{1}{2}}y, \lambda\xi^0 + \lambda^{\frac{1}{2}}\eta\right).$$

4. Show that

$$q_\lambda(y, \eta) = p(x^0, \lambda\xi^0) + r_\lambda(y, \eta),$$

where r_λ satisfies the estimates (6.6).
   Hint: Use Taylor formula with integral reminder at the order 1.

5. Using the inequality (6.7) and the result of Part I show that,

$$\exists c_1 > 0, \ \exists \lambda_0 \geq 1 : |p(x^0, \lambda \xi^0)| \geq c_1 \lambda^{m-\delta}, \quad \forall \lambda \geq \lambda_0.$$

6. Using the fact that, p being classical, $p - p_m \in S^{m-1}$ deduce that,

$$\exists c_2 > 0, \ \exists \lambda_1 \geq 1 : |p_m(x^0, \lambda \xi^0)| \geq c_2 \lambda^{m-\delta}, \quad \forall \lambda \geq \lambda_1.$$

7. Show eventually that,

$$\exists c_3 > 0 \ : |p_m(x, \xi)| \geq c_3 |\xi|^m, \quad \forall x \in \mathbf{R}^d, \ \forall \xi \neq 0.$$

Comments

1. The converse statement of this problem is true. If P is elliptic in the sense given in the beginning of Part 2 then the estimate (6.7) is true with $\delta = 0$ (therefore with all $0 \leq \delta < \frac{1}{2}$). This is proved for instance in [1, 3, 30].
2. The result of the problem is false if $\delta = \frac{1}{2}$ since then P is no more necessarily elliptic. There are characterizations of the operators satisfying (6.7) with $\frac{1}{2} \leq \delta < 1, 1 \leq \delta < \frac{3}{2}$ and so on. These operators are called "subelliptic." Their characterization is more difficult and can be found in Chapter XXVII of the book by L. Hörmander [15].   □

Problem 30 *Local Solvability of Real Vector Fields*

Part 1
We consider, in a neighborhood of zero in $\mathbf{R}^d, d \geq 2$ in which the coordinates are denoted by $(t, x), t \in \mathbf{R}, x \in \mathbf{R}^{d-1}$, a differential operator of order one of the form,

$$X = \frac{\partial}{\partial t} + \sum_{j=2}^{d} a_j(t, x) \frac{\partial}{\partial x_j},$$

where the coefficients a_j are C^1 *real valued* functions.

1. We consider the differential system,

$$\dot{t}(s) = 1, \quad t(0) = 0, \qquad \dot{x}_j(s) = a_j(t(s), x(s)), \quad x_j(0) = y_j, \quad 2 \leq j \leq d.$$

Prove that for all $y = (y_2, \ldots, y_d)$ in a neighborhood of zero in $\mathbf{R}^{d-1}$ this system has a unique solution, denoted by $\varphi(s, y) = (s, x_2(s, y), \ldots, x_d(s, y))$, which is a C^1 function of all its arguments.

2. a) Prove that the Jacobian of φ does not vanish at $(0, 0)$.

   b) Deduce that there exists open neighborhoods U of $(t, x) = (0, 0)$ and V of $(s, y) = (0, 0)$ such that φ is a diffeomorphism from V to U.

   c) For $u \in C^1(U)$ we set $\tilde{u}(s, y) = u(\varphi(s, y))$. Show that,

$$\frac{\partial \tilde{u}}{\partial s}(s, y) = (Xu)(\varphi(s, y)).$$

3. Let b and f be two C^1 functions in U with values in $\mathbf{C}$ and let u_0 be a C^1 function in a neighborhood of $x = 0$. Show that the problem

$$Xu + bu = f, \quad u|_{t=0} = \widetilde{u_0}(x)$$

has a unique C^1 solution u in U.

Part 2

Let $x^0 \in \mathbf{R}^d$ and let V be a neighborhood of x^0. Let $b_j \in C^1(V, \mathbf{R})$, $1 \le j \le d$ and $c, g \in C^1(V, \mathbf{C})$. We assume that $\sum_{j=1}^{d} |b_j(x^0)| \ne 0$ and set

$$Y = \sum_{j=1}^{d} b_j(x)\frac{\partial}{\partial x_j}.$$

For $j_0 \in \{1, \ldots, d\}$ we set,

$$V^0 = \left\{ (x_1, \ldots, x_{j_0-1}, x_{j_0+1}, \ldots, x_d) : (x_1, \ldots, x_{j_0-1}, x_{j_0}^0, x_{j_0+1}, \ldots, x_d) \in V \right\}.$$

1. Prove that there exists $j_0 \in \{1, \ldots, d\}$ such that for all $u_0 \in C^1(V_{x_0})$ the problem,

$$Yu + cu = g, \quad u|_{\{x_{j_0}=x_{j_0}^0\}} = u_0(x_1, \ldots, x_{j_0-1}, x_{j_0+1}, \ldots, x_d)$$

has a unique C^1 solution u in a neighborhood $W \subset V$ of x^0.

Problem 31 *A Necessary Condition for Hypoellipticity*

Let Ω be an open subset of $\mathbf{R}^d$ and let,

$$P(x, D) = \sum_{j,k=1}^{d} a_{jk}(x)D_j D_k + \sum_{k=1}^{d} b_k(x)D_k + c(x), \quad D_j = \frac{1}{i}\frac{\partial}{\partial x_j} \qquad (6.8)$$

be a second order differential operator with C^∞ coefficients in Ω such that for all $x \in \Omega$ the matrix $A(x) = (a_{jk}(x))$ is symmetric with **real** coefficients. Let

$$p(x,\xi) = \sum_{j,k=1}^{d} a_{jk}(x)\xi_j\xi_k = \langle A(x)\xi, \xi \rangle$$

be the principal symbol of P. We assume that,

$$(H) \quad \exists (x^0, \xi^0) \in \Omega \times \mathbf{R}^d \setminus \{0\} : \quad p(x^0, \xi^0) = 0, \quad \sum_{\ell=1}^{d} \left| \frac{\partial p}{\partial \xi_\ell}(x^0, \xi^0) \right| \neq 0.$$

We will denote by $\mathcal{V}_{x^0}$ the set of open neighborhoods of x^0 in Ω.

We will admit that (H) implies,

$$(\star) \quad \exists V \in \mathcal{V}_{x^0}, \quad \exists \varphi \in C^\infty(V) : p(x, \operatorname{grad}\varphi(x))=0, \ \forall x \in V, \quad \operatorname{grad}\varphi(x^0)=\xi^0.$$

Part 1

Our goal is to prove, by contradiction, that P **is not** hypoelliptic in Ω.

1. Give an example of operator on the form (6.8) satisfying (H).

   We shall assume in what follows that P is hypoelliptic.

   Let $V \in \mathcal{V}_{x^0}$. We consider the space $E = \{u \in C^0(V) : Pu \in C^\infty(V)\}$ endowed with the semi-norms $p_{K,N}(u) = \sup_K |u| + \sum_{|\alpha|=N} \sup_K |D^\alpha Pu|$ where K is a compact subset of V and $N \in \mathbf{N}$.

2. Using the closed graph theorem prove that there exists a compact $K \subset V$, $N_0 \in \mathbf{N}$ and $C > 0$ such that,

$$|\operatorname{grad} u(x^0)| \leq C \left(\sup_K |u| + \sum_{|\alpha| \leq N_0} \sup_K |D^\alpha Pu| \right), \quad \forall u \in C^\infty(V). \tag{6.9}$$

3. Let $\varphi \in C^\infty(V)$ be given by $(\star)$ and $w \in C^\infty(V)$. Show that,

$$P(e^{i\lambda\varphi}w) = e^{i\lambda\varphi}(-i\lambda Lw + Pw)$$

where

$$L = \sum_{k=1}^{d} \frac{\partial p}{\partial \xi_k}(x, \operatorname{grad}\varphi(x)) \frac{\partial}{\partial x_k} + b, \ b \in C^\infty(V).$$

4. a) Prove that for all $\alpha \in \mathbf{N}^d$ and $f \in C^\infty(V)$ there exist $c_{\alpha k} \in C^\infty(V), 0 \le k \le |\alpha|$ such that,

$$\partial^\alpha (e^{i\lambda\varphi} f) = \sum_{k=0}^{|\alpha|} c_{\alpha k} \lambda^k e^{i\lambda\varphi}.$$

b) Deduce that if $K \subset V$ is compact and $\alpha \in \mathbf{N}^d$ there exists $C_{K,\alpha} > 0$ such that,

$$\sup_K |\partial^\alpha (e^{i\lambda\varphi} f)| \le C_{K,\alpha} \lambda^{|\alpha|}, \quad \forall \lambda \ge 1.$$

5. From (H) and $(\star)$ there exists ℓ_0 such that $\frac{\partial p}{\partial \xi_{\ell_0}}(x^0, \text{grad}\varphi(x^0)) \ne 0$. Let N_0 be defined in question 2. Show that the problems,

$$Lw_0 = 0, \; w_0|_{x_{\ell_0}=x_{\ell_0}^0} = 1, \; Lw_{j+1} = -iPw_j, \; w_{j+1}|_{x_{\ell_0}=x_{\ell_0}^0} = 0, \; 0 \le j \le N_0 - 1$$

have C^∞ solutions $w_0, w_1, \ldots, w_{N_0}$ in $V_1 \in \mathcal{V}_{x^0}, V_1 \subset V$.
   Hint: Use Problem 30.
6. Set $v = \sum_{j=0}^{N_0} \lambda^{-j} e^{i\lambda\varphi} w_j, \quad \lambda \ge 1.$

a) Show that $Pv = \lambda^{-N_0} e^{i\lambda\varphi} P w_{N_0}$.
b) Deduce that if $K \subset V_1$ is compact there exists $C_K > 0$ such that

$$\sup_K |\partial^\alpha Pv| \le C_K \lambda^{|\alpha|-N_0} \quad \text{and} \quad \sup_K |v| \le C_K.$$

c) Prove that there exists $C > 0$ such that $|\text{grad}v(x_0)| \ge C\lambda$ for λ large enough and deduce a contradiction.

Part 2
Let P be a differential operator of the form (6.8) and $x^0 \in \Omega$. Assume that there exists $\xi^1, \xi^2 \in \mathbf{R}^d \setminus \{0\}$ such that,

$$\left(A(x^0)\xi^1, \xi^1\right) > 0, \quad \left(A(x^0)\xi^2, \xi^2\right) < 0. \tag{6.10}$$

The goal is to prove that there exists $\xi^0 \in \mathbf{R}^d \setminus \{0\}$ such that (H) is satisfied.

1. Show that the matrix $A(x^0)$ has d real eigenvalues $\lambda_1, \ldots, \lambda_d$, not all vanishing and that it has at least one strictly positive eigenvalue λ_{j_0} and at least one strictly negative eigenvalue λ_{j_1}.
2. Set $D = \text{diag}(\lambda_j)$ and $q(\eta) = \langle D\eta, \eta \rangle$. Show that there exists $\eta^0 \in \mathbf{R}^d \setminus \{0\}$ such that $q(\eta^0) = 0$ and $\sum_{\ell=1}^d \left| \frac{\partial q}{\partial \eta_\ell}(\eta^0) \right| \ne 0$.

3. Deduce that there exists $\xi^0 \in \mathbf{R}^d \setminus \{0\}$ such that,

$$p(x^0, \xi^0) = 0 \quad \text{and} \quad \sum_{\ell=1}^{d} \left| \frac{\partial p}{\partial \xi_\ell}(x^0, \xi^0) \right| \neq 0.$$

4. State the conclusion of Parts 1 and 2 as a theorem.

Comments The result proved in this problem appears in the paper [13] by L. Hörmander. It was the motivation for this author to study the hypoellipticity of second order operators whose principal symbol does not change sign, namely the famous operators "sum of squares": $P = \sum_{j=1}^{r} X_j^2 + X_0 + c$.   □

Problem 32 *Wave Front Set of an Indicator Function*

Let $d \geq 2$. If $x = (x_1, \ldots, x_d) \in \mathbf{R}^d$ we set $x' = (x_1, \ldots, x_{d-1})$, so $x = (x', x_d)$. We consider the function $u \in L^\infty(\mathbf{R}^d)$ defined by,

$$u(x) = 1 \text{ if } x_d > 0, \quad u(x) = 0 \text{ if } x_d < 0.$$

1. Show that $\mathrm{WF}(u) \subset \{(x, \xi) \in \mathbf{R}^d \times (\mathbf{R}^d \setminus \{0\}) : x_d = 0\}$.
2. a) Let $\xi_0 \in \mathbf{R}^d \setminus \{0\}$ be such that $\xi_0' \neq 0$. Show that there exist $\varepsilon > 0$ and $c_0 > 0$ such that,

$$\forall \xi \in \Gamma_{\xi_0} := \left\{ \xi \in \mathbf{R}^d \setminus \{0\} : \left| \frac{\xi}{|\xi|} - \frac{\xi_0}{|\xi_0|} \right| < \varepsilon \right\} \quad \text{we have} \quad |\xi'| \geq c_0 |\xi|.$$

   b) Deduce that in Γ_{ξ_0} we have $|\xi| \approx |\xi'|$.
   c) Let $(x_0, \xi_0) \in \mathbf{R}^d \times (\mathbf{R}^d \setminus \{0\})$ such that $(x_0)_d = 0$, $\xi_0' \neq 0$. Let $\varphi \in C_0^\infty(\mathbf{R}^d)$ be such that $\varphi(x) = 1$ near x_0. Prove by integrating by parts that,

$$\xi \in \Gamma_{\xi_0} \implies \widehat{\varphi u}(\xi) = O(|\xi'|^{-N}) \quad \forall N \in \mathbf{N}, \quad |\xi| \to +\infty.$$

   Deduce that $\mathrm{WF}(u) \subset \{(x, \xi) \in \mathbf{R}^d \times (\mathbf{R}^d \setminus \{0\}) : x_d = 0, \xi' = 0\}$.
3. Let $\rho_0 = (x_0, \xi_0) \in \mathbf{R}^d \times (\mathbf{R}^d \setminus \{0\})$ such that $(x_0)_d = 0$, $\xi_0' = 0$. Show that $\rho_0 \in \mathrm{WF}(u)$. Conclude.

Problem 33

The goal of this problem is to determine the wave front set of the continuous function on $\mathbf{R}$ given by,

$$u(x) = \int_0^{+\infty} \frac{e^{ix\eta}}{1+\eta^2}\, d\eta.$$

1. a) Show that for any $k \in \mathbf{N}$ there exists a polynomial P_k of degree $\leq k$ such that,

$$\left(\frac{d}{d\eta}\right)^k \left(\frac{1}{1+\eta^2}\right) = \frac{P_k(\eta)}{(1+\eta^2)^{k+1}}, \quad \forall \eta \geq 0.$$

   b) Deduce that for any $k \in \mathbf{N}$ there exists $C_k > 0$ such that,

$$\left|\left(\frac{d}{d\eta}\right)^k \left(\frac{1}{1+\eta^2}\right)\right| \leq \frac{C_k}{(1+\eta)^{k+2}}, \quad \forall \eta \geq 0.$$

2. a) Prove that for all $N \in \mathbf{N}^*$ there exist $a_{j,N} \in \mathbf{C}, 0 \leq j \leq N-1$ such that for $x \neq 0$ we have

$$u(x) = \sum_{j=0}^{N-1} \frac{a_{j,N}}{x^j} + \frac{i^{N-1}}{x^{N-1}} \int_0^{+\infty} e^{ix\eta} \left(\frac{d}{d\eta}\right)^{N-1} \left(\frac{1}{1+\eta^2}\right) d\eta.$$

   b) Deduce that $WF(u) \subset \{(0, \xi) : \xi \in \mathbf{R} \setminus \{0\}\}$.
3. Let $\varphi \in C_0^\infty(\mathbf{R})$ be such that $\varphi(x) = 1$ if $|x| \leq \frac{1}{2}$ and $\varphi(x) = 0$ if $|x| \geq 1$.

   a) Show that $\widehat{\varphi u}(\xi) = \int_0^{+\infty} \widehat{\varphi}(\xi - \eta)\frac{1}{1+\eta^2}\, d\eta$.
   b) Show that $\widehat{\varphi u}(\xi) = O(|\xi|^{-N})$, $\forall N \in \mathbf{N}$, $\xi \to -\infty$. Deduce that $WF(u) \subset \{(0, \xi) : \xi > 0\}$.

4. Let $\psi \in C_0^\infty(\mathbf{R})$ with values in $(0 + \infty)$ be even and such that $\int \psi(y)^2\, dy = 1$. Set $\varphi = \psi \star \psi$.

   a) Prove that $\varphi(0) = 1$, $\varphi'(0) = 0$, $\widehat{\varphi} \geq 0$, $\widehat{\varphi}(0) > 0$.
   b) Using question 3a) and noticing that $\widehat{\varphi u}(\xi) \geq \int_A \widehat{\varphi}(\xi - \eta)\frac{1}{1+\eta^2}\, d\eta$ where $A = \{\eta \in (0, +\infty) : |\xi - \eta| \leq \varepsilon\}$, $\varepsilon > 0$ small to be chosen, show that there exists $C > 0$ such that $\xi^2 \widehat{\varphi u}(\xi) \geq C$ for any $\xi > 0$ large enough.
   c) Let $\varphi_1 \in C_0^\infty(\mathbf{R})$ be such that $\varphi_1 = 1$ in a neighborhood of φ. Set $\theta = \varphi - \varphi_1$. Show that $\theta(x) = x^2\theta_1(x)$ where $\theta_1 \in C_0^\infty(\mathbf{R})$. Using questions 3a) and 1b) show that,

$$|\widehat{\theta u}(\xi)| \leq |(\widehat{\theta_1})'(\xi)| + C \int_0^{+\infty} |\widehat{\theta_1}(\xi - \eta)|\frac{1}{(1+\eta)^4}\, d\eta.$$

Deduce that for $\varepsilon > 0$ small enough there exists $C > 0$ such that,

$$\xi^{3-\varepsilon}|\widehat{\theta u}(\xi)| \leq C, \quad \forall \xi \geq 1.$$

d) Deduce that there exists $C > 0$ such that $\xi^2 \widehat{\varphi_1 u}(\xi) \geq C$ for $\xi > 0$ large enough, and then that $WF(u) = \{(0, \xi) : \xi > 0\}$.

Problem 34 *A Problem on the Wave Front Set*

Let $u \in \mathcal{D}'(\mathbf{R}^2)$ be such that $x_1 u \in C^\infty(\mathbf{R}^2)$, $x_2 u \in C^\infty(\mathbf{R}^2)$ but $u \notin C^\infty(\mathbf{R}^2)$. We want to determine the wave front set of u.

1. Show that $WF(u) \subset \{(x, \xi) \in \mathbf{R}^2 \times \mathbf{R}^2 \setminus \{0\} : x = 0\}$.
2. We assume in what follows that there exists $\xi_0 \in \mathbf{R}^2 \setminus \{0\}$ such that $(0, \xi_0) \notin WF(u)$. We set $\xi_0 = \lambda \omega_0$, $\lambda \neq 0$, $\omega_0 \in \mathbf{S}^1$. Let $\varphi \in C_0^\infty(\mathbf{R}^2)$ be equal to 1 near zero.

   a) Show that $\frac{\partial}{\partial \xi_j}(\widehat{\varphi u}) \in \mathcal{S}(\mathbf{R}^2)$, $j = 1, 2$.
   b) Let $\xi = \lambda \omega$, $\lambda > 0$, $\omega \in \mathbf{S}^1$ be such that the angle $\theta = (\omega, \omega_0)$ is strictly less than π. Show that there exists $c > 0$ such that $|t\omega + (1 - t)\omega_0| \geq c$.
      Hint: Compute the minimum of the function $t \mapsto |t\omega + (1 - t)\omega_0|^2$ from $[0, 1]$ to $\mathbf{R}$.
   c) Applying the Taylor formula at the order one to the function $\varphi(\lambda \omega')$ where ω' is close to ω, show that $(0, \xi) \notin WF(u)$.
   d) Deduce that u is C^∞ in a neighborhood of zero and determine $WF(u)$.

3. Give an alternative proof of this result using the theorem of propagation of singularities for the operator $P = x_1 + \lambda x_2$ (or $P = \mu x_1 + x_2$) for an appropriate λ (or μ) in $\mathbf{R}$.

Problem 35 *On Propagation of Singularities*

This problem is an application of the theorem on propagation of singularities for a concrete operator.
   We denote by P the differential operator in $\mathbf{R}^2$ given by,

$$P = x\frac{\partial^2}{\partial y^2} - \frac{\partial^2}{\partial x^2}$$

and let $p(x, y, \xi, \eta) = \xi^2 - x\eta^2$ be its symbol. We set $H = \{(x, y) \in \mathbf{R}^2) : x < 0\}$.

1. Let $u \in \mathcal{D}'(\mathbf{R}^2)$ be such that $Pu \in C^\infty(\mathbf{R}^2)$. Prove that $u \in C^\infty(H)$.
2. Give the equations of the bicharacteristics of p. Solve these equations.
3. Let $v \in \mathcal{D}'(\mathbf{R}^2)$ be such that $Pv \in C^\infty(\mathbf{R}^2)$. Prove that if v is C^∞ in a neighborhood of the set $\{(x, y) \in \mathbf{R}^2 : x = 0\}$ then $v \in C^\infty(\mathbf{R}^2)$.

Problem 36 *A Microlocal Defect Measure*

For $\xi_0 \in \mathbf{S}^{d-1}$ and $k \geq 1$ we set $u_k(x) = b(x)e^{ikx \cdot \xi_0}$ where $b \in C_0^\infty(\mathbf{R}^d)$.

1. Prove that the sequence $(u_k)_{k \in \mathbf{N}^*}$ is bounded in $L^2_{\text{loc}}(\mathbf{R}^d)$ and converges to zero weakly in the sense of (5.13).

   The purpose of this problem is to show that this sequence has a defect measure μ given by $\mu = |b(x)|^2 dx \otimes \delta_{\frac{\xi_0}{|\xi_0|}}$.

   Let A be a pseudo-differential operator whose symbol a has compact support in x and is homogeneous of degree zero for $|\xi| \geq 1$.

   Let χ be in $C_0^\infty(\mathbf{R}^d, \mathbf{R})$ equal to one on the support in x of the symbol of A. Set $K = \text{supp } \chi$ and,

$$I_k = (A(\chi u_k), \chi u_k)_{L^2(\mathbf{R}^d)}.$$

2. Prove that we have,

$$I_k = \iint_{\mathbf{R}^d \times \mathbf{R}^d} f_k(x, \eta)\, dx\, d\eta, \quad \text{where,}$$

$$f_k(x, \eta) = (2\pi)^{-d} e^{ix \cdot \eta} a(x, \eta + k\xi_0) \widehat{\chi b}(\eta) \chi(x)\overline{b(x)}.$$

We set

$$J_k = \iint_{|\eta| \geq \frac{1}{2}k} f_k(x, \eta)\, dx\, d\eta, \quad H_k = \iint_{|\eta| \leq \frac{1}{2}k} f_k(x, \eta)\, dx\, d\eta.$$

3. a) Prove that $\lim_{k \to +\infty} J_k = 0$.
   b) Prove that when $k \geq 2$ and $|\eta| \leq \frac{1}{2}k$ we have,

$$a_0(x, \eta + k\xi_0) = a_0\left(x, \xi_0 + \frac{\eta}{k}\right) = a(x, \xi_0) + \frac{1}{k}c_k(x, \eta),$$

   where c_k is such that $|c_k(x, \eta)| \leq C|\eta|$ where C is independent of k.
   c) Prove that when $k \to +\infty$ we have,

$$H_k = (2\pi)^{-d} \iint_{|\eta| \leq \frac{1}{2}k} e^{ix \cdot \eta} a(x, \xi_0)\widehat{\chi b}(\eta)\chi(x)\overline{b(x)}\, d\eta\, dx + o(1).$$

d) Prove that when $k \to +\infty$ we have,

$$(2\pi)^{-d} \iint_{|\eta| \leq \frac{1}{2}k} e^{ix\cdot\eta} a(x, \xi_0) \widehat{\chi b}(\eta) \chi(x) \overline{b(x)} \, d\eta \, dx$$

$$= \int_{\mathbf{R}^d} a(x, \xi_0) |b(x)|^2 \, dx + o(1).$$

e) Conclude.
4. The goal of this question is to prove that the same result holds for $u_k(x) = b(x) e^{ikx\cdot\xi_0}$ where $b \in L^2_{\text{loc}}(\mathbf{R}^d)$. Let $\chi_1 \in C_0^\infty(\mathbf{R}^d)$ be equal to one on a neighborhood of supp χ. We set $\widetilde{b} = \chi_1 b \in L^2(\mathbf{R}^d)$. There exists $(b_n)_{n\in\mathbf{N}} \subset C_0^\infty(\mathbf{R}^d)$ converging to $\widetilde{b}$ in $L^2(\mathbf{R}^d)$. Set,

$$I_k = \left(A(\chi(\widetilde{b} e^{ikx\cdot\xi_0}), \chi \widetilde{b} e^{ikx\cdot\xi_0} \right)_{L^2} - \int_{\mathbf{R}^d} |\widetilde{b}(x)|^2 a(x, \xi_0) |b(x)|^2 \, dx$$

and write $I_k = (1) + (2) + (3) + (4)$ where,

$$(1) = \left(A(\chi(\widetilde{b} - b_n)e^{ikx\cdot\xi_0}), \chi \widetilde{b} e^{ikx\cdot\xi_0} \right)_{L^2},$$

$$(2) = \left(A(\chi b_n e^{ikx\cdot\xi_0}), \chi(\widetilde{b} - b_n)e^{ikx\cdot\xi_0}) \right)_{L^2},$$

$$(3) = \left(A(\chi b_n e^{ikx\cdot\xi_0}), \chi b_n e^{ikx\cdot\xi_0} \right)_{L^2} - \int_{\mathbf{R}^d} |b_n(x)|^2 a(x, \xi_0) \, dx,$$

$$(4) = - \int_{\mathbf{R}^d} |\widetilde{b}(x)|^2 a(x, \xi_0) |b(x)|^2 \, dx + \int_{\mathbf{R}^d} |b_n(x)|^2 a(x, \xi_0) \, dx.$$

a) Prove that there exists a constant $C_0 > 0$ independent of k, n such that,

$$|(1)| + |(2)| + |(4)| \leq C(\|b_n\|_{L^2} + \|\widetilde{b}\|_{L^2})\|b_n - \widetilde{b}\|_{L^2}.$$

b) Prove that $\lim_{k\to+\infty} I_k = 0$ and conclude.

Chapter 7
The Classical Equations

In this chapter, we review some of the main properties of the solutions to the classical partial differential equations. This includes results about equations with analytic coefficients, the Laplace equation, the wave equation, the heat equation, the Schrödinger equation, the Burgers equation, and also the Euler and Navier–Stokes equations in fluid dynamics.

7.1 Equations with Analytic Coefficients

First of all let us recall some definitions concerning the concept of analyticity.

- Let $\Omega \subset \mathbf{R}^d$ be open. A function $f : \Omega \to \mathbf{C}$ is said to be real analytic in Ω if every point $x_0 \in \Omega$ has a neighborhood V_{x_0} in which one can write,

$$f(x) = \sum_{\alpha \in \mathbf{N}^d} a_\alpha (x - x_0)^\alpha \qquad (a_\alpha \in \mathbf{C}),$$

  where the series is absolutely convergent.
- In $\mathbf{C}^d$ denote by $z = (z_1, \ldots, z_d)$ the variable, where $z_j = x_j + i y_j$, $x_j, y_j \in \mathbf{R}$.
  Let $O \subset \mathbf{C}^d$ be open. A function $f : O \to \mathbf{C}$ is said to be holomorphic if it is C^1 with respect to the real variables (x_j, y_j), $1 \leq j \leq d$ and satisfies,

$$\bar{\partial}_j f = 0 \quad \text{in } O \quad \text{for } j = 1, \ldots, d, \quad \text{where } \bar{\partial}_j = \frac{1}{2}\left(\frac{\partial}{\partial x_j} + i \frac{\partial}{\partial y_j} \right).$$

- There is of course a tight link between these two notions. If f is real analytic in a neighborhood V_{x_0} of x_0 in $\mathbf{R}^d$ one can find a neighborhood $\widetilde{V}_{x_0}$ of x_0 in $\mathbf{C}^d$ and

© The Editor(s) (if applicable) and The Author(s), under exclusive license
to Springer Nature Switzerland AG 2020
T. Alazard, C. Zuily, *Tools and Problems in Partial Differential Equations*,
Universitext, https://doi.org/10.1007/978-3-030-50284-3_7

a holomorphic function $\tilde{f}$ in $\tilde{V}_{x_0}$ such that $\tilde{f} = f$ on V_{x_0}. On the other hand, if $\tilde{f}$ is holomorphic in $\tilde{V}_{x_0}$ the restriction of $\tilde{f}$ to $\tilde{V}_{x_0} \cap \mathbf{R}^d$ is real analytic.

- If φ is a C^1 real valued function in a neighborhood V_{x_0} of $x_0 \in \mathbf{R}^d$ we set

$$S = \left\{ x \in V_{x_0} : \varphi(x) = \varphi(x_0) \right\}, \tag{7.1}$$

and we assume that $d\varphi(x) \neq 0$ on S, where $d\varphi$ is the differential of φ. Then S is a C^1 hypersurface.

We shall denote by n the unit normal to S and by $\frac{\partial}{\partial n}$ the normal derivative.

7.1.1 The Cauchy–Kovalevski Theorem

The Cauchy–Kovalevski theorem concerns the question of existence of local solutions to the Cauchy problem for partial differential equations in the framework of real analytic or holomorphic functions.

We assume in this paragraph that the function φ appearing in (7.1) is real analytic.

7.1.1.1 The Linear Version

- Let $m \in \mathbf{N}$, $m \geq 1$ and let $P = \sum_{|\alpha| \leq m} a_\alpha(x) D^\alpha$ be a linear differential operator in a neighborhood of x_0 with real analytic coefficients and principal symbol $p_m = \sum_{|\alpha|=m} a_\alpha(x)\xi^\alpha$. Assume that S is non-characteristic for P that is,

$$p_m(x, d\varphi(x)) \neq 0, \quad \forall x \in S. \tag{7.2}$$

Then for every real analytic functions $u_0, \ldots, u_{m-1}$ on S and every real analytic function f near x_0 the problem,

$$Pu = f, \quad \left(\frac{\partial}{\partial n}\right)^j u \bigg|_S = u_j, \quad 0 \leq j \leq m-1,$$

has a unique real analytic solution u in a neighborhood of x_0.
- Notice that if we make a real analytic change of coordinates from x to $(t, y) \in \mathbf{R} \times \mathbf{R}^{d-1}$ in which x_0 becomes $(0, y_0)$ and S becomes $\{t = 0\}$ (which is always possible locally) then condition (7.2) ensures that the coefficient of D_t^m in the operator P in the new coordinates does not vanish at $(0, y_0)$.

7.1.1.2 The Nonlinear Version

Let $m \in \mathbf{N}$, $m \geq 1$. We set $N = \sharp \{\alpha \in \mathbf{N}^d : |\alpha| \leq m\}$ and $p = (p_\alpha)_{|\alpha| \leq m} \in \mathbf{R}^N$.

Then we have the following result.

- Let $p^0 \in \mathbf{R}^N$ and let $F : V_{x_0} \times \mathbf{R}^N \to \mathbf{R}$ be a real analytic function in a neighborhood of (x_0, p^0) in $\mathbf{R}^d \times \mathbf{R}^N$. Assume that,

$$(i) \quad F(x_0, p^0) = 0, \qquad (ii) \quad \sum_{|\alpha|=m} \frac{\partial F}{\partial p_\alpha}(x_0, p^0)(d\varphi(x_0))^\alpha \neq 0.$$

Let $\tilde{u}$ be any real valued real analytic function near x_0 such that $\partial^\alpha \tilde{u}(x_0) = p_\alpha^0$ for $|\alpha| \leq m$.

Then the Cauchy problem,

$$F(x, (\partial^\alpha u(x))) = 0, \qquad \left(\frac{\partial}{\partial n}\right)^j u \bigg|_S = \left(\frac{\partial}{\partial n}\right)^j \tilde{u} \bigg|_S, \qquad 0 \leq j \leq m-1,$$

has a unique real valued real analytic solution u near x_0.

Comments

i) If we want to consider non-real solutions one has to assume holomorphy for F and then the same result holds where *real analytic* is replaced by *holomorphic*.

ii) Nevertheless if F is a polynomial with respect to the variable p then the above result still holds in the category of real analytic functions even for non-real data.

iii) A short abstract proof of the nonlinear Cauchy-Kovalevski theorem is given by T. Nishida in [23]. $\square$

7.1.2 The Holmgren Uniqueness Theorem

We assume in this paragraph that the function φ appearing in (7.1) is C^1.

- Let $P = \sum_{|\alpha| \leq m} a_\alpha(x) D^\alpha, m \geq 1$, be a linear differential operator with real analytic coefficients in V_{x_0}. Assume that,

$$p_m(x_0, d\varphi(x_0)) \neq 0,$$

where p_m is the principal symbol of P.

Then there exists a neighborhood $W_{x_0} \subset V_{x_0}$ of x_0 such that if $u \in \mathcal{D}'(V_{x_0})$ satisfies,

$$Pu = 0 \text{ in } V_{x_0}, \qquad u = 0 \text{ when } \varphi(x) < \varphi(x_0),$$

then $u = 0$ in W_{x_0}.

- The above result implies the same conclusion when $\varphi \in C^m$ and u is a C^m solution of the problem,

$$Pu = 0 \text{ in } V_{x_0}, \quad \left(\frac{\partial}{\partial n}\right)^j u\bigg|_S = 0, \quad 0 \le j \le m - 1.$$

Comments A proof of the Holmgren Theorem can be found in Chapter 3 of the book by F. John [16].   □

7.2   The Laplace Equation

The Laplace operator (or the "Laplacian") is given by

$$\Delta = \sum_{j=1}^d \frac{\partial^2}{\partial x_j^2}.$$

7.2.1   The Mean Value Property

Recall that a harmonic function is a solution of $\Delta u = 0$.

Let Ω be an open subset of $\mathbf{R}^d$ and $u \in C^2(\Omega)$. Then u is harmonic if and only if u satisfies the mean value property: for any $x \in \Omega$ and any $r > 0$ such that $B(x, r) \subset \Omega$,

$$u(x) = \frac{1}{|B(x,r)|} \int_{B(x,r)} u(y)\,dy.$$

where $|B|$ denotes the Lebesgue measure of the set B.

7.2.2   Hypoellipticity: Analytic Hypoellipticity

Recall that a fundamental solution of a constant coefficient differential operator P is a distribution $E \in \mathcal{D}'(\mathbf{R}^d)$ satisfying $PE = \delta_0$.

- A fundamental solution of the Laplacian is given by,

$$E = \frac{1}{2\pi}\mathrm{Log}\,|x| \quad \text{if } d = 2, \quad E = \frac{|x|^{2-d}}{|S^{d-1}|(2-d)} \quad \text{if } d \ge 3.$$

- The Laplacian is *hypoelliptic* (resp. *analytic hypoelliptic*). This means the following. Let u be a distribution in an open set $\Omega \subset \mathbf{R}^d$. For every open set $\omega \subset \Omega$ if Δu belongs to $C^\infty(\omega)$ (resp. is real analytic in ω) then u belongs to $C^\infty(\omega)$ (resp. is real analytic in ω).

These properties are shared by all elliptic operators with C^∞ (resp. real analytic) coefficients.

7.2.3 The Maximum Principles

- **The weak maximum principle.** Let Ω be a **bounded** open subset of $\mathbf{R}^d$. Let u be a real valued function belonging to $C^0(\overline{\Omega}) \cap C^2(\Omega)$ such that $\Delta u(x) \geq 0$ (resp. $\Delta u(x) \leq 0$) for all $x \in \Omega$. Then for all $x \in \Omega$ we have,

$$u(x) \leq \sup_{y \in \partial\Omega} u(y), \quad (\text{resp. } u(x) \geq \inf_{y \in \partial\Omega} u(y)).$$

- **The strong maximum principle.** Let Ω be a **bounded** and **connected** open subset of $\mathbf{R}^d$. Let u be a real valued function belonging to $C^0(\overline{\Omega}) \cap C^2(\Omega)$ such that $\Delta u(x) \geq 0$ (resp. $\Delta u(x) \leq 0$) for all $x \in \Omega$. Then if u reaches its maximum (resp. its minimum) at a point $x \in \Omega$ then u is constant.
- **The Hopf Lemma.** Let $x_0 \in \mathbf{R}^d$ and $R > 0$. Set $B = \left\{ x \in \mathbf{R}^d : |x - x_0| < R \right\}$. Denote by ∂B its boundary, by n the unitary normal to the boundary and by $\frac{\partial}{\partial n}$ the normal derivative.

   Let u be a C^2 function in a neighborhood of $\overline{B}$ such that $\Delta u(x) \geq 0$, $\forall x \in B$. Assume that there exists a point $y \in \partial B$ such that $u(y) > u(x)$ for all $x \in B$. Then

$$\frac{\partial u}{\partial n}(y) > 0.$$

7.2.4 The Harnack Inequality

- Let Ω be an open subset of $\mathbf{R}^d$ and let u be a nonnegative $C^2(\Omega)$ solution of $\Delta u = 0$. Then for any bounded open subset $O \subset \Omega$ there exists $C > 0$ depending only on d, O, Ω such that,

$$\sup_O u \leq C \inf_O u.$$

7.2.5 The Dirichlet Problem

A relevant problem for the Laplacian is the *Dirichlet problem*. Roughly speaking the question is the following.

Given an open subset Ω in $\mathbf{R}^d$, whose boundary is denoted by Γ, given a distribution f on Ω and u_0 defined on Γ, solve the problem,

$$\Delta u = f \text{ in } \Omega, \quad u = g \text{ on } \Gamma. \tag{7.3}$$

Of course the spaces where f and u_0 are to be taken should be specified. Here are some classical results.

7.2.5.1 Case $g = 0$

- Let $\lambda \geq 0$ and let Ω be an open subset of $\mathbf{R}^d$ ($\lambda > 0$ if Ω is not bounded). Then the operator $P = -\Delta + \lambda$ is a linear isomorphism from $H_0^1(\Omega)$ on $H^{-1}(\Omega)$.
- Assume that Ω is a C^∞ open set and let $\lambda \geq 0$ ($\lambda > 0$ if Ω is not bounded). For any $k \in \mathbf{N}$, the operator $-\Delta + \lambda$ is a linear isomorphism from $H^{k+2}(\Omega) \cap H_0^1(\Omega)$ on $H^k(\Omega)$.
- Under the above conditions, $-\Delta + \lambda$ is a linear isomorphism from $\bigcap_{k \in \mathbf{N}} H^k(\Omega) \cap H_0^1(\Omega)$ to $\bigcap_{k \in \mathbf{N}} H^k(\Omega)$.
- If Ω is a C^∞ bounded open set we have $\bigcap_{k \in \mathbf{N}} H^k(\Omega) = C^\infty(\overline{\Omega})$. Otherwise we have $C_0^\infty(\overline{\Omega}) \subset \bigcap_{k \in \mathbf{N}} H^k(\Omega) \subset C^\infty(\overline{\Omega})$.

7.2.5.2 Case $g \not\equiv 0$.

- Let Ω be a C^1 open subset of $\mathbf{R}^d$. Let $\lambda \geq 0$ ($\lambda > 0$ if Ω is not bounded). Let $f \in H^{-1}(\Omega), g \in H^{\frac{1}{2}}(\partial\Omega)$. There exists a unique $u \in H^1(\Omega)$ such that,

$$(-\Delta + \lambda) u = f \text{ in } \mathcal{D}'(\Omega), \quad \gamma_0 u = g \text{ on } \Gamma.$$

7.2.6 Spectral Theory

- Let Ω be a bounded open subset of $\mathbf{R}^d$. The problem

$$-\Delta u = \lambda u, \quad u \in H_0^1(\Omega) \tag{7.4}$$

is solvable with $u \neq 0$ if and only if λ belongs to a sequence $(\lambda_j)_{j \in \mathbf{N}^*}$ of strictly positive real numbers (called *eigenvalues*) which tends to $+\infty$ when $j \to +\infty$. We can order them as,

$$0 < \lambda_1 \leq \lambda_2 \leq \cdots \leq \lambda_j \leq \cdots$$

where each eigenvalue is repeated according to its finite multiplicity.

- There exists an orthonormal basis $(e_j)_{j \geq 1}$ of $L^2(\Omega)$ such that,

$$- \Delta e_j = \lambda_j e_j, \quad e_j \in H_0^1(\Omega). \tag{7.5}$$

The $e'_j s$ are called *eigenfunctions*. The eigenspace corresponding to each eigenvalue is finite dimensional.

By the hypoellipticity of the Laplacian we have $e_j \in C^\infty(\Omega)$ and, if Ω has a C^∞ boundary, we have $e_j \in C^\infty(\overline{\Omega})$.

- (e_j) is also an orthogonal basis of $H_0^1(\Omega)$ and we have $\|e_j\|_{H_0^1(\Omega)} = \sqrt{\lambda_j}$. More generally, for every $k \in \mathbf{N}$ we have $\|e_j\|_{H^k(\Omega)} \leq C \lambda_j^{\frac{k}{2}}$.

7.2.6.1 Weyl Law

Let Ω be a bounded open subset of $\mathbf{R}^d$ with C^∞ boundary. For $\lambda > 0$ set,

$$N(\lambda) = \# \{ j \geq 1 : \lambda_j \leq \lambda \}.$$

Then we have,

$$N(\lambda) \sim (2\pi)^d C_d \text{vol}\,(\Omega) \lambda^{\frac{d}{2}}, \quad \lambda \to +\infty, \tag{7.6}$$

where vol (Ω) is the Lebesgue measure of Ω and C_d is the Lebesgue measure of the unit ball in $\mathbf{R}^d$.

- As a corollary we have,

$$\lambda_j \sim \frac{(2\pi)^2}{[C_d \text{vol}\,(\Omega)]^{\frac{2}{d}}} j^{\frac{2}{d}}, \quad j \to +\infty.$$

7.2.6.2 Estimates of the Eigenfunctions

- Let $(e_j)_{j \in \mathbf{N}^*}$ be an orthonormal basis of $L^2(\Omega)$ made of eigenfunctions. Then by the estimates of the H^k norms of e_j and the Sobolev embedding one can see that $\|e_j\|_{L^\infty(\Omega)} \leq C \lambda_j^{\frac{d}{4}+\varepsilon}$, $\varepsilon > 0$. However, this is not the best result. Indeed we have the following better bound:

$$\exists C > 0 : \forall j \geq 1, \quad \|e_j\|_{L^\infty(\Omega)} \leq C\lambda_j^{\frac{d-1}{4}}, \tag{7.7}$$

and these estimates are optimal in general. (See Problem 39 for the case where Ω is the unit a ball in $\mathbf{R}^3$ and Problem 40 for the case of the sphere.)
- If we call E_j the eigenspace corresponding to the eigenvalue λ_j then there exists $C > 0$ independent of j such that,

$$\dim E_j \leq C\lambda_j^{\frac{d-1}{2}}.$$

(See Problem 39.)

Comments A comprehensive introduction to the study of the Laplace equation can be found in Chapters 1–9 of D. Gilbarg and N.S. Trudinger [11] and in L.C. Evans [9]. The spectral theory for the Laplace operator, including a proof of the Weyl law, can be found in Chapter 13 of the book by C. Zuily [33]. More advanced results concerning the spectral theory can be found in the book by C.D. Sogge [27].

□

7.3   The Heat Equation

The heat operator in $\mathbf{R}_t \times \mathbf{R}_x^d$ is given by $P = \frac{\partial}{\partial t} - \Delta_x$ where Δ_x is the Laplacian.
- A fundamental solution of this operator is given by the function,

$$E(t, x) = \frac{H(t)}{(4\pi t)^{\frac{d}{2}}} e^{-\frac{|x|^2}{4t}},$$

where $H(t)$ is the Heaviside function, $H(t) = 1$ if $t > 0$, $H(t) = 0$ if $t \leq 0$.
- The heat operator is hypoelliptic.

7.3.1   The Maximum Principle

- Let Ω be an open bounded subset of $\mathbf{R}_x^d$ and $T > 0$. Set $Q = \Omega \times (0, T)$ and denote by ∂Q the boundary of Q.
  If $u \in C^2(Q) \cap C^0(\overline{Q})$ is such that $\frac{\partial u}{\partial t} - \Delta_x u \leq 0$ in Q then,

$$\sup_{\overline{Q}} u = \sup_{\partial Q} u.$$

7.3.2 The Cauchy Problem

Roughly speaking, when $\Omega \neq \mathbf{R}^d$, the problem is the following. Given f defined in $Q = (0, T) \times \Omega$, u_0 defined in Ω and g defined on $(0, T) \times \partial\Omega$ (belonging to some spaces) find u defined on Q satisfying,

$$\frac{\partial u}{\partial t} - \Delta_x u = f, \quad u|_{t=0} = u_0, \quad \gamma_0 u(t, \cdot) = g(t, \cdot), \ t \in (0, T), \qquad (7.8)$$

where γ_0 denotes the trace on the boundary of Ω (when $\Omega = \mathbf{R}^d$ we remove this last condition on $\gamma_0 u$). For instance if $g = 0$ we require $u(t, \cdot) \in H_0^1(\Omega)$.

Here is an example of result which holds. Let Ω be an open bounded subset of $\mathbf{R}_x^d$ and $T > 0$. Set $Q = \Omega \times (0, T)$.

- Let $f \in L^2(Q)$, $u_0 \in L^2(\Omega)$, $g = 0$. Then the Cauchy problem (7.8) has a unique solution u in $C^0([0, T], L^2(\Omega)) \cap L^2((0, T), H_0^1(\Omega))$.
- Assume Ω bounded and let $(e_n)_{n \in \mathbf{N}}$, $e_n \in H_0^1(\Omega)$, be an orthonormal basis on $L^2(\Omega)$ consisting of eigenfunctions of the Laplacian, which means $-\Delta e_n = \lambda_n e_n$ where $\lambda_n > 0$. Then the solution in the above result has the following form. Write,

$$u(t, x) = \sum_{n \geq 0} u_n(t) e_n(x), \quad u_0 = \sum_{n \geq 0} u_{0n} e_n(x), \quad f(t, x) = \sum_{n \geq 0} f_n(t) e_n(x).$$

Then,

$$u_n(t) = e^{-\lambda_n t} u_{0n} + \int_0^t e^{-\lambda_n(t-s)} f_n(s) \, ds.$$

7.4 The Wave Equation

The wave operator on $\mathbf{R}_t \times \mathbf{R}_x^d$ is given by $\Box = \partial_t^2 - \sum_{j=1}^d \partial_{x_j}^2$.

A relevant problem for this operator is the *Cauchy problem*.

7.4.1 Homogeneous Cauchy Problem

Given u_0, u_1 defined on $\mathbf{R}^d$ (in convenient spaces) we have to solve,

$$\Box u = 0, \quad u|_{t=0} = u_0, \quad \partial_t u|_{t=0} = u_1. \qquad (7.9)$$

- In any dimension the solution is given by the explicit formula,

$$u(t, x) = \cos\left(t\sqrt{-\Delta}\right)u_0 + \frac{\sin\left(t\sqrt{-\Delta}\right)}{\sqrt{-\Delta}}u_1.$$

- In dimension $d = 3$ the above formula simplifies to,

$$u(t, x) = \frac{t}{4\pi}\int_{S^2} u_1(x - t\omega)\,d\omega + \frac{1}{4\pi}\int_{S^2} u_0(x - t\omega)\,d\omega$$

$$+ \frac{t}{4\pi}\sum_{i=1}^{3}\int_{S^2}\frac{\partial u_0}{\partial x_i}(x - t\omega)\omega_i\,d\omega, \qquad (7.10)$$

where S^2 is the unit sphere in $\mathbf{R}^3$.

7.4.1.1   Properties of the Solution

7.4.1.2   Decay at Infinity

- If u_0, u_1 belong to $C_0^\infty(\mathbf{R}^d)$ there exists $C = C(u_0, u_1)$ such that for $t \geq 1$,

$$\sup_{x\in\mathbf{R}^d}|u(t, x)| \leq \frac{C}{t^{\frac{d-1}{2}}}.$$

- In dimension $d = 3$ we have the following more precise estimate,

$$\sup_{x\in\mathbf{R}^3}|u(t, x)| \leq \frac{1}{4\pi t}\left(\sum_{i=1}^{3}\left\|\frac{\partial u_1}{\partial x_i}\right\|_{L^1(\mathbf{R}^3)} + \sum_{1\leq|\alpha|\leq 2}\|\partial^\alpha u_0\|_{L^1(\mathbf{R}^3)}\right), \qquad t \geq 1.$$

7.4.1.3   Finite Speed of Propagation

- If $u_0, u_1 \in C_0^\infty(\mathbf{R}^d)$ are such that $\operatorname{supp} u_0$ and $\operatorname{supp} u_1$ are included in the set $\{x \in \mathbf{R}^d,\ |x| \leq R\}$ then for all $t > 0$, $\operatorname{supp} u(t, \cdot) \subset \{x \in \mathbf{R}^d : |x| \leq R + t\}$.

7.4.1.4   Huygens Principle

- Let d be **odd** and consider initial data $u_0, u_1 \in C_0^\infty(\mathbf{R}^d)$ such that $\operatorname{supp} u_0$ and $\operatorname{supp} u_1$ are contained in the set $\{x \in \mathbf{R}^d : |x| \leq R\}$. Then u vanishes in $\{(t, x) : t > R,\ |x| < t - R\}$.

7.4.1.5 Influence Domain

- Let $(t_0, x_0) \in \mathbf{R}_+ \times \mathbf{R}^d$. The value of the solution u at (t_0, x_0) depends only on the values of the data u_0, u_1 on the intersection of the hyperplane $t = 0$ with the backward cone $C(t_0, x_0) = \{(t, x) : |x - x_0| < t_0 - t, \ t < t_0\}$.

7.4.1.6 Conservation of the Energy

- Let $u_0, u_1 \in C_0^\infty(\mathbf{R}^d)$ and let u be the solution of (7.9). For $t \in \mathbf{R}$ set,

$$E(t) = \frac{1}{2} \int_{\mathbf{R}^d} \left(\left| \frac{\partial u}{\partial t}(t, x) \right|^2 + |\nabla_x u(t, x)|^2 \right) dx,$$

$$E(0) = \frac{1}{2} \int_{\mathbf{R}^d} \left(|u_1(x)|^2 + |\nabla_x u_0(x)|^2 \right) dx,$$

where $|\nabla_x v|^2 = \sum_{i=1}^d |\frac{\partial v}{\partial x_i}|^2$. Then for all $t \in \mathbf{R}$ we have $E(t) = E(0)$.

The quantity $E(t)$ is called *the energy of the solution at time* t; the above property says that the energy is conserved and therefore is equal to the energy of the data.

7.4.1.7 Strichartz Estimates

- Let $T > 0, s \geq 0$ and $d \geq 2$. There exists $C > 0$ such that for every solution of (7.9) we have,

$$\|u\|_{L^q((0,T),L^r(\mathbf{R}^d))} \leq C \left(\|u_0\|_{\dot{H}^s(\mathbf{R}^d)} + \|u_1\|_{\dot{H}^{s-1}(\mathbf{R}^d)} \right), \tag{7.11}$$

whenever the following conditions are satisfied,

$$r < +\infty, \quad \frac{2}{q} + \frac{d-1}{r} \leq \frac{d-1}{2}, \quad \frac{1}{q} + \frac{d}{r} = \frac{d}{2} - s. \tag{7.12}$$

7.4.2 Inhomogeneous Cauchy Problem

This is the problem

$$\Box u = f, \quad u|_{t=0} = u_0, \quad \partial_t u|_{t=0} = u_1.$$

- In dimension $d = 3$ the solution is given by the formula,

$$u(t, x) = \frac{t}{4\pi} \int_{S^2} u_1(x - t\omega)\, d\omega + \frac{1}{4\pi} \int_{S^2} u_0(x - t\omega)\, d\omega +$$

$$+ \frac{t}{4\pi} \sum_{i=1}^{3} \int_{S^2} \frac{\partial u_0}{\partial x_i}(x - t\omega)\omega_i\, d\omega + \frac{1}{4\pi} \int_0^t \tau \int_{S^2} f(t-\tau, x-\tau\omega)\, d\omega\, d\tau.$$

7.4.2.1   Finite Speed of Propagation

- If for $j = 0, 1$ we have $u_j \in C_0^\infty(\mathbf{R}^d)$ with $\operatorname{supp} u_j \subset \{x \in \mathbf{R}^d,\ |x| \le R\}$ and if for $t > 0$ $\operatorname{supp} f(t, \cdot) \subset \{x \in \mathbf{R}^d,\ |x| \le R + t\}$ then for all $t > 0$, $\operatorname{supp} u(t, \cdot) \subset \{x \in \mathbf{R}^d : |x| \le R + t\}$.

7.4.2.2   Strichartz Estimates

- We also have an inhomogeneous Strichartz estimate where in the right-hand side of (7.11) we add,

$$\|f\|_{L^{a'}((0,T),L^{b'}(\mathbf{R}^d))},$$

where,

$$b < +\infty, \qquad \frac{2}{b} + \frac{d-1}{a} \le \frac{d-1}{2}, \qquad \frac{1}{a'} + \frac{d}{b'} - 2 = \frac{d}{2} - s.$$

Here a', b' denote the conjugate exponents of a, b.

Notice that the pair (a, b) is completely independent of (q, r) which appears in the left-hand side of (7.11).

7.4.3   The Mixed Problem

When the variable x lies in a bounded open set $\Omega \subset \mathbf{R}^d$, then one of the relevant problem for the wave equation is the Cauchy–Dirichlet problem, also called *a mixed problem*. The homogeneous one can be stated as follows. Given $0 < T \le +\infty$, u_0, u_1 defined on Ω and g defined on $(-T, T) \times \partial\Omega$, find u satisfying,

$$\Box u = 0 \text{ in } (-T, T) \times \Omega, \quad u|_{t=0} = u_0, \quad \partial_t u|_{t=0} = u_1, \quad u|_{(-T,T)\times\partial\Omega} = g. \tag{7.13}$$

Here are some classical results on this problem. Set $I = (-T, T)$.

- For all $(u_0, u_1) \in L^2(\Omega) \times H^{-1}(\Omega)$, $g \in L^2_{loc}(I \times \partial\Omega)$ the problem (7.13) has a unique solution $u \in C^0(I, L^2(\Omega)) \cap C^1(I, H^{-1}(\Omega))$.
- If $(u_0, u_1) \in H_0^1(\Omega) \times L^2(\Omega)$ and $g = 0$ the problem (7.13) has a unique solution $u \in C^0(I, H_0^1(\Omega)) \cap C^1(I, L^2(\Omega))$.

Comments For the basic results concerning the wave equation we refer to the books by S. Alinhac [2] and by F. John [16]. □

7.5   The Schrödinger Equation

The Schrödinger operator on $\mathbf{R}_t \times \mathbf{R}_x^d$ is given by $P = i\partial_t + \Delta_x$ where $\Delta_x = \sum_{j=1}^d \partial_{x_j}^2$.

7.5.1   The Cauchy Problem

7.5.1.1   The Homogeneous Cauchy Problem

We have to solve the problem

$$i\partial_t u + \Delta_x u = 0, \quad u|_{t=0} = u_0. \tag{7.14}$$

- If $u_0 \in S'(\mathbf{R}^d)$ the problem (7.14) has a unique solution $u \in S'(\mathbf{R} \times \mathbf{R}^d)$.
- If $u_0 \in S(\mathbf{R}^d)$ the solution u belongs to $C^\infty(\mathbf{R}, S(\mathbf{R}^d))$. It is given by the formula,

$$u(t, x) = (2\pi)^{-d} \int e^{ix\cdot\xi - it|\xi|^2} \widehat{u}_0(\xi)\, d\xi. \tag{7.15}$$

We set $u(t, \cdot) = S(t)u_0(\cdot)$.
- If $u_0 \in H^s(\mathbf{R}^d)$, where $s \in \mathbf{R}$, the solution $u = S(t)u_0$ belongs to $C^0(\mathbf{R}, H^s(\mathbf{R}^d))$. For $k \in \mathbf{N}$, $(\partial_t^k u)$ belongs to $C^0(\mathbf{R}, H^{s-2k}(\mathbf{R}^d))$ and,

$$\|S(t)u_0\|_{H^s} = \|u_0\|_{H^s} \quad \forall t \in \mathbf{R},$$
$$\|\partial_t^k S(t)u_0\|_{H^{s-2k}} \le \|u_0\|_{H^s} \quad \forall t \in \mathbf{R}, \ \forall k \in \mathbf{N}^*. \tag{7.16}$$

7.5.2   Properties of the Solution

7.5.2.1   Expression of the Solution

- For $u_0 \in S'(\mathbf{R}^d)$ the unique solution $u = S(t)u_0$ of the problem (7.14) can be written for $t \neq 0$,

$$S(t)u_0(\cdot) = \frac{1}{(4\pi |t|)^{d/2}}\, e^{-id\frac{\pi}{4}\,\mathrm{sgn}\,t}\, e^{i\frac{|\cdot|^2}{4t}} \left[\mathcal{F}\left(e^{i\frac{|\cdot|^2}{4t}}\,u_0 \right) \right]\!\left(\frac{\cdot}{2t}\right), \qquad (7.17)$$

where sgn t denotes the sign of t.

7.5.2.2   Infinite Speed of Propagation

By this we mean the following fact: the smoothness of the solution at a time $t \neq 0$ does not depend on the smoothness of the initial data u_0 but depends on its behavior at infinity. This can be made more precise but we shall give here only two examples describing this phenomena.

- If $u_0 \in \mathcal{E}'(\mathbf{R}^d)$ and $S(t)u_0$ is the solution of problem (7.14) then, for all $t \neq 0$, $u(t, \cdot) \in C^{\infty}(\mathbf{R}^d)$.
- Let $u_0(x) = e^{-i\lambda|x|^2}$, where $\lambda > 0$. Then

$$S\left(\frac{1}{4\lambda}\right)u_0 = (4\pi\lambda)^{\frac{d}{2}}\, e^{-in\frac{\pi}{4}}\,\delta_0.$$

Summarizing, an irregular datum but vanishing at infinity gives rise to a C^{∞} solution for $t \neq 0$ while a perfectly smooth datum but oscillating at infinity gives rise to a singular solution.

7.5.2.3   Decay at Infinity of the Solution

- If $u_0 \in L^1(\mathbf{R}^d)$, the solution $S(t)u_0$ of the problem (7.14) satisfies

$$\|S(t)u_0\|_{L^{\infty}(\mathbf{R}^d)} \leq \frac{\|g\|_{L^1}}{(4\pi)^{d/2}}\, |t|^{-\frac{d}{2}}, \quad \text{for } t \neq 0.$$

7.5.2.4   Strichartz Inequality

Let (q, r) be two real numbers. We say that (q, r) is admissible if,

$$\frac{2}{q} = \frac{d}{2} - \frac{d}{r} \quad \text{with} \quad 2 \leq q \leq +\infty \text{ if } d \geq 3 \quad \text{and} \quad 2 < q \leq +\infty \text{ if } d = 2.$$

$$(7.18)$$

- Let (q, r) be an admissible couple. There exists $C > 0$ such that,

$$\|S(t)u_0\|_{L^q(\mathbf{R}, L^r(\mathbf{R}^d))} \leq C\|u_0\|_{L^2(\mathbf{R}^d)}, \quad \forall u_0 \in L^2(\mathbf{R}^d). \tag{7.19}$$

Remark

(i) The above condition on (q, r) comes from the scale invariance of the Schrödinger equation. Indeed if u is a solution of (7.14) with datum u_0, then for all $\lambda > 0$ the function $u_\lambda(t, x) = u(\lambda^2 t, \lambda x)$ is also a solution of (7.14) with datum $u_{\lambda 0}(x) = u_0(\lambda x)$. The estimate (7.19) should therefore be satisfied by u_λ for all $\lambda > 0$ with the same constant C. This is possible only if (7.18) holds.

(ii) If $d = 2$ the above inequality with $q = 2, r = +\infty$ is false.   □

7.5.2.5 The Inhomogeneous Problem

We have to solve the problem

$$i\partial_t u + \Delta_x u = f, \quad u|_{t=0} = u_0. \tag{7.20}$$

The solution of this problem is given by

$$u(t, \cdot) = S(t)u_0(\cdot) + \frac{1}{i} \int_0^t S(t-s)[f(s, \cdot)]\, ds.$$

7.5.2.6 Nonhomogeneous Strichartz Inequality

- Let (q_j, r_j), $j = 1, 2$, be two admissible couples and (q_2', r_2') be the conjugate exponents of (q_2, r_2). Then there exists $C > 0$ such that for any $f \in L^{q_2'}(\mathbf{R}, L^{r_2'}(\mathbf{R}^d))$ we have,

$$\left\| \int_0^t S(t-s)[f(s, \cdot)]\, ds \right\|_{L^{q_1}(\mathbf{R}, L^{r_1}(\mathbf{R}^d))} \leq C\|f\|_{L^{q_2'}(\mathbf{R}, L^{r_2'}(\mathbf{R}^d))}.$$

It is important to notice that the couples (q_1, r_1) and (q_2, r_2) are independent.

Comments For the Strichartz estimates and applications to the nonlinear wave and Schrödinger equations we refer to the book by T. Tao [29].   □

7.6 The Burgers Equation

This is a quasi-linear first order partial differential equation which appears in the study of various domains of applied mathematics. It has been introduced by J. M. Burgers in 1948.

We will be interested in solving the Cauchy problem for this equation, that is, given a **real valued** function u_0 find u satisfying,

$$\partial_t u + u\, \partial_x u = 0, \quad (t,x) \in \mathbf{R} \times \mathbf{R}, \quad u|_{t=0} = u_0. \tag{7.21}$$

- **Case where $u_0 \in C^1(\mathbf{R})$**

  - Let $u_0 \in C^1(\mathbf{R})$ be real valued satisfying $\inf_{x \in \mathbf{R}} u_0' > -\infty$ where $u_0' = \partial_x u_0$. Set,

  $$T^* = +\infty \text{ if } \inf_{x \in \mathbf{R}} u_0' \geq 0 \quad \text{and} \quad T^* = \frac{1}{-\inf_{x \in \mathbf{R}} u_0'} \text{ if } \inf_{x \in \mathbf{R}} u_0' < 0.$$

    Then the problem (7.21) has a unique solution $u \in C^1([0, T^*) \times \mathbf{R})$.
  - In particular if $u_0' \in L^\infty(\mathbf{R})$ and $T > 0$ is such that $T\|u_0'\|_{L^\infty(\mathbf{R})} < 1$ then $T < T^*$. Thus the solution exists on the time interval $[0, T]$.
  - Notice that if $u_0 \in C_0^1(\mathbf{R})$ (that is has compact support), $u_0 \not\equiv 0$, then we have necessarily $T^* < +\infty$.
  - If $u_0 \in C^k(\mathbf{R}), k \geq 2$ with $\inf_{x \in \mathbf{R}} u_0' > -\infty$ then the above solution belongs to $C^k([0, T^*) \times \mathbf{R})$.
  - One can show that, when it is finite, T^* is the right maximal time of existence. For instance assume $u_0 \in C^2(\mathbf{R})$. Let x_0 be a point where $u_0'(x_0) < 0$ then we can show that $\lim_{t \to \frac{1}{-u_0'(x_0)}} |\partial_x u(t, x_0)| = +\infty$. However, notice that the function u itself remains bounded.

- **Case where $u_0 \in H^2(\mathbf{R})$**

  - Let $u_0 \in H^2(\mathbf{R}) \subset W^{1,\infty}(\mathbf{R})$ and let $T > 0$ be such that $T\|u_0'\|_{L^\infty(\mathbf{R})} < 1$. We can apply the previous result and infer that (7.21) has a unique C^1 solution. Then this solution belongs to $C^0([0, T], H^2(\mathbf{R}))$.
  - More generally if $u_0 \in H^k(\mathbf{R})$ for $k \geq 2$ then the solution belongs to $C^0([0, T], H^k(\mathbf{R}))$.

    (See Problem 65 for the above claims.)
  - **Continuity of the flow.** The map $u_0 \mapsto u$ is continuous from $H^2(\mathbf{R})$ to $C^0([0, T], H^2(\mathbf{R}))$.
  - **Nonuniform continuity of the flow.** The map $u_0 \mapsto u$ from $H^2(\mathbf{R})$ to $C^0([0, T], H^2(\mathbf{R}))$ is not uniformly continuous (see Problem 65).

- **Case where** $u_0 \in L^\infty(\mathbf{R})$. When the data is poorly regular (in particular discontinuous) we have to weaken the notion of solution. We write the Burgers problem as,

$$\partial_t u + \frac{1}{2}\partial_x(u^2) = 0, \quad u|_{t=0} = u_0. \tag{7.22}$$

Let $I = (0, +\infty)$. We shall say that u is a weak solution of (7.22) if for all $\varphi \in C_0^1([0, +\infty) \times \mathbf{R})$ we have,

$$\iint_{I \times \mathbf{R}} \left(u\partial_t\varphi + \frac{1}{2}u^2\partial_x\varphi\right)(t, x)\,dx\,dt + \int_{\mathbf{R}} u_0(x)\varphi(0, x)\,dx = 0.$$

Then we have the following result.

- Let $u_0 \in L^\infty(\mathbf{R})$. Then the problem (7.22) has a unique weak solution $u \in L^\infty(I \times \mathbf{R})$ satisfying,

  (i) $\|u\|_{L^\infty(I \times \mathbf{R})} \leq \|u_0\|_{L^\infty(\mathbf{R})}$,

  (ii) there exists $E > 0$ such that for every $a > 0, t \in I, x \in \mathbf{R}$,

  $$\frac{u(t, x + a) - u(t, x)}{a} < \frac{E}{t}.$$

- Condition (ii) above is called *the entropy condition*. This condition permits to select the physically acceptable solution and is essential for the uniqueness. The solution u above is called *the entropy solution*.
- Assume that $u_0 \in L^\infty(\mathbf{R})$ has compact support, then the entropy solution satisfies,

$$\|u(t, \cdot)\|_{L^\infty(\mathbf{R})} \leq Ct^{-\frac{1}{2}}, \quad t \geq 1.$$

Comments For further information on the Burgers equation we refer the reader to the book by J. Smoller [25], Chapters 15 and 16.   □

7.7   The Euler Equations

7.7.1   The Incompressible Euler Equations

These are equations which describe the dynamic of a fluid without viscosity submitted to exterior forces. If we denote by $X(t) = (x_1(t), \ldots, x_d(t)) \in \mathbf{R}^d$ a particle of fluid, its velocity at time t at this point is given by,

$$v(t, X(t)) = \dot{X}(t) := \frac{dX}{dt}(t). \qquad (7.23)$$

It is a vector in $\mathbf{R}^d$. The Euler equations are the mathematical expression of the Newton law which says that the acceleration is proportional to the forces which act on the particle. In our case the acceleration is given by $\ddot{X}(t)$. According to (7.23) we have $\ddot{x}_j(t) = \frac{\partial v_j}{\partial t} + \sum_{k=1}^d \dot{x}_k(t) \frac{\partial v_j}{\partial x_k} = F$. Using again (7.23) we obtain,

$$\frac{\partial v_j}{\partial t} + \sum_{k=1}^d v_k \frac{\partial v_j}{\partial x_k} = F, \quad j = 1, \ldots, d. \qquad (7.24)$$

In the case of basic fluid dynamics, the force which acts comes from the gradient of pressure, so that the Euler system reads,

$$\frac{\partial v_j}{\partial t} + \sum_{k=1}^d v_k \frac{\partial v_j}{\partial x_k} = -\frac{\partial P}{\partial x_j}.$$

We can write these equations in a shorter manner. Set $\sum_{k=1}^d v_k \frac{\partial}{\partial x_k} = v \cdot \nabla_x$, then we have,

$$\frac{\partial v}{\partial t} + (v \cdot \nabla_x)v := -\nabla_x P. \qquad (7.25)$$

7.7.1.1   Incompressibility

It is a physical notion which in fluid dynamics expresses the fact that over time the volume of the fluid remains constant under the action of the forces which act on it. This fact can be expressed mathematically. The *incompressibility* is equivalent to

$$\operatorname{div} v =: \sum_{j=1}^d \frac{\partial v_j}{\partial x_j} = 0. \qquad (7.26)$$

(See Problem 59.)

In the incompressible case, the pressure P appearing in the equations (7.25) is determined by the velocity v. Indeed let us take the divergence of both members of (7.25). Since $\operatorname{div} \frac{\partial v}{\partial t} = \frac{\partial}{\partial t}(\operatorname{div} v) = 0$, using the fact that $\operatorname{div} \nabla_x = \Delta = \sum_{j=1}^d \frac{\partial^2}{\partial x_j^2}$ (the Laplacian) we obtain $-\Delta P = \operatorname{div}(v \cdot \nabla_x)v$, which can be written as

$-P = \Delta^{-1} \mathrm{div}\left((v \cdot \nabla_x)v\right)$, so that the equations (7.25) are equivalent to

$$\frac{\partial v}{\partial t} + (v \cdot \nabla_x)v = \nabla_x \Delta^{-1} \mathrm{div}\left((v \cdot \nabla_x)v\right). \tag{7.27}$$

Thus the Cauchy problem for the incompressible Euler equation consists in finding (v, P) satisfying,

$$\begin{cases} \dfrac{\partial v}{\partial t} + (v \cdot \nabla_x)v = -\nabla_x P, \\[2mm] \mathrm{div}\, v = 0, \\[2mm] v|_{t=0} = v_0. \end{cases} \tag{7.28}$$

Define $\mathrm{div}(v \otimes v)$ as the vector whose i^{th} coordinates is $\sum_{j=1}^{d} \partial_{x_j}(v_i v_j)$. Then it is easy to see that the system (7.28) is equivalent to the system,

$$\begin{cases} \dfrac{\partial v}{\partial t} + \mathrm{div}\,(v \otimes v) = -\nabla_x P, \\[2mm] \mathrm{div}\, v = 0, \\[2mm] v|_{t=0} = v_0. \end{cases} \tag{7.29}$$

7.7.1.2 The Vorticity

If v is the velocity, the *vorticity*, is defined by the skew-symmetric matrix

$$\omega = (\omega_{ij})_{1 \le i, j \le d}, \quad \omega_{ij} = \partial_{x_i} v_j - \partial_{x_j} v_i, \quad 1 \le i, j \le d.$$

- Notice that a fluid for which $\omega = 0$ is called *irrotational*. If v is a solution of the Euler system (7.28) such that $\omega|_{t=0} = 0$ then for every $t \ge 0$ we have $\omega(t, \cdot) = 0$. That means that the *irrotationality* is propagated by the flow. (See Problem 60.)
- If $d = 3$, ω may be identified with the vector $\nabla \times v$. Then, if v is a sufficiently regular solution of the Euler system (7.28), the vorticity satisfies the vectorial equation,

$$\partial_t \omega + (v \cdot \nabla_x)\omega = (\omega \cdot \nabla_x)v, \tag{7.30}$$

where $u \cdot \nabla_x = \sum_{i=1}^{3} u_i \partial_{x_i}$.
- If $d = 2$ then the vorticity may be identified with the scalar function $\omega = \partial_{x_1} v_2 - \partial_{x_2} v_1$. In this case the vorticity equation reduces to

$$\partial_t \omega + (v \cdot \nabla_x)\omega = 0. \tag{7.31}$$

Moreover, (still when $d = 2$) since v satisfies the incompressibility condition div $v = 0$ one can formally recover v from ω as follows.

There exists a function ψ such that $v = \nabla^\perp \psi = {}^t(-\partial_{x_2}\psi, \partial_{x_1}\psi)$. Taking the curl of both members we find that $\Delta_x \psi = \omega$. Let $E = \frac{1}{2\pi}\mathrm{Ln}\,|x|$ be the fundamental solution of the Laplacian in $\mathbf{R}^2$. Then if ω decays sufficiently at infinity we get,

$$\psi(t, x) = (E \star \omega(t, \cdot))(x),$$

which implies that,

$$v(t, x) = (K \star \omega(t, \cdot))(x) = \int_{\mathbf{R}^2} K(x - y)\omega(t, y)\,dy,$$

where $K(x) = \nabla_x^\perp E(x) = \frac{1}{2\pi}{}^t\left(-\frac{x_2}{|x|^2}, \frac{x_1}{|x|^2}\right)$.

This is the so-called *Biot–Savart law*.

- Conversely if v, ω satisfy the vorticity equation (7.31) and div $v = 0$ then,

$$\mathrm{curl}\,(\partial_t v + (v \cdot \nabla_x)v) = \partial_t \omega + (v \cdot \nabla_x)\omega = 0.$$

Therefore, there exists P such that,

$$\partial_t v + v \cdot \nabla_x v = -\nabla_x P.$$

Thus (v, P) satisfies the system (7.28).

Moreover, taking the divergence of both members we find that P is a solution of the equation,

$$-\Delta_x P = \sum_{i,j=1}^{2} \partial_{x_i} v_j \partial_{x_j} v_i.$$

Concerning the existence of solutions for the system (7.28) there is a huge literature and thousands of papers. We just give here a few examples of such results.

7.7.1.3 Classical Solutions: The Lichtenstein Theorem

- Let $d \geq 2$ and $0 < \alpha < 1$. Set $\rho = 1 + \alpha$. Let $v_0 \in W^{\rho,\infty}(\mathbf{R}^d, \mathbf{R}^d)$ be such that div $v_0 = 0$. There exists a unique $T^* > 0$ and a unique solution v of (7.28) on $[0, T^*) \times \mathbf{R}^d$ belonging to the space $L^\infty_{\mathrm{loc}}([0, T^*), W^{\rho,\infty}(\mathbf{R}^d, \mathbf{R}^d))$. Moreover,

$$T^* < +\infty \;\Rightarrow\; \int_0^{T^*} \|v(t, \cdot)\|_{W^{\rho,\infty}}\,dt = +\infty.$$

- Notice that under the above hypothesis on v_0 the map $v_0 \mapsto v$ need not be continuous from $W^{\rho,\infty}(\mathbf{R}^d, \mathbf{R}^d)$ to $L^\infty([0, T], W^{\rho,\infty}(\mathbf{R}^d, \mathbf{R}^d))$ for $T < T^*$ (see Problem 64).

7.7.1.4 Weak Solutions: The Yudovitch Theorem

It is a classical result on the existence and uniqueness of a weak solution for the vorticity equation in dimension $d = 2$.

- Set $I = (0, +\infty)$. Let $\omega_0 \in L^1(\mathbf{R}^2) \cap L^\infty(\mathbf{R}^2)$.
  There exists a unique couple (ω, v) with $\omega \in L^\infty(I, L^1(\mathbf{R}^2) \cap L^\infty(\mathbf{R}^2))$, $v = K \star \omega$ satisfying,

$$\iint_{I \times \mathbf{R}^2} \omega(t, x) \left(\partial_t \varphi + (v \cdot \nabla_x)\varphi\right)(t, x)\, dt\, dx + \int_{\mathbf{R}^2} \varphi(0, x)\omega_0(x)\, dx = 0,$$

for every $\varphi \in C_0^1([0, +\infty) \times \mathbf{R}^2)$.
- Notice that by the above discussion this gives a result on existence of a weak solution for the Euler system (7.29) in dimension $d = 2$.
  Let v_0 be a vector field such that $\operatorname{div} v_0 = 0$ which belongs to $L^2(\mathbf{R}^2, \mathbf{R}^2)$. Assume moreover that $\omega_0 = \operatorname{curl} v_0 \in L^1(\mathbf{R}^2) \cap L^\infty(\mathbf{R}^2)$. Then there exists a unique weak solution v of problem (7.29) belonging to $C^0([0, +\infty), L^2(\mathbf{R}^2, \mathbf{R}^2))$ such that $\omega \in L^\infty(I \times \mathbf{R}^2) \cap L^\infty(I, L^1(\mathbf{R}^2))$.

7.7.2 The Compressible Euler Equations

These are also classical equations. We have to introduce another quantity, the *fluid density*, which will change with time. If we denote by ρ this density, the compressible Euler equations are,

$$\frac{\partial \rho}{\partial t} + \sum_{j=1}^{d} v_j \frac{\partial \rho}{\partial x_j} + \left(\sum_{j=1}^{d} \frac{\partial v_j}{\partial x_j}\right)\rho = 0, \quad \rho|_{t=0} = \rho_0,$$

$$\rho \left(\frac{\partial v_k}{\partial t} + \sum_{j=1}^{d} v_j \frac{\partial v_k}{\partial x_j}\right) + \frac{\partial \rho^\gamma}{\partial x_k} = 0, \quad v_k|_{t=0} = v_{0k}, \quad k = 1, \ldots, d, \quad \gamma > 0.$$

$$(7.32)$$

7.8   The Navier–Stokes Equations

These are partial differential equations describing the dynamics of viscous fluids. As for the Euler equations u denotes the velocity of a particle of fluid, P denotes the pressure, and here we add a parameter μ which stands for the viscosity of the media.

The Cauchy problem for these equations consists in finding $u = (u_1, \ldots, u_d)$ and P solutions of,

$$\begin{cases} \dfrac{\partial u}{\partial t} + \text{div}\,(u \otimes u) - \mu \Delta_x u = -\nabla_x P, \\[2mm] \text{div}\, u = 0, \\[2mm] u|_{t=0} = u_0. \end{cases} \tag{7.33}$$

Recall that $\text{div}(u \otimes u)$ is the vector whose i^{th} coordinate is $\sum_{j=1}^{d} \partial_{x_j}(u_i u_j)$.

7.8.1   Weak Solutions

Let $0 < T \le +\infty$ and $I = (0, T)$. We introduce the space,

$$W_{\text{loc}}^{1,1} = \left\{ u \in L_{\text{loc}}^1(I \times \mathbf{R}^d, \mathbf{R}^d) : \partial_{x_k} u \in L_{\text{loc}}^1(I \times \mathbf{R}^d, \mathbf{R}^d), 1 \le k \le d \right\}.$$

Then $u \in W_{\text{loc}}^{1,1}$ is called a weak solution of (7.33) on $I \times \mathbf{R}^d$ if $\text{div}\, u = 0$ and,

$$\iint_{I \times \mathbf{R}^d} u(t,x) \cdot \partial_t \phi(t,x)\, dx\, dt + \sum_{i,j=1}^{d} \iint_{I \times \mathbf{R}^d} u_i(t,x) u_j(t,x) \partial_{x_j} \phi_i(t,x)\, dx\, dt$$

$$- \mu \iint_{I \times \mathbf{R}^d} \nabla_x u(t,x) \cdot \nabla_x \phi(t,x)\, dx\, dt - \int_{\mathbf{R}^d} u_0(x) \cdot \phi(0,x)\, dx = 0, \tag{7.34}$$

for every $\phi = (\phi_1, \ldots, \phi_d) \in C_0^\infty([0,T) \times \mathbf{R}^d, \mathbf{R}^d)$ such that $\text{div}\, \phi = 0$.

As for the Euler equations there are thousands of papers dealing with these equations. We will just mention some of the most famous results on it.

7.8.2   The Leray Theorem (1934)

- Let $I = (0, +\infty)$ and $d \ge 2$. Consider a vector field $u_0 \in L^2(\mathbf{R}^d, \mathbf{R}^d)$ such that $\text{div}\, u_0 = 0$. Then the Cauchy problem for the Navier–Stokes equation (7.33) has a weak solution $u \in L^\infty(I, L^2(\mathbf{R}^d, \mathbf{R}^d))$ with $\nabla_x u \in L^2(I, L^2(\mathbf{R}^d, \mathbf{R}^d))$. Moreover, this solution satisfies,

$$\|u(t, \cdot)\|_{L^2}^2 + 2\mu \int_0^t \|\nabla_x u(s, \cdot)\|_{L^2}^2 \, ds \leq \|u_0\|_{L^2}^2.$$

- If $d = 2$ the solution belongs to $C^0(I, L^2(\mathbf{R}^d, \mathbf{R}^d))$ and is unique in the set,

$$\left\{ u \in L^\infty(I, L^2(\mathbf{R}^d, \mathbf{R}^d)) : \nabla_x u \in L^2(I, L^2(\mathbf{R}^d, \mathbf{R}^d)) \text{ and } \operatorname{div} u = 0 \right\}.$$

Moreover, it satisfies, for all time $t > 0$, the equality,

$$\|u(t, \cdot)\|_{L^2}^2 + 2\mu \int_0^t \|\nabla_x u(s, \cdot)\|_{L^2}^2 \, ds = \|u_0\|_{L^2}^2.$$

- Notice that when $d \geq 3$ the uniqueness of the weak solution above is, since 1934, an open problem. Moreover, since there are other results showing the uniqueness of the solution in spaces of more regular functions, the smoothness of these weak solutions is as well an open problem.

7.8.3 Strong Solutions: Theorems of Fujita–Kato and Kato

We restrict ourselves to the physical dimension $d = 3$.

- **Fujita–Kato (1964):** let $u_0 \in \dot{H}^{\frac{1}{2}}(\mathbf{R}^3, \mathbf{R}^3)$ be such that $\operatorname{div} u_0 = 0$. There exists a maximal time $T^* > 0$ such that the problem (7.33) has a *unique* solution u such that,

$$u \in C^0\left([0, T^*), \dot{H}^{\frac{1}{2}}(\mathbf{R}^3, \mathbf{R}^3)\right) \cap L_{\text{loc}}^2\left((0, T^*), \dot{H}^{\frac{3}{2}}(\mathbf{R}^3, \mathbf{R}^3)\right).$$

Moreover, there exists $\varepsilon_0 > 0$ such that if $\|u_0\|_{\dot{H}^{\frac{1}{2}}(\mathbf{R}^3, \mathbf{R}^3)} \leq \varepsilon_0$ then $T^* = +\infty$.
- **Kato (1984):** Let $u_0 \in L^3(\mathbf{R}^3, \mathbf{R}^3)$ be such that $\operatorname{div} u_0 = 0$. Then there exists a maximal time $T^* > 0$ such that the problem (7.33) has a *unique* solution u such that,

$$u \in C^0\left([0, T^*), L^3(\mathbf{R}^3, \mathbf{R}^3)\right) \quad \text{and} \quad t^{\frac{1}{2}} \|u(t)\|_{L^\infty(\mathbf{R}^3, \mathbf{R}^3)} \in L^\infty((0, T^*)).$$

Moreover, there exists $\varepsilon_0 > 0$ such that if $\|u_0\|_{L^3(\mathbf{R}^3, \mathbf{R}^3)} \leq \varepsilon_0$, then $T^* = +\infty$.

Comments Concerning the equations involved in fluid mechanics we refer to the book by H. Bahouri, J.-Y. Chemin, and R. Danchin [5]. $\square$

Chapter 8
Statements of the Problems of Chap. 7

This chapter contains problems about the classical partial differential equations.

Problem 37 *Monge–Ampère Equation*

Part 1

In $\mathbf{R}^2$, where the variable is denoted by (x_1, x_2), if u is a C^2 function we shall set $u_{jk} = \partial_{x_j} \partial_{x_k} u$. The purpose of this part is to show that, using the Cauchy–Kovalevski theorem, for every given real analytic and real valued functions K, g, h near the origin one can solve locally the Cauchy problem,

$$u_{11} u_{22} - u_{12}^2 = K(x_1, x_2), \quad u|_{x_2=0} = g(x_1), \quad \frac{\partial u}{\partial x_2}\Big|_{x_2=0} = h(x_1) \tag{8.1}$$

in the analytic framework, under the hypothesis $g''(0) \neq 0$.

1. Show that there exists a constant $A \in \mathbf{C}$ such that if we set,

$$\widetilde{u}(x_1, x_2) = g(x_1) + h(x_1)x_2 + \frac{1}{2}Ax_2^2, \qquad \varphi(x_1, x_2) = x_2,$$

$$p^0 = \left(\widetilde{u}(0,0), (\partial_{x_j}\widetilde{u}(0,0))_{j=1,2}, (\partial_{x_j}\partial_{x_k}\widetilde{u}(0,0))_{j,k=1,2} \right),$$

then all the conditions in the nonlinear Cauchy–Kovalevski theorem are satisfied.
2. Conclude.

Part 2

We want to extend this result to higher dimensions.

© The Editor(s) (if applicable) and The Author(s), under exclusive license to Springer Nature Switzerland AG 2020
T. Alazard, C. Zuily, *Tools and Problems in Partial Differential Equations*, Universitext, https://doi.org/10.1007/978-3-030-50284-3_8

If $d \geq 3$ we denote by $(x', x_d) \in \mathbf{R}^{d-1} \times \mathbf{R}$ the variable in $\mathbf{R}^d$. For $u \in C^2$ we denote by $\mathrm{Hess}(u) = (\partial_{x_j} \partial_{x_k} u)_{1 \leq j,k \leq d}$ its Hessian matrix.

Given real analytic and real valued functions K, g, h near the origin we consider the problem,

$$\det(\mathrm{Hess}(u)) = K, \quad u|_{x_d=0} = g(x'), \quad \frac{\partial u}{\partial x_d}|_{x_d=0} = h(x'), \tag{8.2}$$

under the condition

$$\det(\mathrm{Hess}(g)(0)) \neq 0, \tag{8.3}$$

where $\mathrm{Hess}(g)(0) = (\partial_{x_j} \partial_{x_k} g(0))_{1 \leq j,k \leq d-1}$ is the Hessian matrix of g at zero.

1. **Case 1**. Assume that in a neighborhood of zero in $\mathbf{R}^{d-1}$ we have,

$$g(x') = a_0 + \sum_{j=1}^{d-1} a_j x_j + \sum_{j=1}^{d-1} \lambda_j x_j^2 + O(|x'|^3),$$

where $\lambda_j \neq 0$ for $j = 1, \ldots, d-1$.
Show that one can find $A \in \mathbf{C}$ such that if we set,

$$\tilde{u}(x) = g(x') + h(x')x_d + \frac{1}{2} A x_d^2, \quad \varphi(x) = x_d,$$

$$p^0 = \left(\tilde{u}(0), (\partial_{x_j}\tilde{u}(0))_{j=1,\ldots,d}, (\partial_{x_j}\partial_{x_k}\tilde{u}(0))_{j,k=1,\ldots,d}\right),$$

then we can apply the nonlinear Cauchy–Kovalevski theorem.
2. **Case 2**. Assume (8.3) true.
   For $k \in \mathbf{N}, k \geq 2$, denote by $O(k)$ the set or $k \times k$ orthogonal matrices.

a) Prove that one can find $M' = (m'_{jk}) \in O(d-1)$ such that

$${}^t M' \, (\mathrm{Hess}(g)(0)) \, M' = \mathrm{diag}(\lambda_j),$$

where $\lambda_j \neq 0$ for $j = 1, \ldots, d-1$ and ${}^t M'$ is the transposed matrix of M'.
b) Set,

$$m_{jk} = m'_{jk}, \text{ for } 1 \leq j,k \leq d-1, m_{jd} = m_{dj} = 0, \text{ for } 1 \leq j \leq d-1, m_{dd} = 1.$$

Prove that $M = (m_{jk}) \in O(d)$ and that $My = (M'y', y_d)$ if $y = (y', y_d)$.
c) If $u \in C^2(\mathbf{R}^d)$ we set $v(y) = u(My)$. Prove that,

$$\mathrm{Hess}(v)(y) = {}^t M \mathrm{Hess}(u)(My) M.$$

Hint: Apply the Taylor formula at the order two to $v(y+hy_0)$ and $u(My+hMy_0)$ where y, $y_0 \in \mathbf{R}^d$ and $h \in \mathbf{R}$ is small.

d) Deduce that $\det(\mathrm{Hess}(u)(My)) = \det(\mathrm{Hess}(v)(y))$.

e) We set $\widetilde{g}(y') = g(M'y')$. Prove that in a neighborhood of the origin of $\mathbf{R}^{d-1}$ we have,

$$\widetilde{g}(y') = a_0 + \sum_{j=1}^{d-1} a_j y_j + \frac{1}{2} \sum_{j=1}^{d-1} \lambda_j y_j^2 + O(|y'|^3).$$

f) Applying the result proved in Case 1 show that the problem (8.2) has a unique real analytic solution in a neighborhood of the origin.

Comments The Monge–Ampère equations appear in geometry since, for instance, the Gauss curvature of a surface in $\mathbf{R}^3$ given by a graph $z = u(x, y)$ is given by the formula $K = \frac{u_{xx} u_{yy} - u_{xy}^2}{(1+u_x^2+u_y^2)^{\frac{1}{2}}}$. They appear also in recent works concerning the theory of optimal transportation (see the Lecture Notes [4]).   □

Problem 38 *An Application of the Harnack Inequality*

For $n \geq 1$, $y^0 \in \mathbf{R}^n$ and $\rho > 0$ we shall denote in what follows,

$$B_n(y^0, \rho) = \left\{ y \in \mathbf{R}^n : |y - y^0| := \max_{1 \leq j \leq n} |y_j - y_j^0| \leq \rho \right\}$$

the closed ball with center y^0 and radius ρ.

1. Let Ω be an open subset of $\mathbf{R}^n$ and u be a C^2 nonnegative harmonic function in Ω.

Let $X_0 \in \Omega$ and $r > 0$ be such that $B_n(X_0, r) \subset \Omega$. The purpose of this question is to prove that there exists a constant M depending only on the dimension n such that,

$$\sup_{B_n(X_0, \frac{r}{2})} u \leq M \inf_{B_n(X_0, \frac{r}{2})} u. \tag{8.4}$$

a) Let $X_1, X_2 \in B_n(X_0, \frac{r}{2})$. Using the fact that $B_n(X_2, r) \subset B_n(X_1, 2r)$ and the mean value formula prove that there exists $C = C(n) > 0$ such that, $u(X_1) \geq Cu(X_2)$.

b) Deduce (8.4).

2. Let $\omega \subset \mathbf{R}^d, d \geq 1$, be open and let $\varphi \in C^2(\omega) \cap C^0(\overline{\omega})$, $\varphi \neq 0$, be a solution of the problem,

$$-\Delta\varphi = \lambda\varphi \text{ in } \omega, \quad \varphi|_{\partial\omega} = 0, \quad \lambda > 0. \tag{8.5}$$

a) We set for $t \in \mathbf{R}$, $\psi(t, x) = e^{t\sqrt{\lambda}}\varphi(x)$. Show that ψ is a C^2 harmonic function in $\mathbf{R}_t \times \omega$.

Let $Z(\varphi) = \{x \in \overline{\omega} : \varphi(x) = 0\}$. Let $x_0 \in \omega$ be such that $x_0 \notin Z(\varphi)$ and let $r > 0$ be such that $B_d(x_0, 4r) \cap Z(\varphi) = \emptyset$. Changing φ to $-\varphi$, if necessary, we may assume that $\varphi(x) > 0$ for $x \in B_d(x_0, 4r)$.

Let $X_0 = (\frac{3r}{2}, x_0)$ so that $(r, 2r) \times B_d(x_0, \frac{r}{2}) = B_{d+1}(X_0, \frac{r}{2})$.

b) Prove that,

$$\sup_{B_{d+1}(X_0, \frac{r}{2})} \psi \geq e^{r\sqrt{\lambda}} \inf_{B_{d+1}(X_0, \frac{r}{2})} \psi.$$

c) Using question 1 with $n = d+1$ prove that there exists a constant C depending only on the dimension d such that $r \leq \frac{C}{\sqrt{\lambda}}$, which is equivalent to say that,

$$r > \frac{C}{\sqrt{\lambda}} \Rightarrow B_d(x_0, 4r) \cap Z(\varphi) \neq \emptyset.$$

3. Using the conclusion of question 2c) prove by contradiction that there exists a constant M depending only on d such that,

$$\forall \lambda > 0, \quad \forall x \in \omega, \quad d(x, Z(\varphi)) \leq \frac{M}{\sqrt{\lambda}}, \tag{8.6}$$

where $d(x, Z(\varphi)) = \inf_{z \in Z(\varphi)} |x - z|$.

Problem 39

Part 1

The goal of this part is to show that the estimate (7.7) is optimal in the case where Ω is a ball in $\mathbf{R}^3$.

Let $B = \{x \in \mathbf{R}^3 : |x| < 1\}$. For $k \in \mathbf{Z}, k \neq 0$, we set $u_k(x) = \frac{\sin(k\pi|x|)}{|x|}$.

1. Show that u_k is an eigenfunction of the operator $-\Delta$ for the Dirichlet problem in the ball B.
2. Compute the $L^2(B)$ norm of u_k. Compute the $L^\infty(B)$ norm of u_k.
3. Conclude.

Part 2

In this part we give an application of the estimate (7.7).

Let Ω be a bounded open subset of $\mathbf{R}^d$ with C^∞ boundary. Let λ be an eigenvalue of the Dirichlet problem for the Laplacian and let V be the eigenspace corresponding to λ endowed with the $L^2(\Omega)$ norm.

We recall (see Sect. 7.2.6) that $V \subset C^\infty(\overline{\Omega})$ and that $m := \dim V < +\infty$.

Our goal is to show that there exists $C > 0$ independent of λ such that,

$$m \le C\lambda^{\frac{d-1}{2}}. \tag{8.7}$$

We shall use for that the estimate (7.7).

1. Let $(v_1, \ldots, v_m)$ be an orthonormal basis of V. For $y \in \overline{\Omega}$ we set,

$$u_y(x) = \sum_{i=1}^m \overline{v_i(y)} v_i(x), \qquad a(y) = u_y(y) = \sum_{i=1}^m |v_i(y)|^2.$$

Prove that there exists y_0 in $\overline{\Omega}$ such that,

$$m \le |\Omega|\, a(y_0),$$

where $|\Omega|$ denotes the Lebesgue measure of Ω.

2. Compute $\|u_{y_0}\|^2_{L^2(\Omega)}$ and show that $\|u_{y_0}\|_{L^\infty(\Omega)} \ge a(y_0)$.
3. Deduce that,

$$\sup_{\substack{u \in V \\ \|u\|_{L^2(\Omega)}=1}} \|u\|_{L^\infty(\Omega)} \ge |\Omega|^{-\frac12} m^{\frac12}.$$

4. Using the estimate (7.7) prove (8.7).

Problem 40 *An Estimate of the Eigenfunctions of the Laplacian on the Sphere*

For $d \ge 2$ let $\mathbf{S}^d$ be the unit sphere of $\mathbf{R}^{d+1}$ and $\Delta_{\mathbf{R}^{d+1}} = \sum_{j=1}^{d+1} \frac{\partial^2}{\partial x_j^2}$.

The operator Δ_ω on $\mathbf{S}^d$ (the Laplace–Beltrami operator) is such that for $f \in C^2(\mathbf{R}^{d+1})$, $x = r\omega, r > 0, \omega \in \mathbf{S}^d$ and $F(r, \omega) = f(r\omega)$,

$$(\Delta_{\mathbf{R}^{d+1}} f)(r\omega) = \left(\frac{\partial^2 F}{\partial r^2} + \frac{d}{r}\frac{\partial F}{\partial r} + \frac{1}{r^2}\Delta_\omega F\right)(r, \omega). \tag{8.8}$$

1. For $j \in \mathbb{N}$ and $\omega = (\omega_1, \ldots, \omega_{d+1}) \in S^d$ we set $e_j(\omega) = (\omega_1 + i\omega_2)^j$. Prove that we have,

$$-\Delta_\omega e_j = j(j + d - 1)e_j.$$

Hint: Use the holomorphy of the function $f(x) = (x_1 + ix_2)^j$ in $\mathbb{C}$.

2. Show that there exists $c_0 > 0$ such that,

$$\lim_{j \to +\infty} j^{\frac{d-1}{2}} \|e_j\|^2_{L^2(S^d)} = c_0.$$

Hint: Write $S^d = \{\omega = (\omega_1, \omega_2, \omega') \in \mathbb{R} \times \mathbb{R} \times \mathbb{R}^{d-2} : |\omega| = 1\}$, parametrize S^d by $|\omega'| \leq 1$, $\omega_1 = (1 - |\omega'|^2)^{\frac{1}{2}} \cos \alpha$, $\omega_2 = (1 - |\omega'|^2)^{\frac{1}{2}} \sin \alpha$, $0 < \alpha < 2\pi$, then set $\omega' = \rho \, \Theta$, $0 < \rho < 1$, $\Theta \in S^{d-2}$ and eventually $\rho = \frac{t}{\sqrt{j}}$.

3. Set $u_j = \frac{e_j}{\|e_j\|_{L^2(S^d)}}$. Deduce that there exists $c_1 > 0$ such that,

$$\|u_j\|_{L^\infty(S^d)} \sim c_1 j^{\frac{d-1}{4}} \quad j \to +\infty.$$

Problem 41 *Spectral Theory for the Harmonic Oscillator*

This problem uses heavily the results of Problem 1.

The goal of this problem is to study in $\mathbb{R}^d, d \geq 1$ the harmonic oscillator, $P = -\Delta + |x|^2$ and its spectral theory.

In all what follows we shall denote $E = E(\mathbb{R}^d)$ if $E = L^2, H^s, C_0^\infty, S, S'$.

Let $V = \{u \in L^2 : \nabla_x u \in L^2 : \langle x \rangle u \in L^2\}$ endowed with the scalar product,

$$(u, v)_V = (\nabla_x u, \nabla_x v)_{L^2} + (\langle x \rangle u, \langle x \rangle v)_{L^2}, \quad \langle x \rangle = (1 + |x|^2)^{\frac{1}{2}},$$

and with the corresponding norm $\|u\|_V = (u, u)_V^{\frac{1}{2}}$. Then V is a Hilbert space. We shall denote by V' its dual.

Part 1

Study of the operator $P = -\Delta + \langle x \rangle^2$.

1. Prove that the operator $P = -\Delta + \langle x \rangle^2$ sends V to V', that it is bijective, continuous and that P^{-1} is continuous that is,

$$\exists C > 0 : \|w\|_V \leq C \|(-\Delta + \langle x \rangle^2)w\|_{V'}, \quad \forall w \in V. \tag{8.9}$$

Let $f \in L^2 \subset V'$. Our aim is to show that,

$$\left(u \in V, \ -\Delta u + \langle x \rangle^2 u = f \right) \Longrightarrow u \in H^2, \ \langle x \rangle \nabla_x u \in L^2, \ \langle x \rangle^2 u \in L^2.$$
$$(8.10)$$

Let e_k be the kth vector of the canonical basis of $\mathbf{R}^d$. For $0 < |h| \leq 1$ and $v \in V$ set $D_h v = \frac{1}{h}(v(x - he_k) - v(x))$.

2. a) Prove that we have,

$$(\nabla_x D_{-h} u, \nabla_x v)_{L^2} + (\langle x \rangle D_{-h} u, \langle x \rangle v)_{L^2} = \left(D_{-h} \overline{f}, v \right)_{L^2}$$
$$- \int_{\mathbf{R}^d} (2y_k + h) u(y + he_k) \overline{v(y)} \, dy.$$

b) Taking $v = D_{-h} u \in V$ prove that there exists $C > 0$ independent of u, h such that,

$$\| \nabla_x D_{-h} u \|_{L^2} + \| \langle x \rangle D_{-h} u \|_{L^2} \leq C \left(\| f \|_{L^2} + \| \langle x \rangle u \|_{L^2} \right).$$

c) Prove that, $\Delta u \in L^2$, $\langle x \rangle \nabla_x u \in L^2$, $\langle x \rangle^2 u \in L^2$ and deduce (8.10).
   Hint: Use the fact that a ball in L^2 is weakly compact and Problem 1.

d) We set $D = \left\{ u \in H^2 : \langle x \rangle^2 u \in L^2 \right\}$ endowed with the norm $\| u \|_D = \| u \|_{H^2} + \| \langle x \rangle^2 u \|_{L^2}$. Prove that the operator $P = -\Delta + \langle x \rangle^2$ is an isomorphism from D to L^2.

3. We consider the operator P^{-1} from L^2 to L^2 as follows.

$$P^{-1} : L^2 \to V' \to V \to L^2, \quad u \mapsto u \mapsto P^{-1} u \mapsto P^{-1} u.$$

Prove that Ker $(P^{-1})_{L^2 \to L^2} = \{0\}$, that $P^{-1}(L^2) = D$ and that P^{-1} is a positive compact and self-adjoint operator.
   According to the general theory (see Chap. 1, Sect. 1.2.9) $\sigma(P^{-1})$ (the spectrum of P^{-1}) consists in $\{0\}$ and a sequence of positive eigenvalues (μ_n) which tends to zero when $n \to +\infty$ and $H = \oplus_{n=1}^{+\infty} \text{Ker}(P^{-1} - \mu_n Id)$.
   Therefore, there exists an orthonormal basis $(e_n)_{n \in \mathbf{N}}$ of L^2 such that $P^{-1} e_n = \mu_n e_n$.

4. Show that $e_n \in D$ and that $(-\Delta + |x|^2) e_n = \lambda_n e_n$ where $\lambda_n = \frac{1}{\mu_n} - 1$.

5. a) Let $u \in D, u \neq 0$, be a solution of the equation $-\Delta u + |x|^2 u = \lambda u$ where $\lambda \geq 0$.
   Prove that for every $k \in \mathbf{N}, k \geq 2$ we have, $u \in H^k$, $\langle x \rangle u \in H^{k-1}$, $\langle x \rangle^2 u \in H^{k-2}$.
   Hint: Use (8.10) proved in question 2 and an induction on k.

   b) Prove that,

$$\| \nabla_x u \|_{L^2}^2 + \| |x| u \|_{L^2}^2 = \lambda \| u \|_{L^2}^2.$$

c) Prove that,

$$d \, \|u\|_{L^2}^2 \le \|\nabla_x u\|_{L^2}^2 + \| \, |x| u \|_{L^2}^2 .$$

Hint: Compute the quantity $2\mathrm{Re} \int_{\mathbf{R}^d} \partial_{x_j} u(x) x_j \overline{u(x)} \, dx$.

d) Deduce that $\lambda \ge d$.

6. a) Prove that the family $(e_n)_{n \in \mathbf{N}}$ is orthogonal in V and that $\|e_n\|_V^2 = \lambda_n + 1$.

b) Prove that $V = \left\{ u = \sum_{n \in \mathbf{N}} u_n e_n : \sum_{n \in \mathbf{N}} \lambda_n |u_n|^2 < +\infty \right\}$.

c) Prove that $D = \left\{ u = \sum_{n \in \mathbf{N}} u_n e_n : \sum_{n \in \mathbf{N}} \lambda_n^2 |u_n|^2 < +\infty \right\}$.

7. a) Let $\lambda \ne \lambda_n$ for every $n \in \mathbf{N}$. Prove that the operator $-\Delta + |x|^2 - \lambda : D \to L^2$ is invertible.

Hint: To prove the surjectivity if $f = \sum_n f_n e_n \in L^2$ set $u = \sum_n \frac{f_n}{\lambda_n - \lambda} e_n$.

b) Deduce that if we define the spectrum of $-\Delta + |x|^2$ as the set $S = \left\{ \lambda \in \mathbf{C} : -\Delta + |x|^2 - \lambda : D \to L^2 \text{ is not invertible} \right\}$ then

$$S = \cup_{n \in \mathbf{N}} \{\lambda_n\} .$$

Part 2

The goal of this part is to determine explicitly the spectrum of $\mathcal{H} = -\Delta + |x|^2$.

Case 1. $d = 1$ We shall set in what follows,

$$L^+ = -\frac{d}{dx} + x, \quad L^- = \frac{d}{dx} + x, \quad \mathcal{H} = -\frac{d^2}{dx^2} + x^2.$$

1. Prove the following identities,

$$\mathcal{H} = L^+ L^- + 1, \quad L^-(L^+)^n = (L^+)^n L^- + 2n(L^+)^{n-1}, \quad n \in \mathbf{N}, n \ge 1.$$

Hint: Use an induction.

2. a) Let $\phi_0(x) = \exp(-\frac{1}{2} x^2) \in S(\mathbf{R})$. Prove that $\mathcal{H}\phi_0 = \phi_0$.

b) We set $\phi_n = (L^+)^n \phi_0$. Prove that $\phi_n(x) = P_n(x)\exp(-\frac{1}{2} x^2)$ where P_n is a polynomial of order n whose coefficient of x^n is 2^n.

c) Prove that $\mathcal{H}\phi_n = (2n + 1)\phi_n$ for all $n \in \mathbf{N}$.

d) Prove that the family (ϕ_n) is orthogonal in L^2.

Hint: Prove by induction on $n \ge 1$ that for all $m < n$ we have $(\phi_m, \phi_n)_{L^2} = 0$ and use question 1.

e) Prove that the family (ϕ_n) is an orthonormal basis of L^2.

Hint: Prove that

$$\left[f \in L^2 \text{ and } (f, \phi_n)_{L^2} = 0 \forall n \in \mathbf{N} \right] \Longrightarrow \left(\int x^n f(x) e^{-\frac{x^2}{2}} \, dx = 0, \, \forall n \in \mathbf{N} \right)$$

$$\Longrightarrow f = 0.$$

Let $e_n = \frac{\phi_n}{\|\phi_n\|_{L^2}}$. Then (e_n) is an orthonormal basis of L^2 and $\mathcal{H}e_n = (2n+1)e_n$.

3. Let $u \in D$, $u \not\equiv 0$, $\lambda \in \mathbf{R}$ such that $\mathcal{H}u = \lambda u$. We want to prove that there exists $k \in \mathbf{N}$ such that $\lambda = 2k+1$.

a) Prove that $u = \sum_{n \in \mathbf{N}} (u, e_n)_{L^2} e_n$ where the series converges in L^2.
b) Prove that $(u, (\mathcal{H} - \lambda)\varphi)_{L^2} = 0$ for all $\varphi \in D$.
c) Taking $\varphi = e_k$, $k \in \mathbf{N}$ prove that there exists $k \in \mathbf{N}$ such that $\lambda = 2k+1$.

4. For any $k \in \mathbf{N}$ let $E_k = \mathrm{Ker}\,(\mathcal{H} - (2k+1))$. We want to show that we have $\dim E_k = 1$.

Let $k \in \mathbf{N}$ be fixed. Let $u \in D$ be a nontrivial solution of $(\mathcal{H} - (2k+1))u = 0$. Let $n \in \mathbf{N}$, $n \neq k$. Show that $(u, e_n)_{L^2} = 0$. Conclude.

Case 2. $d \geq 2$

1. Extend the results of questions 1 to 4 to higher dimensions.
   Hint: For $j = 1, \ldots, d$ set

$$L_j^+ = -\frac{\partial}{\partial x_j} + x_j, \qquad L_j^- = \frac{\partial}{\partial x_j} + x_j.$$

Introduce also $\phi_0(x) = \exp(-\frac{1}{2}|x|^2)$ and, for $\alpha \in \mathbf{N}^d$, set $\phi_\alpha = \sum_{k=1}^{d}(L_k^+)^{\alpha_k}\phi_0$. Prove that if $\mathcal{H} = -\Delta + |x|^2$ then $\mathcal{H}\phi_\alpha = (2|\alpha| + d)\phi_\alpha$. Eventually prove that the multiplicity of $\lambda = 2n + d$ is equal to $\sum_{|\alpha|=n} 1$.

Problem 42 *Ground State for the Hydrogen Atom*

This problem uses Problem 13.
   For $\psi \in H^1(\mathbf{R}^3)$ we set,

$$F(\psi) = \|\nabla_x \psi\|_{L^2(\mathbf{R}^3)}^2 - \left\| \frac{\psi}{|x|^{\frac{1}{2}}} \right\|_{L^2(\mathbf{R}^3)}^2, \tag{8.11}$$

$$E = \inf\left\{ F(\psi) : \psi \in H^1(\mathbf{R}^3), \|\psi\|_{L^2(\mathbf{R}^3)} = 1 \right\} \geq -\infty. \tag{8.12}$$

The goal of this problem is to show that $E = -\frac{1}{4}$, that the infimum is reached for $\widetilde{\psi} = \frac{1}{\sqrt{8\pi}}e^{-\frac{1}{2}|x|}$ and that the operator $-\Delta - \frac{1}{|x|} - E$ from $H^2(\mathbf{R}^3)$ to $L^2(\mathbf{R}^3)$ has a kernel of dimension one generated by $\widetilde{\psi}$.

This is linked with the study of the hydrogen atom in quantum mechanics.

Unless otherwise stated we shall denote in what follows $X = X(\mathbf{R}^3)$ if $X = L^2, H^k, C_0^\infty, \mathcal{D}'$.

Part 0
1. Prove that if $\psi \in H^1$ then $\frac{\psi}{|x|} \in L^2$. Prove also that $\frac{\psi}{|x|^{\frac{1}{2}}} \in L^2$ and that we have

$$\left\| \frac{\psi}{|x|^{\frac{1}{2}}} \right\|_{L^2} \leq C \|\psi\|_{L^2}^{\frac{1}{2}} \|\nabla_x \psi\|_{L^2}^{\frac{1}{2}}.$$

Hint: Use the inequality of Hardy (4.9) (see Problem 13).

Part 1. *Existence of a Minimizer*
An element $u \in H^1$ will be called a minimizer if $\|u\|_{L^2} = 1$ and $F(u) = E$.

1. Prove that $E < 0$.
    Hint: Consider $\psi(x) = \varepsilon^{\frac{3}{2}} \chi(\varepsilon x)$ where $\chi \in C_0^\infty$.
2. a) Prove that,

$$\exists (\psi_n)_{n \in \mathbf{N}} \subset H^1 \text{ with } \|\psi_n\|_{L^2} = 1 \text{ such that } \lim_{n \to +\infty} F(\psi_n) = E.$$

   b) Prove that the sequence (ψ_n) is bounded in H^1.
   c) Deduce that there is a subsequence $(\psi_{\sigma(n)})$ which converges weakly in L^2 to $\widetilde{\psi}$ and $(\nabla_x \psi_{\sigma(n)})$ converges weakly in L^2 to $\nabla_x \widetilde{\psi}$. Deduce that,

$$\|\widetilde{\psi}\|_{L^2} \leq 1, \quad \|\nabla_x \widetilde{\psi}\|_{L^2}^2 \leq \liminf_{n \to +\infty} \|\nabla_x \psi_{\sigma(n)}\|_{L^2}^2.$$

   d) Prove that $\lim_{n \to +\infty} \left\| \frac{\psi_{\sigma(n)} - \widetilde{\psi}}{|x|^{\frac{1}{2}}} \right\|_{L^2} = 0$.
    Hint: Cut the integral into $\{|x| \leq R\}$ and $\{|x| > R\}$ and use questions a), b), and c).
3. a) Prove that $F(\widetilde{\psi}) \leq E$.
    Hint: Use questions 2c) and d).
   b) Prove that $F(\widetilde{\psi}) \geq E \|\widetilde{\psi}\|_{L^2}^2$.
   c) Deduce that $\widetilde{\psi}$ is a minimizer.
4. a) Let $\psi \in H^1$ and $(\varphi_n) \subset H^2$ such that $(\varphi_n) \to \psi$ in H^1. Prove that $\lim_{n \to +\infty} F(\varphi_n) = F(\psi)$.
    Hint: Use question 1.
   b) For $j = 1, 2$ we set $E_j = \inf \{ F(\psi) : \psi \in H^j, \|\psi\|_{L^2} = 1 \}$. Prove that $E_1 = E_2$.

Part 2
A minimizer is an eigenfunction of $\mathcal{H} = -\Delta - \frac{1}{|x|}$ with eigenvalue E.
    Let $\widetilde{\psi}$ be a minimizer.

1. a) Let $\varepsilon > 0$ and $\varphi \in C_0^\infty$. Set $f = \widetilde{\psi} \pm \varepsilon\varphi$. Using the fact that $F(f) \geq E\|f\|_{L^2}^2$ prove that,

$$\text{Re} \left(-\Delta\widetilde{\psi} - \frac{\widetilde{\psi}}{|x|} - E\widetilde{\psi}, \overline{\varphi} \right) = 0,$$

where $\langle \cdot, \cdot \rangle$ denotes the bracket between $\mathcal{D}'$ and C_0^∞.

b) Changing φ to $i\varphi$ prove that, $\mathcal{H}\tilde{\psi} = E\tilde{\psi}$ in $\mathcal{D}'$.

c) Prove that $\tilde{\psi} \in D = \left\{ u \in H^2 : \frac{\psi}{|x|} \in L^2 \right\}$.

2. Let Ker $(\mathcal{H} - E) = \{u \in D : (\mathcal{H} - E)u = 0\}$. Prove that if $u \in$ Ker $(\mathcal{H} - E)$, $u \neq 0$ then $\frac{u}{\|u\|_{L^2}}$ is a minimizer.

Part 3. *Properties of the Minimizers*

Let $\tilde{\psi}$ be a minimizer. We show that $\tilde{\psi}(x) \neq 0$ for all $x \in \mathbf{R}^3$.   (8.13)

1. Let $u \in H^1$ be a real valued function. Prove that $|u| \in H^1$ and that $\nabla_x |u| = 1_{\{x:u(x)\neq 0\}} \frac{u}{|u|} \nabla_x u$.

   Hint: Consider for $\varepsilon > 0$, $u_\varepsilon = \sqrt{u^2 + \varepsilon^2} - \varepsilon$.

2. a) Prove that if $\tilde{\psi}$ is a minimizer then $|\tilde{\psi}|$ is also a minimizer.

   b) Deduce that $|\tilde{\psi}| \in D$ and,

$$(-\Delta + \omega^2)|\tilde{\psi}| = \frac{|\tilde{\psi}|}{|x|}, \quad \text{where } E = -\omega^2 \neq 0.$$

   c) Prove that,

$$|\tilde{\psi}(x)| = \frac{1}{4\pi} \int_{\mathbf{R}^3} \frac{e^{-\omega|x-y|}}{|x-y|} \frac{|\tilde{\psi}(y)|}{|y|} dy.$$

   Hint: Let $G = \frac{1}{4\pi} \frac{e^{-\omega|x|}}{|x|}$. Then $(-\Delta + \omega^2)G = \delta_0$.

   d) Prove that $|\tilde{\psi}(x)| > 0$ for every $x \in \mathbf{R}^3$.

We show that $|x|^k |\tilde{\psi}| \in L^2 \cap L^\infty$, $\forall k \in \mathbf{N}$.   (8.14)

3. Prove by induction on $k \in \mathbf{N}$ that $|x|^k |\tilde{\psi}| \in L^2 \cap L^\infty$ for all $k \in \mathbf{N}$.

   Hint: Use question c) and $L^1 \star L^2 \subset L^2$, $L^2 \star L^2 \subset L^\infty$.

We show that dim Ker$(\mathcal{H} - E) = 1$.   (8.15)

4. a) Prove that if $u \in$ Ker$(\mathcal{H} - E)$ then $u \equiv 0$ or $u \neq 0$ everywhere.

   b) Let $u_1, u_2 \in$ Ker$(\mathcal{H} - E)$ with $u_2 \neq 0$. Let x_0 be such that $u_2(x_0) \neq 0$. Let $u = u_1 - \frac{u_1(x_0)}{u_2(x_0)} u_2$. Prove that $u \equiv 0$ and conclude.

We show that if $\tilde{\psi} > 0$ then $\tilde{\psi}$ is radial.   (8.16)

5. Let A be a $d \times d$ orthogonal matrix. Set $\tilde{\psi}_A(x) = \tilde{\psi}(Ax)$. Show that $\tilde{\psi}_A \in$ Ker $(\mathcal{H} - E)$ and deduce that $\tilde{\psi}$ is radial, that means $\tilde{\psi}(x) = u(|x|)$.
   Hint: Use the fact that $\Delta \tilde{\psi}_A(x) = (\Delta \tilde{\psi})(Ax)$.

Part 4
The goal of this part is to determine E and $\tilde{\psi}$.
   Let $\theta \in D$ be a solution of $\mathcal{H}\theta = \lambda\theta$, with $\lambda = -\mu^2$, $\mu > 0$ such that,

$$\theta > 0, \quad |x|^k \theta \in L^2 \cap L^\infty \quad \forall k \in \mathbf{N}, \quad \theta \text{ is radial.}$$

1. Prove that $F(\theta) = \lambda \|\theta\|_{L^2}^2$ where F is defined in (8.11).
2. a) We set $|x| = r$ and $\theta(x) = u(r)$. Prove that u is continuous on $[0, +\infty)$ and satisfies the equation $-u''(r) - \frac{2}{r}u' - \frac{1}{r}u(r) = \lambda u(r)$ in $\mathcal{D}'((0, +\infty))$.
   b) Let $v(r) = ru(r)$. Prove that $r^k v \in L^\infty((0, +\infty))$ for all $k \in \mathbf{N}$. Deduce that $v \in L^1((0, +\infty))$.
   c) Prove that v satisfies $-v''(r) - \frac{1}{r}v(r) = \lambda v(r)$ in $\mathcal{D}'((0, +\infty))$.
3. a) Set $w(r) = 1_{\{r>0\}}v(r) \in L^1(\mathbf{R})$ and $T = -rw'' - w - \lambda rw \in \mathcal{S}'(\mathbf{R})$. Show that $T = \sum_{k=0}^{N} c_k \delta_0^{(k)}$, $c_k \in \mathbf{C}$.
   b) Let $\varphi \in C_0^\infty(\mathbf{R})$. For $\varepsilon > 0$ set $\varphi_\varepsilon(r) = \varphi(\frac{r}{\varepsilon})$. Prove that $\lim_{\varepsilon \to 0} \langle T, \varphi_\varepsilon \rangle = 0$.
      Hint: Use the fact that $v \in L^\infty((0, +\infty)) \cap C^0([0, +\infty))$ and $v(0) = 0$.
   c) Using question a) prove that $T = 0$.
4. a) Let $f(\xi) = \widehat{w}(\xi) = \int_0^{+\infty} e^{-ir\xi} v(r) \, dr$. Prove that f is a C^∞ function on $\mathbf{R}$ which can be extended as a holomorphic function (still denoted by f) in the set $\{z \in \mathbf{C} : \text{Im } z < 0\}$.
   b) Show that f satisfies the differential equation,

$$(\xi^2 + \mu^2)f'(\xi) + (2\xi + i)f(\xi) = 0, \quad \xi \in \mathbf{R}, \quad \lambda = -\mu^2, \mu > 0.$$

5. a) Let G be a holomorphic function in the open set $\{z \in \mathbf{C} : \text{Im } z < 0\}$ and continuous in the set $\{z \in \mathbf{C} : \text{Im } z \leq 0.\}$ Assume that $G|_{\text{Im } z=0} = 0$. Prove that $G = 0$.
      Hint: Use the Holmgren theorem and the principle of isolated zeros.
   b) Deduce that,

$$(z^2 + \mu^2)f'(z) + (2z + i)f(z) = 0, \quad z \in \mathbf{C}, \text{Im } z < 0.$$

6. a) Let $\varepsilon > 0$ small and $B = \{z \in \mathbf{C} : |z + i\mu| \leq \varepsilon\}$. Set $I = \int_B \frac{f'(z)}{f(z)} \, dz$. Show that I is an integer $n \in \mathbf{N}$.
   b) Using the residue formula prove that $I = \frac{1}{2\mu} - 1$ and then that $\mu = \frac{1}{2(n+1)}$.
   c) Deduce that the minimum energy E is equal to $-\frac{1}{4}$.
   d) Using the form of $\frac{f'}{f}$ deduce that $f(z) = \frac{C}{(z - \frac{i}{2})^2}$.

7. Recall that $\frac{C}{(\xi - \frac{i}{2})^2} = f(\xi) = \widehat{w}(\xi)$ so $w(r) = \mathcal{F}_{\xi \to r}^{-1} \left(\frac{C}{(\xi - \frac{i}{2})^2} \right)$ since w and f
are in $L^1(\mathbf{R})$. Our goal is to compute w.

a) Let $\varphi(r) = 1_{r>0} e^{-\frac{r}{2}} \in L^1(\mathbf{R})$. Compute $\widehat{\varphi}$ and deduce $\mathcal{F}(r\varphi)$.

b) Prove that $w(r) = 1_{r>0} Cre^{-\frac{r}{2}}$ and then that the set of minimizers is generated by $\widetilde{\psi}(x) = \frac{1}{\sqrt{8\pi}} e^{-\frac{1}{2}|x|}$.

Comments We would like to warmly thank our colleague P. Gérard for sharing his notes on the subject of this problem.   □

Problem 43 *On the Dirichlet–Neuman Operator*

This problem uses Problem 5.

The goal of this problem is to define the Dirichlet–Neuman operator in a particular geometry and to prove its continuity in the Sobolev spaces.

Notations Let $h > 0$ and $\eta \in W^{1,\infty}(\mathbf{R})$ be such that $\|\eta\|_{L^\infty(\mathbf{R})} \leq \frac{h}{2}$. We set,

$$\Omega = \left\{ (x, y) \in \mathbf{R}^2 : -h < y < \eta(x) \right\}, \quad \Sigma = \left\{ (x, y) \in \mathbf{R}^2 : y = \eta(x) \right\},$$

$$\widetilde{\Omega} = \left\{ (x, z) \in \mathbf{R}^2 : -h < z < 0 \right\}.$$

We shall denote by $C^{\infty,0}(\Omega)$ the space of C^∞ functions on Ω which vanish near Σ and by $\mathcal{H}^{1,0}(\Omega)$ the closure of $C^{\infty,0}(\Omega)$ in $H^1(\Omega)$, endowed with the $H^1(\Omega)$ norm. We will admit that,

$$\mathcal{H}^{1,0}(\Omega) = \left\{ u \in H^1(\Omega) : \gamma_0(u) = 0 \right\},$$

where γ_0 denotes the trace on Σ.

$C_0^\infty(\Omega)$ will denote the space of C^∞ functions with compact support in Ω.

We will set eventually $\Delta_{xy} = \frac{\partial^2}{\partial x^2} + \frac{\partial^2}{\partial y^2}$.

Part 1

Case $\eta = 0$. Assume $\eta \equiv 0$, that is, $\Omega = \left\{ (x, y) \in \mathbf{R}^2 : -h < y < 0 \right\}$. Let $\psi \in \mathcal{S}(\mathbf{R})$. We consider the problem,

$$\Delta_{xy} u = 0 \text{ in } \Omega, \quad u|_{y=0} = \psi, \quad (\partial_y u)|_{y=-h} = 0. \tag{8.17}$$

1. Prove that this problem has at most one solution belonging to $C^2([-h, 0],$ $H^s(\mathbf{R}))$ for all $s \in \mathbf{R}$.

   Hint: If u_1, u_2 are two solutions compute $-\left(\Delta_{xy}u, u\right)_{L^2(\Omega)}$ where $u = u_1 - u_2$.

2. a) Show that $u = \frac{\mathrm{ch}((y+h)|D_x|)}{\mathrm{ch}(h|D_x|)}\psi$, where $\mathrm{ch}(t) = \frac{1}{2}(e^t + e^{-t})$ is the hyperbolic cosine, is the unique solution of (8.17) in $C^2([-h, 0], H^s(\mathbf{R}))$ for every $s \in \mathbf{R}$.

      Hint: Use the Fourier transform in x.

   b) Set $G(0)\psi = (\partial_y u)|_{y=0}$. Deduce that $G(0)\psi = a(D_x)\psi$ where $a(\xi) = |\xi|\,\mathrm{th}\,(h|\xi|)$ and th denotes the hyperbolic tangent.

   c) Show that $a \in S^1$ and that for all $s \in \mathbf{R}$ we have,

$$\|G(0)\psi\|_{H^{s-\frac{1}{2}}(\mathbf{R})} \le \|\psi\|_{H^{s+\frac{1}{2}}(\mathbf{R})}, \qquad \forall \psi \in H^{s+\frac{1}{2}}(\mathbf{R}).$$

Part 2

Case $\eta \ne 0$.

1. Prove the inequality, $\|u\|_{L^2(\Omega)} \le \frac{3h}{2}\|\partial_y u\|_{L^2(\Omega)}$, for all $u \in \mathcal{H}^{1,0}(\Omega)$.

2. a) Let $\psi \in H^{\frac{1}{2}}(\mathbf{R})$. For $(x, z) \in \widetilde{\Omega}$ let $\widetilde{\psi}(x, z) = \chi(z)e^{z|D_x|}\psi(x)$ where $\chi \in C^\infty$, $\chi(z) = 1$ if $z \ge -\frac{h}{4}$, $\chi(z) = 0$ if $z \le -\frac{h}{3}$.

      Show that $\underline{\widetilde{\psi}}|_{z=0} = \psi$, $\underline{\widetilde{\psi}}|_{z=-h} = 0$ and that there exists $C > 0$ such that,

$$\|\underline{\widetilde{\psi}}\|_{H^1(\widetilde{\Omega})} \le C\|\psi\|_{H^{\frac{1}{2}}(\mathbf{R})}.$$

   b) Deduce that there exists $\underline{\psi} \in H^1(\Omega)$ such that,

$$\underline{\psi}|_\Sigma = \psi, \qquad \underline{\psi}(x, y) = 0 \text{ if } y \le -\frac{5}{6}h, \text{ and,}$$

$$\exists C > 0 : \|\underline{\psi}\|_{H^1(\Omega)} \le C(1 + \|\eta\|_{W^{1,\infty}(\mathbf{R})})\|\psi\|_{H^{\frac{1}{2}}(\mathbf{R})}.$$

3. We consider the sesquilinear form,

$$a(u, v) = \iint_\Omega \partial_x u(x, y)\overline{\partial_x u(x, y)}\,dy\,dx + \iint_\Omega \partial_y u(x, y)\overline{\partial_y u(x, y)}\,dy\,dx.$$

   Show that it is coercive on $\mathcal{H}^{1,0}(\Omega)$.

4. a) Let $\psi \in H^{\frac{1}{2}}(\mathbf{R})$. Prove that there exists a unique $u \in \mathcal{H}^{1,0}(\Omega)$ such that,

$$\left(-\Delta_{xy}u, \varphi\right) = \left(\Delta_{xy}\underline{\psi}, \varphi\right) \qquad \forall \varphi \in C_0^\infty(\Omega).$$

b) Let $\phi = u + \psi$. Prove that ϕ is the unique solution of the problem,

$$\left(-\Delta_{xy}u, \varphi\right) = 0 \quad \forall \varphi \in C_0^\infty(\Omega), \quad \gamma_0(\phi) = \psi \text{ on } \Sigma \qquad (8.18)$$

and that there exists $C > 0$ such that,

$$\|\partial_x\phi\|_{L^2(\Omega)} + \|\partial_y\phi\|_{L^2(\Omega)} \le C(1 + \|\eta\|_{W^{1,\infty}(\mathbf{R})})\|\psi\|_{H^{\frac{1}{2}}(\mathbf{R})}.$$

5. Prove that the map,

$$(x, z) \mapsto (x, \rho(x, z)), \quad \text{where} \quad \rho(x, z) = \frac{1}{h}(z + h)\eta(x) + z,$$

is a diffeomorphism from $\widetilde{\Omega}$ to Ω.

6. Let $\widetilde{\phi}(x, z) = \phi(x, \rho(x, z))$. Prove that $\widetilde{\phi}$ is the solution of problem,

$$(\Lambda_1^2 + \Lambda_2^2)\widetilde{\phi} = 0 \text{ in } \mathcal{D}'(\widetilde{\Omega}), \quad \widetilde{\phi}|_{z=0} = \psi,$$

where

$$\Lambda_1 = \frac{1}{\partial_z\rho}\partial_z, \quad \Lambda_2 = \partial_x - \frac{\partial_x\rho}{\partial_z\rho}\partial_z.$$

7. Let $U = (\Lambda_1 - (\partial_x\rho)\Lambda_2)\widetilde{\phi}$. Show that,

$$\partial_z U = -\partial_x\left((\partial_z\rho)\Lambda_2 U\right).$$

8. a) Prove that there exists $C > 0$ such that,

$$\|U\|_{L^2((-h,0),L^2(\mathbf{R}))} + \|\partial_z U\|_{L^2((-h,0),H^{-1}(\mathbf{R}))} \le C(1 + \|\eta\|_{W^{1,\infty}(\mathbf{R})})\|\psi\|_{H^{\frac{1}{2}}(\mathbf{R})}.$$

b) Deduce from Problem 5 that U is continuous on $[-h, 0]$ with values in $H^{-\frac{1}{2}}(\mathbf{R})$ and that there exists $C > 0$ such that,

$$\|U|_{z=0}\|_{H^{-\frac{1}{2}}(\mathbf{R})} \le C(1 + \|\eta\|_{W^{1,\infty}(\mathbf{R})})\|\psi\|_{H^{\frac{1}{2}}(\mathbf{R})}.$$

9. Show that

$$U|_{z=0} = \sqrt{1 + (\partial_x\eta)^2}\left(\frac{\partial\phi}{\partial n}\right)|_\Sigma,$$

where $\frac{\partial}{\partial n}|_\Sigma$ is the normal derivative on Σ.

Comments The map which to ψ (the Dirichlet data of the problem (8.18)) gives $\sqrt{1 + (\partial_x \eta)^2} \left(\frac{\partial \phi}{\partial n} \right) |_\Sigma$ (the normalized Neumann data of the same problem) is called the Dirichlet–Neuman operator. It is denoted by $G(\eta)$. Therefore, we have shown that

$$\|G(\eta)\psi\|_{H^{-\frac{1}{2}}(\mathbf{R})} \leq C(1 + \|\eta\|_{W^{1,\infty}(\mathbf{R})}) \|\psi\|_{H^{\frac{1}{2}}(\mathbf{R})},$$

for all $\psi \in H^{\frac{1}{2}}(\mathbf{R})$. This operator plays a role in establishing the water wave equations. □

Problem 44

Let $u \in H^1(\mathbf{R}^3)$ be a solution of the equation,

$$-\Delta u + u = u^3, \tag{8.19}$$

where $\Delta = \sum_{j=1}^{3} \frac{\partial^2}{\partial x_j^2}$ is the Laplacian.

The goal of the problem is to prove that u decays exponentially at infinity.

Part 1
1. Using the Sobolev embedding show that $u^3 \in L^2(\mathbf{R}^3)$. Deduce that $u \in H^2(\mathbf{R}^3)$.
2. Using (3.3) prove by induction on k that $u \in H^k(\mathbf{R}^3)$ for every $k \geq 2$.

Part 2. *Preliminaries*
1. Compute the Fourier transform of the $L^1(\mathbf{R}^3)$ function $E(x) = \frac{e^{-|x|}}{4\pi|x|}$.

   Hint: Use the fact that $\widehat{E}$ is radial, for fixed ξ use an orthogonal matrix A sending ξ to $(|\xi|, 0, \ldots, 0)$ then use the polar coordinates in $\mathbf{R}^3$.
2. Deduce that E is the unique solution in $\mathcal{S}'(\mathbf{R}^3)$ of the equation $E - \Delta E = \delta_0$ where δ_0 is the Dirac distribution at zero.

Part 3
Let $u \in H^1(\mathbf{R}^3)$ be a solution of (8.19).

1. Show that we have,

$$u(x) = (E \star u^3)(x) = \int_{\mathbf{R}^3} f(x, y) \, dy, \quad f(x, y) = \frac{e^{-|x-y|}}{4\pi|x-y|} u(y)^3.$$

   For $n \in \mathbf{N}$ large, set $\omega_n = \{x \in \mathbf{R}^3 : 2^n \leq |x| \leq 2^{n+1}\}$ and $A_n = \sup_{x \in \omega_n} |u(x)|$.
2. Prove that $\lim_{n \to +\infty} A_n = 0$.

3. Prove that there exists $C = C(u) > 0$ independent of n such that for $x \in \omega_n$,

$$I_1 := \int_{|y| \leq 2^{n-1}} f(x, y) \, dy \leq C e^{-2^{n-1}}, \quad I_2 := \int_{|y| \geq 2^{n+2}} f(x, y) \, dy \leq C e^{-2^{n-1}},$$

$$I_3 := \int_{y \in \omega_{n-1}} f(x, y) \, dy \leq C A_{n-1}^3, \quad I_4 := \int_{y \in \omega_n} f(x, y) \, dy \leq C A_n^3,$$

$$I_5 := \int_{y \in \omega_{n+1}} f(x, y) \, dy \leq C A_{n+1}^3,$$

and deduce that,

$$A_n \leq C(e^{-2^{n-1}} + A_{n-1}^3 + A_n^3 + A_{n+1}^3). \tag{8.20}$$

4. Prove that there exists $C' > 0$, $N_0 \in \mathbf{N}$ such that for $N \geq N_0$ we have,

$$\sum_{n=N}^{+\infty} A_n \leq C'(e^{-2^{N-2}} + A_{N-1}^3).$$

Hint: Using (8.20) estimate $\sum_{n=N}^{N+M} A_n$; then take N large enough and use question 2. Then let M go to $+\infty$.

5. a) Set $B_N = e^{-2^{N-2}} + A_N$. Show that there exists $N_1 \in \mathbf{N}$ such that $B_{N+1} \leq (1 + C')B_N^2$ for $N \geq N_1$ and deduce that for $N \geq N_1$ we have,

$$B_N \leq \frac{1}{1+C'} \left((1 + C')^2 B_{N_1} \right)^{2^{N-N_1}}$$

b) Prove that if N_1 is large enough one can find $K > 0$ and $\delta > 0$ such that $B_N \leq K e^{-\delta 2^N}$ for $N \geq N_1$.

c) Prove that there exist $C > 0$, $R > 0$, $\delta_1 > 0$ such that $|u(x)| \leq C e^{-\delta_1 |x|}$ for $|x| \geq R$.

Problem 45 *On the Smoothing Effect for the Heat Equation*

The goal of this problem is to study the smoothing effect for the heat equation. Consider the Cauchy problem,

$$\begin{cases} \partial_t u - \Delta u = 0, \\ u_{|t=0} = g. \end{cases} \tag{8.21}$$

We first prove a classical result: if $g \in H^s(\mathbf{R}^d)$ then $u \in L^2((0, T); H^{s+1}(\mathbf{R}^d))$ for any $T > 0$. We then prove a refined version which, in dimension $d = 2$ for instance, states that if $g \in L^2(\mathbf{R}^2)$ then u belongs to $L^2((0, T); L^\infty(\mathbf{R}^2))$.

Part 1

Let $d \geq 1$ and $s \in \mathbf{R}$. Given $T > 0$ and g in $H^s(\mathbf{R}^d)$, we denote by u the unique function in $C^0([0, T]; H^s(\mathbf{R}^d)) \cap C^1([0, T]; H^{s-2}(\mathbf{R}^d))$ solution to the Cauchy problem (8.21).

1. Given an integer $\ell \in \mathbf{N}^*$, introduce the Fourier multiplier P_ℓ defined by,

$$\widehat{P_\ell f}(\xi) = \hat{f}(\xi) \quad \text{if } |\xi| \leq \ell, \qquad \widehat{P_\ell f}(\xi) = 0 \quad \text{if } |\xi| > \ell,$$

where $\hat{v}(\xi) = \int v(y)e^{-iy\xi} \, dy$ is the Fourier transform of v.

Define u_ℓ by $u_\ell(t, \cdot) = P_\ell(u(t, \cdot))$. Show that $u_\ell \in C^1([0, T], H^s(\mathbf{R}^d))$ and that, for any time $t \in (0, T)$,

$$\frac{1}{2}\frac{d}{dt} \|u_\ell(t, \cdot)\|_{H^s}^2 + \|\nabla u_\ell(t, \cdot)\|_{H^s}^2 = 0.$$

2. Deduce that u belongs to $L^2((0, T), H^{s+1}(\mathbf{R}^d))$.

Part 2

We now take $g \in H^{\frac{d}{2}-1}(\mathbf{R}^d)$ and we denote by u the unique solution of (8.21). It follows from the previous part that,

$$\|u\|_{L^2((0,T),H^{\frac{d}{2}}(\mathbf{R}^d))} \leq C \|g\|_{H^{\frac{d}{2}-1}(\mathbf{R}^d)}.$$

The goal of this part is to improve this to,

$$\|u\|_{L^2((0,T),L^\infty(\mathbf{R}^d))} \leq C \|g\|_{H^{\frac{d}{2}-1}(\mathbf{R}^d)}. \tag{8.22}$$

1. Consider the Littlewood–Paley decomposition in $\mathbf{R}^d$, $I = \sum_{j \geq -1} \Delta_j$. Set $(\Delta_j u)(t, \cdot) = \Delta_j(u(t, \cdot))$. Prove that $\Delta_j u$ is the unique solution of the problem,

$$(\partial_t - \Delta)\Delta_j u = 0, \qquad (\Delta_j u)_{|t=0} = \Delta_j g.$$

2. Show that there exists $C > 0$ such that, for any time t in $[0, T]$ and any $j \geq -1$,

$$\|\Delta_j u(t, \cdot)\|_{L^\infty} \leq C \sum_{j \geq -1} 2^{j\frac{d}{2}} \|\Delta_j u(t, \cdot)\|_{L^2}.$$

Deduce that, for some $c > 0$,

$$\|u(t, \cdot)\|_{L^\infty} \leq C \sum_{j \geq -1} 2^{j\frac{d}{2}} e^{-ct2^{2j}} \|\Delta_j g\|_{L^2}.$$

3. a) Consider two sequences $(f_n)_{n\in\mathbf{Z}} \in \ell^1(\mathbf{Z})$ and $(g_n)_{n\in\mathbf{Z}} \in \ell^2(\mathbf{Z})$. Prove that the sequence $(h_n)_{n\in\mathbf{Z}}$ defined by,

$$h_n = \sum_{k\in\mathbf{Z}} f_{n-k} g_k$$

belongs to $\ell^2(\mathbf{Z})$.

b) Consider two sequences (a_j) and (a_k) indexed by $j, k = -1, 0, 1, \dots$. Prove that,

$$\sum_{j=-1}^{\infty} \sum_{k=-1}^{+\infty} \frac{a_j a_k}{2^{|k-j|}} \leq \sum_{n=-1}^{\infty} a_n^2.$$

c) Set, $a_j = 2^{j\left(\frac{d}{2}-1\right)} \|\Delta_j g\|_{L^2}$. Prove that,

$$\|u\|_{L^2((0,T);L^\infty)}^2 \leq \sum_{j=-1}^{\infty} \sum_{k=-1}^{\infty} \frac{a_j a_k}{c 2^{|k-j|}},$$

and then deduce (8.22).

Problem 46 *Entropy Notions and the Heat Equation*

In this problem, we are interested in finding quantities which are nonincreasing along the flow of the heat equation.

We shall consider functions $h = h(t, x) : [0, +\infty) \times \mathbf{R}^d \to (0, +\infty)$ which are 2π-*periodic* with respect to x_j for any $1 \leq j \leq d$, belong to $C^\infty([0, +\infty) \times \mathbf{R}^d)$ and satisfy the heat equation,

$$\partial_t h - \Delta h = 0, \quad \text{in } (0, +\infty) \times \mathbf{R}^d,$$

where $\Delta = \sum_{j=1}^{d} \partial_j^2$, $\partial_j = \frac{\partial}{\partial x_j}$.

Notations We shall set in what follows, $\nabla h = (\partial_j h)_{1\leq j\leq d}$ and,

$$\nabla^2 h = (\partial_i \partial_j h)_{1\leq i,j\leq d}, \quad \nabla h \otimes \nabla h = ((\partial_i h)(\partial_j h))_{1\leq i,j\leq d}, \quad |(a_{ij})|^2 = \sum_{i,j=1}^{d} a_{ij}^2.$$

1. Show that,

$$(\partial_t - \Delta)\mathrm{Log}\, h = \frac{|\nabla h|^2}{h^2},$$

$$(\partial_t - \Delta)(h\,\mathrm{Log}\, h) = -\frac{|\nabla h|^2}{h},$$

$$(\partial_t - \Delta)\frac{|\nabla h|^2}{h} = -2h\left|\frac{\nabla^2 h}{h} - \frac{\nabla h \otimes \nabla h}{h^2}\right|^2.$$

2. Introduce the function $t \mapsto H(t)$, known as the *Boltzmann's entropy*, defined by,

$$H(t) = \int_{[0,2\pi]^d} h(t,x)\mathrm{Log}\,(h(t,x))\,dx.$$

Prove that the function H decays in a convex manner that is,

$$\frac{d}{dt}H \le 0 \quad \text{and} \quad \frac{d^2}{dt^2}H \ge 0.$$

(The quantity H is also known as the Gibbs' entropy or also as the Shannon's entropy.)

3. In this question, we further assume that the initial data $h_0(x) = h(0,x)$ is a probability density which means that,

$$\int_{[0,2\pi]^d} h_0(x)\,dx = 1.$$

a) Verify that $\int_{[0,2\pi]^d} h(t,x)\,dx = 1$ for all time.
b) Introduce the functions $t \mapsto F(t)$ and $t \mapsto J(t)$ defined by,

$$F(t) = \int_{[0,2\pi]^d} \frac{|\nabla h(t,x)|^2}{h(t,x)}\,dx,$$

$$J(t) = \int_{[0,2\pi]^d} h(t,x)\left|\frac{\nabla^2 h(t,x)}{h(t,x)} - \frac{\nabla h(t,x) \otimes \nabla h(t,x)}{h(t,x)^2}\right|^2 dx.$$

Show that,

$$\frac{d}{dt}F + 2J = 0.$$

(The quantity F is called the *Fisher's information*).

c) Given a real number λ introduce the quantity,

$$A(\lambda) = \sum_{i,j=1}^{d} \int_{[0,2\pi]^d} h \left(\frac{\partial_{ij} h}{h} - \frac{(\partial_i h)(\partial_j h)}{h^2} + \lambda \delta_{ij} \right)^2 dx,$$

where $\delta_{ij} = 1$ when $i = j$ and 0 when $i \neq j$. Verify that,

$$A(\lambda) = J - 2\lambda F + \lambda^2 d.$$

Then by choosing λ appropriately deduce that,

$$J \geq \frac{1}{d} F^2.$$

d) Consider the function $t \mapsto N(t)$ defined by,

$$N(t) = \exp\left(-\frac{2}{d} H(t) \right).$$

(This function, introduced by Shannon, is called the *entropy power*). Prove that the function $t \mapsto N(t)$ is concave that is,

$$\frac{d^2}{dt^2} N \leq 0.$$

e) Let $u : [0, +\infty) \to (0, +\infty)$ be is a C^1 function satisfying for some $K > 0$ the inequality,

$$\partial_t u + K u^2 \leq 0.$$

Prove that,

$$u(t) \leq \frac{u(0)}{1 + K u(0) t}.$$

Hint: Consider the function $1/u$.

f) Conclude that there exists a constant $C > 0$ such that, for any time $t \geq 1$,

$$\int_{[0,2\pi]^d} \frac{|\nabla h(t, x)|^2}{h(t, x)} dx \leq \frac{C}{t}.$$

Problem 47 *Lyapunov Functions for the Mean-Curvature Equation*

In this problem, we consider functions $h = h(t, x) : [0, +\infty) \times \mathbf{R}^d \to \mathbf{R}$ which are 2π-*periodic* with respect to x_j for any $1 \leq j \leq d$, belong to $C^\infty([0, +\infty) \times \mathbf{R}^d)$ and are solutions to the mean-curvature equation,

$$\partial_t h + \sqrt{1 + |\nabla h|^2}\kappa = 0 \quad \text{where} \quad \kappa = -\operatorname{div}\left(\frac{\nabla h}{\sqrt{1 + |\nabla h|^2}}\right).$$

Here ∇ and div denote the derivatives and the divergence with respect to the spatial variable $x = (x_1, \ldots, x_d)$.

We are interested in finding quantities which are nonincreasing along the flow of this equation.

1. Show that,

$$\frac{d}{dt}\int_{[0,2\pi]^d}\sqrt{1 + |\nabla h|^2}\,dx = \int_{[0,2\pi]^d}(\partial_t h)\kappa\,dx.$$

Deduce that

$$\frac{d}{dt}\int_{[0,2\pi]^d}\sqrt{1 + |\nabla h|^2}\,dx \leq 0.$$

2. In the rest of the problem we assume that the space dimension d is equal to 1 and we use the notations,

$$h_x = \partial_x h, \quad h_{xx} = \partial_{xx} h.$$

Verify that,

$$\partial_t h + \sqrt{1 + h_x^2}\kappa = \partial_t h - \frac{\partial_{xx} h}{1 + h_x^2},$$

and deduce that,

$$\frac{d}{dt}\int_0^{2\pi} h_x^2\,dx \leq 0.$$

3. a) Verify that,

$$\partial_t h + \sqrt{1 + h_x^2}\kappa = \partial_t h - \partial_x \arctan(\partial_x h).$$

   b) Prove that,

$$\frac{d}{dt}\int_0^{2\pi} h^2\,dx \leq 0.$$

c) Compute $\partial_t(h_x \arctan(h_x))$ and then prove that,

$$\frac{d^2}{dt^2}\int_0^{2\pi} h^2\, dx \geq 0.$$

(Together with the previous result this shows that $\int_0^{2\pi} h^2\, dx$ decays in a convex manner.)

d) Prove that $\dot{h} = \partial_t h$ satisfies the equation,

$$\partial_t \dot{h} = -\partial_x\left(\frac{\dot{h}_x}{1+h_x^2}\right).$$

e) Prove that,

$$\frac{d}{dt}\int_0^{2\pi} (\partial_t h)^2\, dx \leq 0,$$

and then deduce that,

$$\frac{d}{dt}\int_0^{2\pi} (1+h_x^2)\kappa^2\, dx \leq 0.$$

Problem 48 *The Boussinesq Equation*

In this problem, we consider functions $h = h(t,x) : [0,+\infty) \times \mathbf{R}^d \to (0,+\infty)$ which are 2π-*periodic* with respect to x_j for any $1 \leq j \leq d$, belong to $C^\infty([0,+\infty) \times \mathbf{R}^d)$ and are solutions to the following Boussinesq's equation,

$$\partial_t h - \operatorname{div}(h\nabla h) = 0,$$

where ∇ and div denote the derivatives and the divergence with respect to the spatial variable $x = (x_1, \ldots, x_d)$.

As in the previous problems we are interested in finding quantities which are nonincreasing along the flow of this equation.

Part 1. *A Functional Inequality*
The goal of this part is to prove a Sobolev type inequality which will be used to study the Boussinesq equation.

We want to prove that, for any $d \geq 1$ and any positive function $\theta \in C^\infty(\mathbf{R}^d)$ which is 2π-*periodic* with respect to x_j for any $1 \leq j \leq d$ we have,

$$\int_{[0,2\pi]^d} |\nabla \theta^{1/2}|^4\, dx \leq \frac{9}{16}\int_{[0,2\pi]^d} (\Delta\theta)^2\, dx.$$

1. Verify that,

$$\int_{[0,2\pi]^d} (\Delta\theta)^2 \, dx = \int_{[0,2\pi]^d} \left|\nabla^2\theta\right|^2 dx,$$

where $\nabla^2\theta = (\partial_j\partial_k\theta)_{1\le j,k\le d}$.

2. Show that,

$$I := 16 \int_{[0,2\pi]^d} \left|\nabla\theta^{1/2}\right|^4 dx,$$

satisfies,

$$I = -\int_{[0,2\pi]^d} \left(\nabla\theta^{-1}\cdot\nabla\theta\right) |\nabla\theta|^2 \, dx.$$

3. Deduce that I can be written under the form,

$$I = \int_{[0,2\pi]^d} \theta^{-1}\Delta\theta|\nabla\theta|^2 \, dx + 2\int_{[0,2\pi]^d} \theta^{-1}[(\nabla\theta\cdot\nabla)\nabla\theta]\cdot\nabla\theta \, dx.$$

4. Prove that, $|(\nabla\theta\cdot\nabla)\nabla\theta| \le |\nabla\theta||\nabla^2\theta|$ and deduce that,

$$I \le 3\, I^{1/2}\left(\int_{[0,2\pi]^d} (\Delta\theta)^2 \, dx\right)^{1/2}.$$

Conclude the proof.

Part 2. *Lyapunov Functionals*
Assume that h is a smooth *positive* solution to $\partial_t h - \mathrm{div}(h\nabla h) = 0$ which is 2π-periodic with respect to x_j for any $1 \le j \le d$.

1. Let $m \ge 0$. Show that,

$$\frac{d}{dt}\int_{[0,2\pi]^d} h^{m+1}\, dx + m(m+1)\int_{[0,2\pi]^d} h^m|\nabla h|^2\, dx = 0.$$

2. Show that,

$$\frac{d}{dt}\int_{[0,2\pi]^d} h^2\,|\nabla h|^2 \, dx \le 0.$$

Hint: Multiply the equation by $\partial_t(h^2)$ and integrate by parts.

3. The goal of this question is to prove that,

$$m \in [0, (1+\sqrt{7})/2] \quad \Rightarrow \quad \frac{d}{dt} \int_{[0,2\pi]^d} h^m \, |\nabla h|^2 \, dx \le 0.$$

a) Check that,

$$\frac{d}{dt} \int_{[0,2\pi]^d} h^m \, |\nabla h|^2 \, dx = \int_{[0,2\pi]^d} m h^{m-1} \mathrm{div}(h\nabla h)|\nabla h|^2 \, dx$$

$$+ \int_{[0,2\pi]^d} 2h^m \nabla h \cdot \nabla \mathrm{div}(h\nabla h) \, dx.$$

b) Prove that,

$$\frac{d}{dt} \int_{[0,2\pi]^d} h^m \, |\nabla h|^2 \, dx = \frac{m^2+1}{2} \int_{[0,2\pi]^d} h^{m-1} |\nabla h|^4 \, dx$$

$$+ m \int_{[0,2\pi]^d} h^m \Delta h |\nabla h|^2 \, dx - 2 \int_{[0,2\pi]^d} \left(\mathrm{div} \left(h^{(m+1)/2} \nabla h \right) \right)^2 \, dx.$$

Hint: Use the identity,

$$\mathrm{div}(h^m \nabla h) \, \mathrm{div}(h\nabla h) = \left(\mathrm{div} \left(h^{(m+1)/2} \nabla h \right) \right)^2 - \frac{(m-1)^2}{4} h^{m-1} |\nabla h|^4.$$

c) Prove that,

$$(m+1) \int_{[0,2\pi]^d} h^m \, |\nabla h|^2 \, \Delta h \, dx = - \int_{[0,2\pi]^d} h^{m+1} (\Delta h)^2 \, dx$$

$$+ \int_{[0,2\pi]^d} \mathrm{div} \left(h^{m+1} \nabla h \right) \Delta h \, dx.$$

Hint: Integrate by parts twice.
       Deduce that,

$$\int_{[0,2\pi]^d} h^m \, |\nabla h|^2 \, \Delta h \, dx = - \frac{1}{m+1} \int_{[0,2\pi]^d} h^{m+1} (\Delta h)^2 \, dx$$

$$+ \frac{1}{m+1} \int_{[0,2\pi]^d} \left(\mathrm{div} \left(h^{(m+1)/2} \nabla h \right) \right)^2 \, dx$$

$$- \frac{m+1}{4} \int_{[0,2\pi]^d} h^{m-1} |\nabla h|^4 \, dx.$$

Hint: Use the identity,

$$\operatorname{div}\left(h^{m+1}\nabla h\right)\Delta h = \left(\operatorname{div}\left(h^{(m+1)/2}\nabla h\right)\right)^2 - \frac{(m+1)^2}{4}h^{m-1}|\nabla h|^4.$$

d) Deduce that,

$$\frac{d}{dt}\int_{[0,2\pi]^d} h^m\,|\nabla h|^2\,dx = \frac{m^2-m+2}{4}\int_{[0,2\pi]^d} h^{m-1}|\nabla h|^4\,dx$$

$$-\frac{m}{m+1}\int_{[0,2\pi]^d} h^{m+1}(\Delta h)^2\,dx - \frac{m+2}{m+1}\int_{[0,2\pi]^d}\left(\operatorname{div}\left(h^{(m+1)/2}\nabla h\right)\right)^2\,dx.$$

e) Use the inequality proved in Part 1 to show that,

$$\int_{[0,2\pi]^d}\left(\operatorname{div}\left(h^{(m+1)/2}\nabla h\right)\right)^2\,dx \geq \frac{(m+3)^2}{36}\int_{[0,2\pi]^d} h^{m-1}|\nabla h|^4\,dx.$$

f) Deduce that for any $m \geq 0$,

$$\frac{d}{dt}\int_{[0,2\pi]^d} h^m\,|\nabla h|^2\,dx + I_m \leq 0,$$

with,

$$I_m = \frac{m}{m+1}\int_{[0,2\pi]^d} h^{m+1}(\Delta h)^2\,dx + C_m\int_{[0,2\pi]^d} h^{m-1}|\nabla h|^4\,dx,$$

where

$$C_m = \frac{m+2}{m+1}\cdot\frac{(m+3)^2}{36} - \frac{m^2-m+2}{4}.$$

g) Verify that $C_m \geq 0$ for $m \in [0,(1+\sqrt{7})/2]$.

4. Conclude from the previous questions that the square of the L^2-norm decays in a convex manner that is,

$$\frac{d}{dt}\int_{[0,2\pi]^d} h^2\,dx \leq 0 \quad\text{and}\quad \frac{d^2}{dt^2}\int_{[0,2\pi]^d} h^2\,dx \geq 0.$$

5. We now introduce the Boltzmann's entropy (already considered in Problem 46) that is the function $t \mapsto H(t)$ defined by,

$$H(t) = \int_{[0,2\pi]^n} h(t,x)\mathrm{Log}\,(h(t,x))\,dx.$$

Prove that H decays in a convex manner that is,

$$\frac{d}{dt}H \leq 0 \quad \text{and} \quad \frac{d^2}{dt^2}H \geq 0.$$

Problem 49 *A Morawetz' Inequality*

This problem uses Problem 13.

Let $\square = \partial_t^2 - \sum_{i=1}^3 \partial_{x_i}^2$ be the wave operator in $\mathbf{R} \times \mathbf{R}^3$ and set $I = (0, +\infty)$. Let $f, g \in C_0^\infty(\mathbf{R}^3)$. We recall that the unique solution of the problem,

$$\square u = 0 \text{ in } I \times \mathbf{R}^3, \quad u|_{t=0} = f, \quad \partial_t u|_{t=0} = g$$

can be written as $u = u_1 + u_2 + u_3$, where $u_1(t, x) = \frac{t}{4\pi} \int_{S^2} g(x - t\omega) \, d\omega$,

$$u_2(t, x) = \frac{1}{4\pi} \int_{S^2} f(x - t\omega) \, d\omega, \quad u_3(t, x) = \frac{t}{4\pi} \sum_{i=1}^3 \int_{S^2} \frac{\partial f}{\partial x_i}(x - t\omega)\omega_i \, d\omega.$$

1. Using the Cauchy–Schwarz inequality and the polar coordinates in $\mathbf{R}^3$ prove that there exist constants $C_j > 0$ (independent of f, g) such that for all $x \in \mathbf{R}^3$,

$$\|u_1(\cdot, x)\|_{L^2(I)} \leq C_1 \|g\|_{L^2(\mathbf{R}^3)}, \tag{8.23}$$

$$\|u_2(\cdot, x)\|_{L^2(I)} \leq C_2 \left\| \frac{f(x - \cdot)}{|\cdot|} \right\|_{L^2(\mathbf{R}^3)}, \tag{8.24}$$

$$\|u_3(\cdot, x)\|_{L^2(I)} \leq C_3 \left(\sum_{i=1}^3 \|\partial_{x_i} f\|_{L^2(\mathbf{R}^3)}^2 \right)^{\frac{1}{2}}. \tag{8.25}$$

2. Deduce that there exists $C > 0$ (independent of f, g) such that,

$$\sup_{x \in \mathbf{R}^3} \|u(\cdot, x)\|_{L^2(I)} \leq C \left(\|g\|_{L^2(\mathbf{R}^3)} + \left(\sum_{i=1}^3 \|\partial_{x_i} f\|_{L^2(\mathbf{R}^3)}^2 \right)^{\frac{1}{2}} \right).$$

Hint: Use the Hardy inequality (4.9) (see Problem 13).

Problem 50 *A Nonlinear Wave Equation*

In what follows we will denote $\Box = \partial_t^2 - \sum_{j=1}^3 \partial_{x_j}^2$ the wave operator in $\mathbf{R} \times \mathbf{R}^3$ and $|\nabla_x u|^2 = \sum_{j=1}^3 (\partial_{x_j} u)^2$.

The goal of this problem is to study the existence of a solution to the problem

$$\Box u = |\nabla_x u|^2 - (\partial_t u)^2, \quad u|_{t=0} = 0, \quad \partial_t u|_{t=0} = g(x), \tag{8.26}$$

where g will be specified in the sequel.

Part 1

1. Let u be a smooth solution of (8.26). Set $v = e^u$. Of which problem v is it solution?
2. Show that $v = 1 + w$ where $w(t, x) = \frac{t}{4\pi} \int_{S^2} g(x - t\omega)\, d\omega$.

   We assume in what follows that $g \in L^\infty(\mathbf{R}^3)$, $\nabla_x g \in L^1(\mathbf{R}^3)$ and that $\lim_{|x| \to +\infty} g = 0$.
3. Writing $g(x - t\omega) = -\int_t^{+\infty} \frac{d}{dt}[g(x - s\omega)]\, ds$ show that,

$$|w(t, x)| \le \frac{1}{4\pi t} \|\nabla_x g\|_{L^1(\mathbf{R}^3)},$$

   where $\|\nabla_x g\|_{L^1(\mathbf{R}^3)} = \sum_{j=1}^3 \|\partial_{x_j} g\|_{L^1(\mathbf{R}^3)}$.
4. a) Prove that $|w(t, x)| < 1$ if $t > \frac{1}{4\pi} \|\nabla_x g\|_{L^1(\mathbf{R}^3)}$.
   b) Prove that,

$$|w(t, x)| \le \frac{1}{4\pi} \|\nabla_x g\|_{L^1(\mathbf{R}^3)} \|g\|_{L^\infty(\mathbf{R}^3)} \quad \text{if} \quad t \le \frac{1}{4\pi} \|\nabla_x g\|_{L^1(\mathbf{R}^3)}.$$

5. Assume that $\|\nabla_x g\|_{L^1(\mathbf{R}^3)} \|g\|_{L^\infty(\mathbf{R}^3)} < 4\pi$. Show that the problem (8.26) has a unique solution defined on $\mathbf{R} \times \mathbf{R}^3$.

Part 2

We consider the case where $\lim_{|x| \to +\infty} g = 0$, but $\|\nabla_x g\|_{L^1(\mathbf{R}^3)} \|g\|_{L^\infty(\mathbf{R}^3)} > 4\pi$.

Let $\delta > 0$ be small. Let φ_0 be the piecewise C^1 function on $[0, +\infty)$ defined by,

$$\varphi_0(t) = 1 \qquad \text{for } t \in [0, 1],$$

$$\varphi_0(t) = -\frac{1}{\delta} t + \frac{1+\delta}{\delta} \qquad \text{for } t \in [1, 1+\delta],$$

$$\varphi_0(t) = 0 \qquad \text{for } t \ge 1 + \delta.$$

Let $\varepsilon > 0$ be small and set,

$$\lambda = \frac{1+\varepsilon}{1-\varepsilon}, \quad g(x) = -\varphi_0\left(\frac{|x|}{\lambda}\right).$$

1. Let $|x_0| \leq \varepsilon\lambda$ and $1 < t_0 < (1-\varepsilon)\lambda = (1+\varepsilon)$. Show that $w(t_0, x_0) = -t_0$. Deduce that $v(t_0, x_0) < 0$, and then that the problem (8.26) has no globally defined solution on $\mathbf{R} \times \mathbf{R}^3$.

   We want now to give a lower bound for the quantity $\|\nabla_x g\|_{L^1(\mathbf{R}^3)} \|g\|_{L^\infty(\mathbf{R}^3)}$.

2. Using polar coordinates show that

$$\|\nabla_x g\|_{L^1(\mathbf{R}^3)} \|g\|_{L^\infty(\mathbf{R}^3)} \geq 4\pi \left(1 + \delta + \frac{\delta^2}{3}\right).$$

Part 3

Give a very simple example of function g, which does not tend to zero at infinity, for which the problem (8.26) has no globally defined solution on $\mathbf{R} \times \mathbf{R}^3$.

Comments The equation considered in this problem is the very beginning example of equations of the type $\Box u = f(\partial_t u, \nabla_x u)$ fully studied by S. Klainerman in a series of papers (see the book by C. Sogge [26]). $\Box$

Problem 51 *On the Strichartz Estimate for the Wave Equation*

Consider the Cauchy problem for the wave equation in $\mathbf{R}_t \times \mathbf{R}_x^d$,

$$\Box u = 0, \quad u|_{t=0} = 0, \quad \partial_t u|_{t=0} = u_1. \tag{8.27}$$

Part 1

The goal of this part is to show that in dimension $d = 3$ the end point Strichartz estimate does not hold. More precisely we want to show that for every $T > 0$ there is no positive constant C such that,

$$\|u\|_{L^2((0,T),L^\infty(\mathbf{R}^3))} \leq C\|u_1\|_{L^2(\mathbf{R}^3)}. \tag{8.28}$$

In $\mathbf{R}^3$ we denote by $B(x_0, r)$ the Euclidian ball of center x_0 and radius r and by e_1 the point $(1, 0, 0)$. Let $A = B(2e_1, 2) \setminus B(e_1, 1)$ and $1_A(x)$ be its indicator function.

1. a) Show that $A = \left\{x \in \mathbf{R}^3 : \frac{1}{4}|x|^2 < x_1 \leq \frac{1}{2}|x|^2\right\}$.

   b) Deduce that in the spherical coordinates $(\rho, \theta_1, \theta_2) \in (0, +\infty) \times (0, \pi) \times (0, 2\pi)$ we have,

$$A \subset \left\{(\rho, \theta_1, \theta_2) \in (0, 4) \times (0, \frac{\pi}{2}) \times (0, 2\pi) : \right.$$

$$\left. \arccos\left(\min(\frac{\rho}{2}, 1)\right) \leq \theta_1 \leq \arccos\left(\frac{\rho}{4}\right)\right\}.$$

Hint: We recall the spherical coordinates:

$$x_1 = \rho \cos\theta_1, \quad x_2 = \rho \cos\theta_1 \cos\theta_2, \quad x_3 = \rho \cos\theta_1 \sin\theta_2,$$

$$dx = \rho^2 \sin\theta_1 \, d\rho \, d\theta_1 \, d\theta_2.$$

c) Let $g_A : \mathbf{R}^3 \setminus \{0\} \to \mathbf{R}$ be defined by,

$$g_A(x) = \frac{1_A(x)}{|x|^2 (1 + |\ln|x||)^\alpha}, \quad \frac{1}{2} < \alpha \le 1. \tag{8.29}$$

Show that $g_A \in L^2(\mathbf{R}^3)$.

In what follows we shall take in (8.27) $u_1 = g_A$ and we recall then that the solution u can be written as,

$$u(t, x) = \frac{t}{4\pi} \int_{S^2} g_A(x - t\omega) \, d\omega. \tag{8.30}$$

2. We set for $k \in \mathbf{N}$, $A_k = A \cap \{x \in \mathbf{R}^3 : 2^{-k+1} < |x| \le 2^{-k+2}\}$.

a) Show that

$$u(t, te_1) = \frac{t}{4\pi} \sum_{k=0}^{+\infty} m_k, \quad \text{where } m_k = \int_{S^2} g_{A_k}(te_1 - t\omega) \, d\omega.$$

b) Show that for $1 < t < 2$ and $\omega \in S^2$ we have,

$$te_1 - t\omega \in A_k \iff 2^{-2k+1} < t^2(1 - \omega_1) \le 2^{-2k+3}.$$

c) Show that there exists $c_0 > 0$ such that for large k we have,

$$I_k := \int_{\{\omega \in S^2 : te_1 - t\omega \in A_k\}} d\omega = c_0 \, t^{-2} \, 2^{-2k}.$$

Hint: Use the spherical coordinates recalled in question 1b) with $\rho = 1$.

d) Deduce that there exists $c_1 > 0$ such that, $m_k \ge c_1 t^{-2}/(1 + k)^\alpha$.

3. Show that (8.28) does not hold for this u.

Part 2

The goal of this part is to show that for any $s \ge 0$ there is no $C > 0$ such that,

$$\|u\|_{L^2(\mathbf{R}^+, L^\infty(\mathbf{R}^3))} \le C \|u_1\|_{H^s(\mathbf{R}^3)}, \tag{8.31}$$

for every $u_1 \in H^s(\mathbf{R}^3)$, where u is the solution of (8.27).

Let $g_A \in L^2(\mathbf{R}^3)$ be the function defined in (8.29). Let $(g_n) \in S(\mathbf{R}^3)$ be a sequence which converges to g_A in $L^2(\mathbf{R}^3)$ and denote by u and u_n the corresponding solutions of (8.27) given by (8.30).

1. Our first goal is to prove that (u_n) cannot be bounded in $L^2(\mathbf{R}^+, L^\infty(\mathbf{R}^3))$. We argue by contradiction.

   Assume that there exists $M > 0$ such that $\|u_n\|_{L^2(\mathbf{R}^+, L^\infty(\mathbf{R}^3))} \leq M$.

a) Show that

$$\|u_n - u\|_{L_x^\infty(\mathbf{R}^3,\, L_t^2(\mathbf{R}))} \leq C\|g_n - g_A\|_{L^2(\mathbf{R}^3)}.$$

b) Using the Banach–Alaoglu theorem show that there is a subsequence $(u_{\sigma(n)})$ and $v \in L^2(\mathbf{R}^+, L^\infty(\mathbf{R}^3))$ such that,

$$\langle u_{\sigma(n)}, \varphi \rangle \to \langle v, \varphi \rangle, \quad \forall \varphi \in L^2(\mathbf{R}^+, L^1(\mathbf{R}^3)).$$

c) Prove that $v = u$ and using Part 1 show a contradiction.
d) Deduce that there exists a subsequence u_{n_k} tending to $+\infty$ in $L^2(\mathbf{R}^+, L^\infty(\mathbf{R}^3))$ when n_k goes to $+\infty$.

2. For $\lambda > 0$ we set $g_{n_k}^\lambda(x) = \lambda^{-\frac{3}{2}} g_{n_k}(\frac{x}{\lambda})$ and we denote by $u_{n_k}^\lambda$ the corresponding solution of (8.27) with $u_0 = 0$, $u_1 = g_{n_k}^\lambda$.
a) Prove that

$$(i) \qquad \|g_{n_k}^\lambda\|_{L^2(\mathbf{R}^3)} = \|g_{n_k}\|_{L^2(\mathbf{R}^3)},$$

$$(ii) \qquad \|u_{n_k}^\lambda\|_{L^2(\mathbf{R}^+, L^\infty(\mathbf{R}^3))} = \|u_{n_k}\|_{L^2(\mathbf{R}^+, L^\infty(\mathbf{R}^3))},$$

$$(iii) \qquad \lim_{\lambda \to +\infty} \|g_{n_k}^\lambda\|_{H^s(\mathbf{R}^3)} = (2\pi)^{\frac{3}{2}} \|g_{n_k}\|_{L^2(\mathbf{R}^3)}.$$

b) Prove that (8.31) cannot hold with a fixed C.

Problem 52

For $u_0 \in S'(\mathbf{R}^d)$ we denote by $S(t)u_0$ the solution of the problem,

$$i\partial_t u + \Delta u = 0, \quad u|_{t=0} = u_0.$$

1. Let $2 \leq p \leq +\infty$ and p' its conjugate defined by $\frac{1}{p} + \frac{1}{p'} = 1$.

   Prove that there exists $C > 0$ such that for all $u_0 \in L^{p'}(\mathbf{R}^d)$ and all $t \neq 0$ we have,

$$\|S(t)u_0\|_{L^p(\mathbf{R}^d)} \leq C|t|^{-d(\frac{1}{2} - \frac{1}{p})} \|u_0\|_{L^{p'}(\mathbf{R}^d)}.$$

2. Assume $d \geq 2$ and $2 < p < \frac{2d}{d-2}$. Let $u_0 \in H^1(\mathbf{R}^d)$. Using the density of $C_0^\infty(\mathbf{R}^d)$ in $H^1(\mathbf{R}^d)$, the Sobolev embedding and question 1 show that,

$$\lim_{t \to +\infty} \|S(t)u_0\|_{L^p(\mathbf{R}^d)} = 0.$$

Problem 53 *On the Smoothing Effect for the Schrödinger Equation*

Let $u_0 \in C_0^\infty(\mathbf{R}^d)$. We know that the problem,

$$i\partial_t u + \Delta u = 0, \quad u|_{t=0} = u_0, \quad \text{where } \Delta = \sum_{j=1}^d \partial_{x_j}^2, \qquad (8.32)$$

has a unique solution $u \in C^1(\mathbf{R}, L^2(\mathbf{R}^d))$ such that $\|u(t)\|_{L^2(\mathbf{R}^d)} = \|u_0\|_{L^2(\mathbf{R}^d)}$. The purpose of this problem is to show that, when $d = 1$, for any $\sigma > \frac{1}{2}$ we have

$$\langle x \rangle^{-\sigma} u \in L^2((0, T), H^{\frac{1}{2}}(\mathbf{R})).$$

1. Let $f, g \in C^1(\mathbf{R}, L^2(\mathbf{R}^d))$. Show that the map $t \mapsto (f(t), g(t))_{L^2(\mathbf{R}^d)}$ is C^1 on $\mathbf{R}$ and that,

$$\partial_t (f(t), g(t))_{L^2(\mathbf{R}^d)} = (\partial_t f(t), g(t))_{L^2(\mathbf{R}^d)} + (f(t), \partial_t g(t))_{L^2(\mathbf{R}^d)}.$$

2. Let $B \in \mathrm{Op}(S^0)$ and let u be the solution of the problem (8.32). Show that,

$$\partial_t (Bu(t), u(t))_{L^2(\mathbf{R}^d)} = (i[B, \Delta]u(t), u(t))_{L^2(\mathbf{R}^d)}.$$

3. Deduce that for all $T > 0$ there exists $C > 0$ such that,

$$\left| \int_0^T (i[B, \Delta]u(t), u(t))_{L^2(\mathbf{R}^d)} \, dt \right| \leq C\|u_0\|_{L^2(\mathbf{R}^d)}^2.$$

We assume in what follows that $d = 1$.

Let $b = -\frac{1}{2}\frac{\xi}{\langle\xi\rangle}\int_0^x \frac{dy}{\langle y\rangle^{2\sigma}} dy$ where $\sigma > \frac{1}{2}$ and $\langle a \rangle = (1 + a^2)^{\frac{1}{2}}$. Let $B = \text{Op}(b)$.

4. a) Show that $b \in S^0$.
   b) Prove that $i[B, \Delta] = -\langle x \rangle^{-2\sigma} \Lambda^{-1}\partial_x^2 + R$ where $R \in \text{Op}(S^0)$ and $\Lambda = (Id - \partial_x^2)^{\frac{1}{2}}$.
   c) Deduce that for all $T > 0$ there exists $C > 0$ such that,

$$\int_0^T \|\partial_x \Lambda^{-\frac{1}{2}} \langle x \rangle^{-\sigma} u(t)\|_{L^2(\mathbf{R})}^2 \, dt \leq C\|u_0\|_{L^2(\mathbf{R})}^2,$$

and then that,

$$\int_0^T \| \langle x \rangle^{-\sigma} u(t)\|_{H^{\frac{1}{2}}(\mathbf{R})}^2 \, dt \leq C(1+T)\|u_0\|_{L^2(\mathbf{R})}^2. \tag{8.33}$$

5. Let $u_0 \in L^2(\mathbf{R})$ and $(\varphi_k) \subset C_0^\infty(\mathbf{R})$ be a sequence which converges to u_0 in $L^2(\mathbf{R})$. Denote by u (resp. u_k) the solution of the problem (8.32) with datum u_0 (resp. φ_k).

   a) Prove that $(\langle x \rangle^{-\sigma} u_k)$ converges to an element v in $L^2((0, T), H^{\frac{1}{2}}(\mathbf{R}))$.
   b) Prove that the sequence (u_k) converges to u in $L^2((0, T), L^2(\mathbf{R}))$ and deduce that it converges to u in $\mathcal{D}'((0, T) \times \mathbf{R})$.
   c) Deduce that u satisfies $\langle x \rangle^{-\sigma} u \in L^2((0, T), H^{\frac{1}{2}}(\mathbf{R}))$ together with the inequality (8.33).

Problem 54

Let $d \geq 1$, $s > d/2$ and let $f \in C^\infty(\mathbf{C}, \mathbf{C})$ be such that $f(0) = 0$. Our goal is to study the existence of a solution to the problem,

$$i\partial_t u + \Delta u = f(u), \quad u|_{t=0} = u_0, \tag{8.34}$$

when $u_0 \in H^s(\mathbf{R}^d)$.
   We shall use the following result (see (3.3)):

   (i) if $u \in H^s(\mathbf{R}^d)$ we have $f(u) \in H^s(\mathbf{R}^d)$ and
   (ii) there exists $\mathcal{F} : \mathbf{R}^+ \to \mathbf{R}^+$ nondecreasing such that

$$\|f(u)\|_{H^s(\mathbf{R}^d)} \leq \mathcal{F}\left(\|u\|_{H^s(\mathbf{R}^d)}\right) \|u\|_{H^s(\mathbf{R}^d)}.$$

Notations

- We shall denote by $u(t)$ the function $x \mapsto u(t, x)$ and we shall set $H^s = H^s(\mathbf{R}^d)$.

- $S(t)$ will denote the Schrödinger propagator defined by $S(t) f = \mathcal{F}^{-1} \left(e^{-it|\xi|^2} \hat{f} \right)$.
- We recall that for $a \in \mathbf{R}$ and $f \in S'(\mathbf{R}^d)$ $S(t+a)f$ satisfies,

$$(i \partial_t + \Delta) \, (S(t+a)f) = 0, \quad (S(t+a)f) \, |_{t=0} = S(a)f.$$

Part 1
We study here the existence problem. We fix $u_0 \in H^s$ with $s > d/2$.
Let $\mathcal{A}$ be the operator $u \mapsto \mathcal{A}(u)$ where,

$$\mathcal{A}(u)(t) = S(t)u_0 - i \int_0^t S(t - \sigma)[f(u(\sigma))] \, d\sigma.$$

We set $R = 2\|u_0\|_{H^s}$, $I = (-T, T)$ and,

$$B = \left\{ u \in L^\infty(I, H^s) \cap C^0(I, H^s) : \|u\|_{L^\infty(I, H^s)} \leq R \right\}.$$

endowed with the $L^\infty(I, H^s)$ norm.

1. Show that if T is small enough with respect to $\|u_0\|_{H^s}$ the operator $\mathcal{A}$ sends B to B.
2. Show that if T is small enough with respect to $\|u_0\|_{H^s}$ the operator $\mathcal{A}$ is contractive from B to B.
3. Deduce that the equation $u = \mathcal{A}(u)$ has a solution $u \in B$.
4. Using the linear theory show that this solution u satisfies (8.34).

Part 2
We study here the uniqueness in the space $L^\infty(I, H^s) \cap C^0(I, H^s)$.
   Let u_1, u_2 be two solutions in $L^\infty(I, H^s) \cap C^0(I, H^s)$ of the equation $u = \mathcal{A}(u)$.
We consider the set,

$$F = \left\{ t \in I : u_1(t) = u_2(t) \text{ in } H^s \right\}.$$

1. Show that F is a nonempty closed subset of I.
2. We want to show that F is open in I.

   a) Let $t_0 \in F$. Set $v_j(t) = u_j(t + t_0)$, $j = 1, 2$. Show that v_j satisfies,

   $$v_j(t) = S(t)u_j(t_0) - i \int_0^t S(t - \sigma')[f(v_j(\sigma'))] \, d\sigma', \quad t \in (-T - t_0, T - t_0).$$

   Hint: Use the fact that $S(a)S(b) = S(a + b)$ for $a, b \in \mathbf{R}$.
   b) Let $\varepsilon > 0$ be such that $(t_0 - \varepsilon, t_0 + \varepsilon) \subset I$. Using question a) show that,

   $$\|v_1(t) - v_2(t)\|_{H^s} \leq \int_{-\varepsilon}^\varepsilon \|f(v_1(\sigma)) - f(v_2(\sigma))\|_{H^s} \, d\sigma, \quad t \in (-\varepsilon, \varepsilon).$$

c) Deduce that there exists $\varepsilon > 0$ such that $(t_0 - \varepsilon, t_0 + \varepsilon) \subset F$ and conclude.

Part 3

We show here that if T is small enough, the map $u_0 \mapsto u$ is locally Lipschitz from H^s to $L^\infty(I, H^s)$.

1. Let $u_0, v_0 \in H^s$ be such that $\|u_0\|_{H^s} \leq \frac{R}{2}$, $\|v_0\|_{H^s} \leq \frac{R}{2}$ and let u, v be the corresponding solutions. Prove that there exist $T_0 > 0$ and $C > 0$ such that for $T \leq T_0$ we have,

$$\|u(t) - v(t)\|_{H^s} \leq C \|u_0 - v_0\|_{H^s}, \quad \forall t \in (-T, T).$$

Problem 55 *Infinite Speed Propagation for the Schrödinger Equation*

Let $P = i\partial_t + \Delta_x$ be the Schrödinger operator and for $j = 1, \ldots, d$, set

$$L_j = x_j + 2it\partial_{x_j}.$$

We consider the Cauchy problem,

$$Pu = 0, \quad u|_{t=0} = g.$$

1. Prove that $[P, L_j] = 0$.
2. Set $L^\alpha = L_1^{\alpha_1} \cdots L_d^{\alpha_d}$. Using the fact that $[P, L^\alpha] = 0$ prove that if $g \in L^2(\mathbf{R}^d)$ is such that $x^\alpha g \in L^2(\mathbf{R}^d)$ for all $\alpha \in \mathbf{N}^d$ then for $t \neq 0$, the function $x \mapsto u(t, x)$ belongs to $C^\infty(\mathbf{R}^d)$.

Problem 56 *Soliton for the Benjamin–Ono Equation*

1. Compute the Fourier transform of the function $f(\xi) = e^{-|\xi|}$, $\xi \in \mathbf{R}$.
2. Deduce the Fourier transform of the function $g(\xi) = |\xi| e^{-|\xi|}$.
     Hint: We may consider for $\lambda > 0$ the functions $g_\lambda(\xi) = |\xi| e^{-\lambda|\xi|}$.
3. Set for $x \in \mathbf{R}$, $Q(x) = \frac{1}{1+x^2}$ and for $c \in \mathbf{R} \setminus \{0\}$,

$$u(t, x) = 4cQ(c(x + ct)), \quad t \in \mathbf{R}, x \in \mathbf{R}.$$

a) Show that Q belongs to $H^s(\mathbf{R})$ for any real number s.
b) We define the operator $|D_x|$ on $\mathcal{S}'(\mathbf{R})$ by

$$|D_x|T = \mathcal{F}^{-1}\left(|\xi| \widehat{T}\right).$$

Show that

$$|D_x|u = -cu + \frac{1}{2}u^2.$$

c) Deduce that u satisfies the Benjamin–Ono equation,

$$\partial_t u + \partial_x |D_x| u = \frac{1}{2}\partial_x(u^2).$$

Problem 57 *Averaging Lemma*

Let $(\Omega, \mathcal{T}, \mu)$ be a measured set such that $\mu(\Omega) < +\infty$. We consider a family $(p_\omega(\xi))_{\omega \in \Omega}$ which is L^∞ in ω with values in the set of homogeneous symbols of degree 1 on $\mathbf{R}^d$.

Part 1
We assume that there exists $\delta \in (0, 1)$ and $c_0 > 0$ such that,

$$\forall \theta \in \mathbf{S}^{d-1}, \ \forall a > 0, \quad \mu\{\omega \in \Omega : |p_\omega(\theta)| \leq a\} \leq c_0 \, a^{2\delta}. \qquad (\star)$$

1. Let $u \in L^2(\mathbf{R}^d \times \Omega, dx\, d\mu)$. Show that the function $\omega \to u(x, \omega)$ belongs to $L^1(\Omega, d\mu)$ for almost all $x \in \mathbf{R}^d$.
   Set,

$$U(x) = \int_\Omega u(x, \omega)\, d\mu(\omega).$$

The goal of this part is to show that $(\star)$ implies that there exists $C > 0$ such that,

$$\|U\|_{H^\delta(\mathbf{R}^d)} \leq C \left(\|u\|_{L^2(\mathbf{R}^d \times \Omega, dx\, d\mu)} + \| \operatorname{Op}(p_\omega)u \|_{L^2(\mathbf{R}^d \times \Omega, dx\, d\mu)} \right), \qquad (\star\star)$$

for all $u \in L^2(\mathbf{R}^d \times \Omega, dx\, d\mu)$ such that $\operatorname{Op}(p_\omega)u \in L^2(\mathbf{R}^d \times \Omega, dx\, d\mu)$.
2. a) Prove that $U \in L^2(\mathbf{R}^d)$.
   b) Prove that $\widehat{U}(\xi) = \int_\Omega \widehat{u}(\xi, \omega)d\mu(\omega)$ (where $\widehat{u}$ is the Fourier transform of u with respect to x) and that

$$|\widehat{U}(\xi)|^2 \leq C \int_\Omega |\widehat{u}(\xi, \omega)|^2 d\mu(\omega).$$

For $\xi \in \mathbf{R}^d, \xi \neq 0$ we set,

$$I_1(\xi) = \int_{|p_\omega(\xi)| \leq 1} \widehat{u}(\xi, \omega) \, d\mu(\omega), \quad I_2(\xi) = \int_{|p_\omega(\xi)| \geq 1} \widehat{u}(\xi, \omega) \, d\mu(\omega).$$

3. Show that there exists $C > 0$ such that,

$$|I_1(\xi)|^2 \leq \frac{C}{|\xi|^{2\delta}} \int_\Omega |\widehat{u}(\xi, \omega)|^2 \, d\mu(\omega), \quad \forall \xi \neq 0.$$

4. Prove that for $\xi \neq 0$ we have,

$$|I_2(\xi)|^2 \leq \frac{1}{|\xi|^2} \left(\int_{|p_\omega(\theta)| \geq \frac{1}{|\xi|}} \frac{d\mu(\omega)}{|p_\omega(\theta)|^2} \right) \left(\int_\Omega |p_\omega(\xi) \widehat{u}(\xi, \omega)|^2 \, d\mu(\omega) \right),$$

where $\theta = \frac{\xi}{|\xi|} \in \mathbf{S}^{d-1}$.

5. We fix $\theta \in \mathbf{S}^{d-1}$. Let ν be the measure on $\mathbf{R}^+$ which is the image of the measure μ by the map $\Omega \to \mathbf{R}^+$, $\omega \mapsto |p_\omega(\theta)|$.

   a) How does condition ($\star$) translates on ν?
   b) Show that,

$$\frac{1}{|\xi|^2} \int_{|p_\omega(\theta)| \geq \frac{1}{|\xi|}} \frac{d\mu(\omega)}{|p_\omega(\theta)|^2} = \frac{1}{|\xi|^2} \int_{\frac{1}{|\xi|}}^{+\infty} \frac{d\nu(t)}{t^2} = \lim_{N \to +\infty} \frac{1}{|\xi|^2} \int_{\frac{1}{|\xi|}}^{\frac{N}{|\xi|}} \frac{d\nu(t)}{t^2},$$

$$:= \lim_{N \to +\infty} J_N(\xi).$$

6. Writing $J_N(\xi) = \frac{1}{|\xi|^2} \sum_{k=1}^{N-1} \int_{\frac{k}{|\xi|}}^{\frac{k+1}{|\xi|}} \frac{d\nu(t)}{t^2}$ and using ($\star$) prove that there exists $C > 0$ such that,

$$J_N(\xi) \leq \frac{C}{|\xi|^{2\delta}}.$$

7. Deduce ($\star\star$).

Part 2
We study the converse. We assume ($\star\star$) satisfied. We recall that $U(x) = \int_\Omega u(x, \omega) \, d\mu(\omega)$.

1. Let $g : \mathbf{R}^d \to \mathbf{R}$ be a continuous function such that,

$$\forall \psi \in \mathcal{S}(\mathbf{R}^d), \quad \int_{\mathbf{R}^d} g(\xi) |\widehat{\psi}(\xi)|^2 \, d\xi \leq 0.$$

Show that $g(\xi) \leq 0$ for all $\xi \in \mathbf{R}^d$.

2. Applying $(\star\star)$ to $u(x, \omega) = \psi(x)\varphi(\omega)$ where $\psi \in S(\mathbf{R}^d)$ and $\varphi \in L^2(\Omega)$ is well chosen, using the Fourier transform and the previous question prove that $(\star)$ is satisfied.

Problem 58 *Kinetic Equations*

Part 1

The last question of Part 2 of this problem uses Part 1 in Problem 57.

For $t \in \mathbf{R}, x \in \mathbf{R}^d, v \in \mathbf{R}^d, d \geq 1$, we consider the problem,

$$\frac{\partial f}{\partial t}(t, x, v) + v \cdot \nabla_x f(t, x, v) = 0, \quad f(0, x, v) = f^0(x, v). \tag{8.35}$$

1. Using the method of characteristics prove that for $f^0 \in C^1(\mathbf{R}_x^d \times \mathbf{R}_v^d)$ this problem has a unique solution $f \in C^1(\mathbf{R}_t \times \mathbf{R}_x^d \times \mathbf{R}_v^d)$.

2. Prove that if in addition $f^0 \in L^p(\mathbf{R}_x^d \times \mathbf{R}_v^d)$ then for $t \in \mathbf{R}$ we have $f(t) \in L^p(\mathbf{R}_x^d \times \mathbf{R}_v^d)$, where $f(t)$ denotes the function $(x, v) \mapsto f(t, x, v)$, and,

$$\|f(t)\|_{L^p(\mathbf{R}_x^d \times \mathbf{R}_v^d)} = \|f^0\|_{L^p(\mathbf{R}_x^d \times \mathbf{R}_v^d)}.$$

If for every $(t, x) \in \mathbf{R}_t \times \mathbf{R}_x^d$ the function $v \mapsto f(t, x, v)$ belongs to $L^1(\mathbf{R}^d)$ we define the macroscopic density by,

$$\rho(t, x) = \int_{\mathbf{R}^d} f(t, x, v) \, dv.$$

3. Assume that $f^0 \in L^1(\mathbf{R}_x^d, L^\infty(\mathbf{R}_v^d))$. Prove that we have for $t \neq 0$,

$$|\rho(t, x)| \leq \frac{1}{t^d} \|f^0\|_{L^1(\mathbf{R}_x^d, L^\infty(\mathbf{R}_v^d))}.$$

4. In this question we want to prove a kind of Strichartz estimate for the macro-scopic density. Let a, q, r be positive real numbers such that,

$$1 < r < \frac{d}{d-1}, \quad \frac{2}{q} = d\left(1 - \frac{1}{r}\right), \quad 1 \leq a = \frac{2r}{r+1} < \frac{2d}{2d-1}.$$

Notice then that $q > 1$.

a) Let $\Phi \in C_0^\infty(\mathbf{R}_t \times \mathbf{R}_x^d)$. Prove the estimate,

$$\left| \int_{\mathbf{R}} \int_{\mathbf{R}^d} \rho(t,x) \Phi(t,x) \, dx \, dt \right|$$

$$\leq \|f^0\|_{L^a(\mathbf{R}_x^d \times \mathbf{R}_v^d)} \left\| \int_{\mathbf{R}} |\Phi(t, x+vt)| \, dt \right\|_{L^{a'}(\mathbf{R}_x^d \times \mathbf{R}_v^d)},$$

where a' is the conjugate exponent of a defined by $\frac{1}{a} + \frac{1}{a'} = 1$.

b) We set $I := \left\| \int_{\mathbf{R}} |\Phi(t, x+vt)| \, dt \right\|_{L^{a'}(\mathbf{R}_x^d \times \mathbf{R}_v^d)}^2$. Prove that,

$$I = \left\| \left(\int_{\mathbf{R}} |\Phi(t, x+vt)| \, dt \right)^2 \right\|_{L^{r'}(\mathbf{R}_x^d \times \mathbf{R}_v^d)},$$

where r' is the conjugate exponent of r.

c) Prove that,

$$I \leq \int_{\mathbf{R}} \|\Phi(t, \cdot)\|_{L^{r'}(\mathbf{R}^d)} \left(\int_{\mathbf{R}} \frac{1}{|t-s|^{\frac{d}{r'}}} \|\Phi(s, \cdot)\|_{L^{r'}(\mathbf{R}^d)} \, ds \right) dt.$$

d) Deduce that,

$$I \leq \|\Phi\|_{L^{q'}(\mathbf{R}_t, L^{r'}(\mathbf{R}_x^d))} \left\| \frac{1}{|t|^{\frac{d}{r'}}} \star |\Phi| \right\|_{L^q(\mathbf{R}_t, L^{r'}(\mathbf{R}_x^d))}.$$

e) Using the Hardy–Littlewood–Sobolev inequality prove that one can find $C > 0$ such that,

$$I \leq C \|\Phi\|_{L^{q'}(\mathbf{R}_t, L^{r'}(\mathbf{R}_x^d))}^2.$$

Hint: Show that $\frac{d}{r'} < 1$ and that if b is defined by $\frac{1}{q} = \frac{1}{b} + \frac{d}{r'} - 1$ then $b = q'$.

f) Using the previous questions deduce that,

$$\left| \int_{\mathbf{R}} \int_{\mathbf{R}^d} \rho(t,x) \Phi(t,x) \, dx \, dt \right| \leq \|f^0\|_{L^a(\mathbf{R}_x^d \times \mathbf{R}_v^d)} \|\Phi\|_{L^{q'}(\mathbf{R}_t, L^{r'}(\mathbf{R}_x^d))}.$$

5. Prove that there exists $C(d) > 0$ such that,

$$\|\rho\|_{L^q(\mathbf{R}_t, L^r(\mathbf{R}_x^d))} \leq C(d) \|f^0\|_{L^a(\mathbf{R}_x^d \times \mathbf{R}_v^d)}.$$

Part 2

In this part we want to apply the result obtained in Part 1 of Problem 57 to the following situation.

Let $M > 0$ and $B(0, M) = \{v \in \mathbf{R}^d : |v| \leq M\}$. Let $\chi \in C_0^\infty(\mathbf{R}_v^d)$ with supp $\chi \subset B(0, M)$. We shall take,

$$\Omega = \mathbf{R}^d, \quad \mu = \chi(v)\,dv, \quad p_v(\tau, \xi) = \tau + v \cdot \xi.$$

For $a > 0$ and $(\tau, \xi) \in \mathbf{S}^{d-1}$ we set,

$$E_{\tau,\xi}^a = \left\{v \in \mathbf{R}^d : |v| \leq M, |\tau + v \cdot \xi| \leq a\right\}.$$

1. Let O be a $d \times d$ orthogonal matrix and,

$$O(E_{\tau,\xi}^a) = \left\{w \in \mathbf{R}^d : w = Ov, v \in E_{\tau,\xi}^a\right\}.$$

Prove that,

$$O(E_{\tau,\xi}^a) = \left\{w \in \mathbf{R}^d : |w| \leq M, |\tau + \langle w, O\xi \rangle| \leq a\right\} \quad \text{and,}$$

$$\mu(E_{\tau,\xi}^a) = \mu_1(O\,E_{\tau,\xi}^a), \quad \text{where } \mu_1 = \chi(O^{-1}w)\,dw.$$

2. Prove that one can find a $d \times d$ orthogonal matrix O such that,

$$\mu(E_{\tau,\xi}^a) = \mu_1 \left\{w \in \mathbf{R}^d : |w| \leq M, |\tau + |\xi|w_d| \leq a\right\}.$$

3. Let $a > \frac{1}{2}$. Prove that for any $(\tau, \xi) \in \mathbf{S}^d$ we have,

$$\mu(E_{\tau,\xi}^a) \leq 2\mu(B(0, M))a.$$

4. Let $0 < a \leq \frac{1}{2}$.

   a) Prove that one can find ε_0 depending only on M such that if $(\tau, \xi) \in \mathbf{S}^d$ and $|\xi| \leq \varepsilon_0$ then $|\tau + v \cdot \xi| > a$.

   b) Let $|\xi| > \varepsilon_0$. Prove then that there exists a constant $C(d, M) > 0$ such that,

$$\mu(E_{\tau,\xi}^a) \leq \frac{C(d, M)}{\varepsilon_0}a.$$

Hint: Use question 2.

5. Deduce from Part 1 of Problem 57 that if $f \in L^2(\mathbf{R}_t \times \mathbf{R}_x^d \times \mathbf{R}_v^d,\ dt\, dx\, dv)$ is a solution of the equation,

$$\frac{\partial f}{\partial t}(t, x, v) + v \cdot \nabla_x f(t, x, v) = 0$$

then for every $\chi \in C_0^\infty(\mathbf{R}_v^d)$ we have $\int_{\mathbf{R}^d} \chi(v) f(t, x, v)\, dv \in H^{\frac{1}{2}}(\mathbf{R}_t \times \mathbf{R}_x^d)$.

Problem 59 *Incompressibility*

The purpose of this problem is to give a mathematical interpretation of incompressibility in hydrodynamics.

Part 1
Let $M_d(\mathbf{R})$ be the algebra of $d \times d$ matrices with real coefficients.

Let $I \subset \mathbf{R}$ be an open interval, $A \in C^0(I, M_d(\mathbf{R}))$, $B \in C^1(I, M_d(\mathbf{R}))$, $t_0 \in I$ satisfying,

$$\frac{dB}{dt}(t) = A(t)B(t), \quad t \in I, \tag{8.36}$$

Let $\Delta(t) = \det B(t)$ and let $\mathrm{Tr}(A(t))$ be the trace. The goal of this part is to prove that,

$$\frac{d\Delta}{dt}(t) = \Delta(t)\mathrm{Tr}(A(t)) \quad \text{and therefore,} \quad \Delta(t) = \Delta(t_0)\exp\left(\int_{t_0}^t \mathrm{Tr}(A(s))\, ds\right).$$

Let $B(t) = (b_{ij}(t))_{1 \le i, j \le d}$ and let $\ell_i(t)$ be the ith line of the matrix $B(t)$, so that,

$$\frac{d\Delta}{dt}(t) = \sum_{i=1}^d \det\left(\ell_1(t), \ldots, \ell_i'(t), \ldots, \ell_d(t)\right).$$

1. Show that,

$$\frac{d\Delta}{dt}(t) = \sum_{i=1}^d \frac{db_{jk}}{dt}(t)\mathrm{cof}(b_{jk}(t)),$$

where $\mathrm{cof}(b_{jk})$ is the cofactor of b_{jk}.

2. We set $\text{adj}(B(t)) = {}^t(\text{cof}(b_{ik}(t)))$ the adjoint matrix of $B(t)$. Show that,

$$\text{Tr}\left(\frac{dB}{dt}(t)\text{adj}(B(t))\right) = \frac{d\Delta}{dt}(t).$$

3. Conclude.

Let $v \in C^1(\mathbf{R}_t \times \mathbf{R}_x^d, \mathbf{R}^d)$, $T > 0$ and for $t \in [0, T)$ let $X(t)$ be the solution of the problem,

$$\dot{X}(t) = v(t, X(t)), \quad X(0) = y \in \mathbf{R}^d. \tag{8.37}$$

We shall set $X(t) = X(t, y) = \Phi_t(y)$.

Let $\Omega_0 \subset \mathbf{R}^d$ be a bounded open set and $\Omega_t = \Phi_t(\Omega_0)$. Let μ be the Lebesgue measure.

Our goal is to show that the equality $\mu(\Omega_t) = \mu(\Omega_{t'})$ for all $t, t' \in [0, T)$ is equivalent to the fact that $(\text{div}_x v)(t, x) = 0$ for all $(t, x) \in [0, T) \times \mathbf{R}^d$ where $\text{div}_x v = \sum_{j=1}^d \partial_{x_j} v_j$.

Part 2

We assume $(\text{div}_x v)(t, x) = 0$ for all $(t, x) \in [0, T) \times \mathbf{R}^d$.

1. a) Show that the map $\Phi_t : \Omega_0 \rightarrow \Omega_t$ is injective for all $t \in [0, T)$.
    Hint: Assume $X(t_0, y) = X(t_0, y')$; show that,

$$E = \big\{t \in [0, T) : X(t, y) = X(t, y')\big\}$$

    is a nonempty open and closed subset of $[0, T)$.
  b) Compute its Jacobian determinant $\Delta(t, y)$.
      Hint: Using (8.37) show that the matrix $\left(\frac{\partial X_j}{\partial y_k}(t, y)\right)$ satisfies a differential equation.
  c) Deduce that Φ_t is a C^1-diffeomorphism from $\mathbf{R}^d$ to $\mathbf{R}^d$.
2. a) For $t \in [0, T)$ set $V(t) = \mu(\Omega_t) = \int_{\Omega_t} dx$. Show that $V(t) = \int_{\Omega_0} \Delta(t, y) dy$.
  b) Deduce that,

$$V'(t) = \int_{\Omega_0} (\text{div}_x v)(t, X(t, y))\Delta(t, y) dy.$$

  c) Conclude.

Part 3

We assume that there exists $(t_0, x_0) \in [0, T) \times \mathbf{R}^d$ such that $(\text{div}_x v)(t_0, x_0) \neq 0$.

Let y_0 be the unique point in $\mathbf{R}^d$ such that $X(t_0, y_0) = x_0$. Then there exists Ω_0 a neighborhood of y_0 such that,

$$(\text{div}_x v)(t_0, X(t_0, y)) \neq 0, \quad \forall y \in \Omega_0.$$

Set $\Omega_t = \Phi_t(\Omega_0)$.

1. Prove that there exists $t, t' \in [0, T)$ such that $\mu(\Omega_t) \neq \mu(\Omega_{t'})$.

Problem 60 *Propagation of the Irrotationality*

This problem uses Problem 59.

Let $v \in C^2(\mathbf{R} \times \mathbf{R}^d, \mathbf{R}^d)$ be a solution of the Euler equations on $[0, T) \times \mathbf{R}^d$,

$$\partial_s v + (v \cdot \nabla_x)v = -\nabla_x P, \quad \mathrm{div}_x v = 0, \quad v|_{s=0} = v_0. \tag{8.38}$$

We recall (see Problem 59) that if $X(s, y)$ is the solution on $[0, T)$ of the problem

$$\frac{dX}{ds}(s, y) = v(s, X(s, y)), \quad X(0, y) = y \in \mathbf{R}^d,$$

then for all $s \in [0, T)$, the map $y \mapsto X(s, y)$ is a diffeomorphism from $\mathbf{R}^d$ to $\mathbf{R}^d$.
Set,

$$(\mathrm{curl}\, v)(s, x) =: A(s, x) = (\omega_{ij}(s, x))_{1 \leq i, j \leq d}, \quad \omega_{ij} = \partial_i v_j(s, x) - \partial_j v_i(s, x).$$

Our goal is to show that if $(\mathrm{curl}\, v_0)(\cdot) = 0$ then $(\mathrm{curl}\, v)(s, \cdot) = 0$ for all $s \in [0, T)$.

1. Using (8.38) show that for all $1 \leq i < j \leq d$ we have,

$$\partial_s \omega_{ij} + (v \cdot \nabla_x)\,\omega_{ij} + \sum_{k=1}^{d} \left(\partial_i v_k \partial_k v_j - \partial_j v_k \partial_k v_i\right) = 0. \tag{8.39}$$

2. Prove that,

$$\partial_i v_k \partial_k v_j - \partial_j v_k \partial_k v_i = \partial_i v_k \omega_{kj} + \partial_j v_k \omega_{ki}.$$

3. Denote by $B(s, \cdot)$ the matrix $(\partial_i v_j(s, \cdot))_{1 \leq i, j \leq d}$. Show that $A(s, \cdot)$ satisfies the matrix differential equation,

$$\frac{dA}{ds}(s, \cdot) + (v \cdot \nabla_x)A(s, \cdot) + B(s, \cdot)A(s, \cdot) - A(s, \cdot)^t B(s, \cdot) = 0.$$

4. a) Set $C(s) = A(s, X(s, \cdot))$. Show that $C(s)$ satisfies the differential equation,

$$\frac{dC}{ds}(s) + B(s, X(s, \cdot))C(s) - C(s)^t B(s, X(s, \cdot)) = 0.$$

b) Denoting by $\| \cdot \|$ any matrix norm deduce that,

$$\|C(s)\| \leq \|C(0)\| + 2 \int_0^s \|B(\sigma, X(\sigma, \cdot))\|\,\|C(\sigma)\|\,d\sigma.$$

5. Conclude.

Problem 61

If $v = (v_j)_{1 \leq j \leq 3}$ is a vector field on $\mathbf{R}^3$ and E is a functional space we shall say that $v \in E$ if $v_j \in E$, $1 \leq j \leq 3$. If $v \in C^1(\mathbf{R}^3)$ we shall denote div $v = \sum_{j=1}^3 \partial_j v_j$ its divergence.

1. Let $f \in \mathcal{S}(\mathbf{R}^3)$. Show that there exists $v \in H^{+\infty}(\mathbf{R}^3) = \cap_{s \in \mathbf{R}} H^s(\mathbf{R}^3)$ such that div $v = f$.

   Hint: Consider the quantity $\frac{\xi_j}{|\xi|^2}\widehat{f}$.

2. a) Choosing f appropriately, show that there exists $v \in H^{+\infty}(\mathbf{R}^3)$ such that,

$$\int_{\mathbf{R}^3} \text{div } v(x)\, dx \neq 0.$$

   b) Deduce that there exists v such that div $v \in L^1(\mathbf{R}^3)$ but there exists $j \in \{1, 2, 3\}$ such that $\partial_j v_j \notin L^1(\mathbf{R}^3)$.

3. We set for $j = 1, 2, 3$, $\rho_j(\xi) = \frac{\xi_j}{|\xi|^2} \in L^2(\mathbf{R}^3)$. Let T_j be the operator defined on $\mathcal{S}(\mathbf{R}^3)$ by

$$T_j u = \mathcal{F}^{-1}\left(\rho_j(\xi)\widehat{u}\right).$$

Prove that for $\sigma \in \mathbf{R}$ there exists $C > 0$ such that,

$$\|T_j u\|_{H^{\sigma+1}(\mathbf{R}^3)} \leq C\|u\|_{L^1(\mathbf{R}^3)} + C\|u\|_{H^\sigma(\mathbf{R}^3)}, \quad \forall u \in \mathcal{S}(\mathbf{R}^3).$$

Problem 62 *Nonuniform Continuity of the Euler Flow*

Our goal is to show that the flow of the incompressible Euler system on the torus $\mathbf{T}^2 = \mathbf{R}^2/(2\pi\mathbf{Z})^2$ (flow which is continuous) is not uniformly continuous in any H^s for $s \in \mathbf{R}$.

Notations A function $f : \mathbf{T}^2 \to \mathbf{C}$ is a function $f : \mathbf{R}^2 \to \mathbf{C}$ which is 2π-periodic in (x_1, x_2) that is $f(x_1 + 2\pi, x_2) = f(x_1, x_2) = f(x_1, x_2 + 2\pi)$. To $f \in L^1(\mathbf{T}^2)$ we associate its Fourier coefficient,

$$\widehat{f}(k) = \iint_{(0,2\pi)^2} e^{-ik\cdot x} f(x)\, dx, \quad k \in \mathbf{Z}^2,$$

For $s \in \mathbf{R}$ we define then $H^s(\mathbf{T}^2)$ as the space of f such that,

$$\|f\|^2_{H^s(\mathbf{T}^2)} := \sum_{k \in \mathbf{Z}^2} (1 + |k|^2)^s |\widehat{f}(k)|^2 < +\infty.$$

Eventually if $U = (u_1, u_2)$ we set, $\|U\|^2_{(H^s(\mathbf{T}^2))^2} = \|u_1\|^2_{H^s(\mathbf{T}^2)} + \|u_2\|^2_{H^s(\mathbf{T}^2)}$.

The Euler system on $U = (u_1, u_2)$ is,

$$\frac{\partial U}{\partial t} + (U \cdot \nabla_x) U - \nabla_x \Delta^{-1} \mathrm{div}\,((U \cdot \nabla_x) U) = 0, \qquad t > 0,\ x \in \mathbf{T}^2,$$

$$\mathrm{div}\, U = 0, \qquad U|_{t=0} = U^0, \tag{8.40}$$

where $\Delta^{-1} v = \mathcal{F}^{-1}(\frac{1}{|\xi|^2}\widehat{v})$.

The goal of the following questions is to construct two sequences of data $(U_n^0), (V_n^0)$ such that,

$$\lim_{n \to +\infty} \|U_n^0 - V_n^0\|_{(H^s(\mathbf{T}^2))^2} = 0, \tag{8.41}$$

but, for $t > 0$ arbitrary small and for n large enough,

$$\|U_n(t, \cdot) - V_n(t, \cdot)\|_{(H^s(\mathbf{T}^2))^2} \geq c_0(t) > 0, \tag{8.42}$$

where U_n, V_n denote the solutions of the system (8.40) with data U_n^0, V_n^0 at $t = 0$.

1. Let $s \in \mathbf{R}$. Compute the norms in $H^s(\mathbf{T}^2)$ of the functions $f(x_1, x_2) = C \in \mathbf{C}$, $f(x_1, x_2) = \sin(n x_1)$, $f(x_1, x_2) = \sin(n x_2)$ where $n \in \mathbf{Z}$.
   For $n \in \mathbf{N}$ and $\omega \in \mathbf{R}$ we set $U_{n,\omega} = (u_1, u_2)$ where,

$$u_1(t, x) = \omega n^{-1} + n^{-s} \cos(nx_2 - \omega t), \quad u_2(t, x) = \omega n^{-1} + n^{-s} \cos(nx_1 - \omega t). \tag{8.43}$$

2. a) Compute

$$A = \mathrm{div}\, U_{n,\omega}, \qquad B = \frac{\partial}{\partial t} U_{n,\omega} + (U_{n,\omega} \cdot \nabla_x) U_{n,\omega}$$

and

$$C = \text{div}\left((U_{n,\omega} \cdot \nabla_x)U_{n,\omega}\right).$$

b) Set $w_n = \sin(nx_1 - \omega t)\sin(nx_2 - \omega t)$. Compute $\Delta\left(\frac{1}{2n^2}w_n\right)$.

c) Let Q be a periodic solution of equation $\Delta Q = w_n$ and let $R = Q + \frac{1}{2n^2}w_n$. Prove that $\Delta R = 0$.

d) Writing $R(x) = \frac{1}{(2\pi)^2}\sum_{k\in\mathbf{Z}^2}\widehat{R}(k)e^{ik\cdot x}$ show that $R(x) = \frac{1}{(2\pi)^2}\widehat{R}(0)$ and therefore that $Q = \Delta^{-1}w_n = -\frac{1}{2n^2}w_n + C_n$, $C_n \in \mathbf{C}$.

e) Compute $\nabla_x\Delta^{-1}\text{div}\left((U_{n,\omega} \cdot \nabla_x)U_{n,\omega}\right)$.

f) Deduce that $U_{n,\omega}$ is a solution of (8.40).
   Let,

$$U_{n,\omega}^0 = U_{n,\omega}|_{t=0} = \left(\omega n^{-1} + n^{-s}\cos(nx_2),\ \omega n^{-1} + n^{-s}\cos(nx_1)\right),$$
(8.44)

and

$$U_n^0 = U_{n,1}^0, \quad V_n^0 = U_{n,-1}^0.$$

3. Show that (8.41) is satisfied.

4. Let U_n and V_n be the solutions of system (8.40) corresponding to these data. Show that,

$$\|U_n(t, \cdot) - V_n(t, \cdot)\|_{(H^s(\mathbf{T}^2))^2} \geq C_3|\sin(t)| - \frac{C_4}{n}$$

and conclude.

Comments This problem is inspired by an article by A. Himonas and G. Misiolek [12]. □

Problem 63 *Finite Time Blow-Up for the Euler Equations*

Let $T > 0$. We consider the system on $[0, T) \times \mathbf{R}^3$,

$$\frac{\partial\rho}{\partial t} + \sum_{j=1}^{3}v_j\frac{\partial\rho}{\partial x_j} + \left(\sum_{j=1}^{3}\frac{\partial v_j}{\partial x_j}\right)\rho = 0, \quad \rho|_{t=0} = \rho_0,$$
(8.45)

$$\rho\left(\frac{\partial v_k}{\partial t} + \sum_{j=1}^{3}v_j\frac{\partial v_k}{\partial x_j}\right) + 3\rho^2\frac{\partial\rho}{\partial x_k} = 0, \quad v_k|_{t=0} = v_{0k}, \quad k = 1, 2, 3. \quad (8.46)$$

This system corresponds to the compressible Euler equations in the case where $P = \rho^3$. Set $v = (v_1, v_2, v_3)$.

We shall say that (ρ, v) is a smooth solution if,

(C_0) $\rho \geq 0$,

(C_1) $\rho, v_k \in C^1([0, T) \times \mathbf{R}^3, \mathbf{R})$, $k = 1, 2, 3$,

(C_2) ρ satisfies (8.45), v satisfies (8.46)   if $\rho(t, x) \neq 0$

and $\dfrac{\partial v_k}{\partial t} + \displaystyle\sum_{j=1}^{3} v_j \dfrac{\partial v_k}{\partial x_j} = 0$ if $\rho(t, x) = 0$.

Such solutions exist. We consider in what follows a smooth solution.

Part 1

For $R > 0$, let $B(0, R) = \{x \in \mathbf{R}^3 : |x| < R\}$. The goal is to prove that,

$$\text{supp } \rho_0 \cup \text{supp } v_0 \subset B(0, R) \Rightarrow \text{supp } \rho(t) \cup \text{supp } v(t) \subset B(0, R), \quad \forall t \in [0, T).$$
$$(8.47)$$

1. a) Let $|x_0| > R$ and $T_0 < T$. Set,

$$M = \sup_{t \in [0, T_0]} \left(|\nabla_x \rho(t, x_0)| + \sum_{j=1}^{3} |\nabla_x v_j(t, x_0)| \right).$$

Prove that there exists $C > 0$ independent of x_0 such that for $t \in [0, T_0]$,

$$|\partial_t \rho(t, x_0)| + \sum_{j=1}^{3} |\partial_t v_j(t, x_0)| \leq CM(|\rho(t, x_0)| + \sum_{j=1}^{3} |v_j(t, x_0)|).$$

   b) Set $u(t) = \rho(t, x_0)^2 + \sum_{j=1}^{3} v_j(t, x_0)^2$. Deduce that there exists $C > 0$ independent of x_0 such that for $t \in [0, T_0]$,

$$|\partial_t u(t)| \leq Cu(t).$$

2. Prove (8.47).

Part 2

We assume here that ρ_0 and v_0 have compact support and that $\rho_0 \neq 0$.

Our goal is to prove that the solution $(\rho(t), v(t))$ cannot be global, that is $T < +\infty$.

1. Prove, from the equations (8.45), (8.46), the identities,

$$\frac{\partial \rho}{\partial t} + \sum_{j=1}^{3} \frac{\partial}{\partial x_j}(\rho v_j) = 0,$$

$$\frac{\partial}{\partial t}(\rho v_k) + \sum_{j=1}^{3} \frac{\partial}{\partial x_j}(\rho v_j v_k) + 3\rho^2 \frac{\partial \rho}{\partial x_k} = 0, \quad 1 \leq k \leq 3,$$

$$\frac{\partial}{\partial t}\left(\rho |v|^2 + \rho^3\right) + \sum_{j=1}^{3} \frac{\partial}{\partial x_j}\left(v_j(\rho |v|^2 + 3\rho^3)\right) = 0, \quad |v|^2 = \sum_{j=1}^{3}(v_j)^2.$$

$$\tag{8.48}$$

2. Prove that the quantity,

$$E = \int_{\mathbf{R}^3} \left(\rho(t, x)|v(t, x)|^2 + \rho(t, x)^3\right) dx$$

is independent of t and that $E > 0$.
3. Prove that the quantity $\int_{\mathbf{R}^3} \rho(t, x) \, dx$ is independent of t.
   We set in what follows,

$$H(t) = \int_{\mathbf{R}^3} |x|^2 \rho(t, x) \, dx.$$

4. Prove that there exists $M > 0$ such that $H(t) \leq M, \quad \forall t \in [0, T)$.
5. Prove that H is C^1 on $(0, T)$ and that,

$$H'(t) = 2\sum_{k=1}^{3} \int_{\mathbf{R}^3} \rho(t, x) x_k v_k(t, x) \, dx.$$

6. a) Prove that H' is C^1 on $(0, T)$ and that,

$$H''(t) = \int_{\mathbf{R}^3} \left(\rho(t, x)|v(t, x)|^2 + 3\rho^3(t, x)\right) dx.$$

   b) Deduce that $H''(t) \geq E, \forall t \in [0, T)$.
7. Prove that $H(t) \geq H(0) + tH'(0) + \frac{1}{2}Et^2$.
8. Deduce that T is necessarily bounded.

Comments This problem is inspired by an article by T. Makino, S. Ukai, and S. Kawashima [20].   □

Problem 64 *On the Continuity of the Flow Map of the Incompressible Euler Equation*

Consider the Cauchy problem for the incompressible Euler equations:

$$\begin{cases} \dfrac{\partial u}{\partial t} + (u \cdot \nabla)u = -\nabla P, \qquad \operatorname{div} u = 0, \\[2mm] u|_{t=0} = u_0. \end{cases} \tag{8.49}$$

Let $0 < \alpha < 1$. Set $\rho = 1 + \alpha$. The purpose of this problem is to show that when $u_0 \in W^{\rho,\infty}(\mathbf{R}^3, \mathbf{R}^3)$ although there exists for some $T > 0$ a unique solution $u \in L^\infty(([0, T], W^{\rho,\infty}(\mathbf{R}^3, \mathbf{R}^3))$, the map $u_0 \mapsto u$ is not necessarily continuous in these spaces.

Let $T > 0$ and let f, g, h be three functions belonging to $W^{\rho,\infty}(\mathbf{R}, \mathbf{R})$. For $t \in [0, T]$ and $x \in \mathbf{R}^3$ we set,

$$u(t, x) = (f(x_2), 0, h(x_1 - tf(x_2))), \quad v(t, x) = (g(x_2), 0, h(x_1 - tg(x_2))).$$

1. Show that $u, v \in L^\infty([0, T], W^{\rho,\infty}(\mathbf{R}^3, \mathbf{R}^3))$.
2. Show that u and v are two solutions of (8.49) with an appropriate P.
3. Given $\delta > 0$ show that we can find f, g such that $\|u_0 - v_0\|_{W^{\rho,\infty}} \le \delta$ and $f(s) \ne g(s)$ for $s \in \mathbf{R}$.
   We set $a = T \max(\|f\|_{L^\infty}, \|g\|_{L^\infty})$ and we choose h such that,

$$h'(s) = |s|^\alpha \quad \text{for} \quad -2a \le s \le 2a.$$

4. Prove that,

$$A(t) := \|u(t, \cdot) - v(t, \cdot)\|_{W^{\rho,\infty}} \ge \|h'(\cdot - tf(\cdot)) - h'(\cdot - tg(\cdot))\|_{W^{\alpha,\infty}}.$$

5. Deduce that, $A(t) \ge B(t)$ where, $B(t)$ is equal to,

$$\sup_E \frac{||x_1 - tf(x_2)|^\alpha - |x_1 - tg(x_2)|^\alpha - |y_1 - tf(y_2)|^\alpha + |y_1 - tg(y_2)|^\alpha|}{|x_1 - y_1|^\alpha},$$

where $E = \{(x, y) \in [-a, a]^2 : x_1 \ne y_1\}$.
6. Evaluating the right-hand side at $x_2 = y_2 = c$ with $-a < c < a$ and then at $x_1 = tg(c), y_1 = tf(c)$ prove that $A(t) \ge 2$.
7. Conclude.

Comments This problem is inspired by a paper by G. Misiolek and T. Yoneda [22].

□

Problem 65 *On the Cauchy Problem for the Burgers Equation*

Our goal here is to study the Cauchy problem for the Burgers equation,

$$\partial_t u + u\, \partial_x u = 0, \quad u|_{t=0} = u_0, \quad t \geq 0, x \in \mathbf{R}, \tag{8.50}$$

where u_0 is a *real valued* function.

Part 1. *Necessary Condition, Data in $C^1(\mathbf{R})$*
In this part we assume the existence of a real valued solution u which is C^1 on $[0, +\infty) \times \mathbf{R}$.

1. Consider the characteristic of the equation (8.50) $(t(s), x(s))$ starting at $(0, y)$ defined by,

$$\dot{t}(s) = 1, \quad t(0) = 0, \qquad \dot{x}(s) = u(s, x(s)) \quad x(0) = y \in \mathbf{R}.$$

   We have obviously $t(s) = s$. Prove that, as long as $x(s)$ exists, the solution u is constant on the characteristic. Deduce that $x(s) = y + s u_0(y)$ and therefore that $x(s)$ exists for all $s \geq 0$.
2. Prove that we have necessarily $u(s, y + s u_0(y)) = u_0(y)$.

Part 2. *Sufficient Condition, Data in $C^1(\mathbf{R})$*
In this part assuming that $u_0 \in C^1(\mathbf{R})$, and $\inf_{y \in \mathbf{R}} u_0' > -\infty$, we want to solve the problem (8.50) using the results in Part 1.

1. Let $F(t, y) = y + t u_0(y)$. Prove that for $t \geq 0$,

$$\inf_{y \in \mathbf{R}} u_0' \geq 0, \Longrightarrow \left(\frac{\partial F}{\partial y}(t, y) > 0, \quad \forall (t, y) \in [0, +\infty) \times \mathbf{R} \right),$$

$$\inf_{y \in \mathbf{R}} u_0' < 0 \Longrightarrow \left(\frac{\partial F}{\partial y}(t, y) > 0 \text{ for } 0 \leq t < \frac{1}{- \inf_{y \in \mathbf{R}} u_0'}, y \in \mathbf{R} \right).$$

2. a) We set $T^* = +\infty$ if $\inf_{\mathbf{R}} u_0' \geq 0$ and $T^* = \frac{1}{-\inf_{\mathbf{R}} u_0'}$ if $\inf_{\mathbf{R}} u_0' < 0$. Prove that there exists a unique map $\kappa \in C^1([0, T^*) \times \mathbf{R})$ such that, for all $t \in [0, T^*)$,

$$y + t u_0(y) = x \Longleftrightarrow y = \kappa(t, x), \quad \forall x, y \in \mathbf{R}.$$

   b) Prove that if $T \|u_0'\|_{L^\infty(\mathbf{R})} < 1$ then $T < T^*$.
3. Using question 2 in Part 1. show that the problem (8.50) has a unique solution $u \in C^1([0, T^*) \times \mathbf{R})$.
4. Prove that if u_0 has compact support then necessarily $T^* < +\infty$.
   Hint: Use a contradiction argument.
5. We assume $t \in [0, T]$ where $T \|u_0'\|_{L^\infty(\mathbf{R})} < 1$. Prove that,

$$(\partial_x \kappa)(t, x) = \frac{1}{1 + t u_0'(\kappa(t, x))}, \quad (\partial_t \kappa)(t, x) = \frac{-u_0(\kappa(t, x))}{1 + t u_0'(\kappa(t, x))}.$$

Part 3. *Maximal Time of Existence*

The purpose of this part is to show that when T^* is finite it is the right maximal time of existence. Assume $u_0 \in C^2(\mathbf{R})$ with $\inf_{x \in \mathbf{R}} u_0' < 0$ and let $u \in C^2([0, T^*) \times \mathbf{R})$ be the solution found in Part 2.

1. a) For $x_0 \in \mathbf{R}$ such that $u_0'(x_0) < 0$ set $T_{x_0}^* = \frac{1}{-u_0'(x_0)}$ and for $0 \le t < T_{x_0}^*$
   set $q(t) = (\partial_x u)(t, x_0 + t u_0(x_0))$. Then q is a C^1 function. Prove that q is a solution of the problem,

$$q'(t) = -q(t)^2, \quad q(0) = u_0'(x_0).$$

   b) Deduce that $\lim_{t \to T_{x_0}^*} q(t) = -\infty$ and the maximal time of existence as a C^1
   function of the solution of the problem (8.50) is $T^* = \frac{1}{-\inf_{x \in \mathbf{R}} u_0'}$.

Unless otherwise stated we shall denote in what follows $X = X(\mathbf{R})$ if $X = L^2, L^\infty, H^k, C_0^\infty$.

Part 4. *Data in H^2*

The goal of this part is to show that if $u_0 \in H^2$ and T is such that $T \|u_0'\|_{L^\infty} < 1$ then the solution found in Part 2 belongs to $C^0([0, T], H^2)$.

1. Let $u_0 \in H^2$ and T be such that $T \|u_0'\|_{L^\infty} < 1$. Prove that we may apply the result in Part 2 and that the problem (8.50) has a unique solution $u \in C^1([0, T] \times \mathbf{R})$ given by $u(t, x) = u_0(\kappa(t, x))$.
2. Prove that for fixed $t \in [0, T]$ the function $x \mapsto u(t, x)$ belongs to H^1.
3. Prove that for fixed $t \in [0, T]$ the function $x \mapsto \partial_x^2 u(t, x)$ belongs to L^2.

   The goal of the following questions is to prove the continuity. Let $t_0 \in [0, T]$ and $(t_n) \subset [0, T]$ such that $\lim_{n \to +\infty} t_n = t_0$. Let $\varphi \in C_0^\infty$ with $\operatorname{supp} \varphi \subset \{x : |x| \le R_0\}$.
   We set $I_n = \int_{\mathbf{R}} |\varphi(\kappa(t_n, x)) - \varphi(\kappa(t_0, x))|^2 \, dx$ and we want to prove that $\lim_{n \to +\infty} I_n = 0$.

4. a) Set $R = R_0 + (t_0 + 1) \|u_0\|_{L^\infty(\mathbf{R})}$ and take $n \ge n_0$ so large that $|t_n - t_0| \le 1$ for $n \ge n_0$. Prove that,

$$I_n = \int_{|x| \le R} |\varphi(\kappa(t_n, x)) - \varphi(\kappa(t_0, x))|^2 \, dx.$$

   Hint: Prove that if $|x| \ge R$ we have, $|\kappa(t_n, x)| \ge R_0$ and $|\kappa(t_0, x)| \ge R_0$.
   b) We set $B_R = \{x : |x| \le R\}$ and we denote by $|B_R|$ its volume. Prove that for $n \ge n_0$ we have,

$$I_n \le |B_R| \|\varphi'\|_{L^\infty} \frac{\|u_0\|_{L^\infty}}{1 - T \|u_0'\|_{L^\infty}} |t_n - t_0|$$

and conclude.

5. Let $v \in L^2$ and $J_n = \int_{\mathbf{R}} |v(\kappa(t_n, x)) - v(\kappa(t_0, x))|^2 \, dx$. Prove that $\lim_{n\to+\infty} J_n = 0$.

   Hint: Take $(\varphi_k) \subset C_0^\infty$ such that $(\varphi_k) \to v$ in L^2 and use question 4b) above.

6. Let $f = f(t, x)$ be a bounded continuous function on $[0, T] \times \mathbf{R}$ and $v \in L^2$. Prove that the function $w(t, x) = f(t, x)v(\kappa(t, x))$ belongs to $C^0([0, T), L^2)$.

7. Let u be the solution considered in question 1. Prove that $\partial_x^k u \in C^0([0, T), L^2)$ for $k = 0, 1, 2$.

Part 5. *Nonuniform Continuity of the Flow*

The goal of this part is to show that, for $T > 0$ fixed, the map $u_0 \mapsto u$ from $H^2(\mathbf{R})$ to $C([0, T], H^2(\mathbf{R}))$ is not uniformly continuous.

1. Prove that it is sufficient to find $\varepsilon_0 > 0$ and to construct two sequences of data (u_n^0), $(v_n^0) \subset H^2$ such that $\lim_{n\to+\infty} \|u_n^0 - v_n^0\|_{H^2} = 0$ and if (u_n), (v_n) denote the corresponding solutions of (8.50) then $\sup_{t\in[0,T]} \|u_n(t) - v_n(t)\|_{H^2} > \varepsilon_0$ for $n \geq N$.

   Let $\chi \in C_0^\infty$, $\chi(x) = 1$ if $|x| \leq \frac{1}{2}$, $\chi(x) = 0$ if $|x| \geq 1$. Let (λ_n), (ε_n) be two sequences of positive real numbers such that,

$$\lim_{n\to+\infty} \lambda_n = +\infty, \qquad \lim_{n\to+\infty} \varepsilon_n = 0, \qquad \lim_{n\to+\infty} \lambda_n \varepsilon_n = +\infty. \qquad (8.51)$$

Set

$$u_n^0(x) = \lambda_n^{-\frac{3}{2}} \chi(\lambda_n x), \qquad v_n^0(x) = u_n^0(x) + \varepsilon_n \chi(x). \qquad (8.52)$$

Then obviously,

$$\lim_{n\to+\infty} \|u_n^0 - v_n^0\|_{H^2} = 0. \qquad (8.53)$$

2. a) Prove that $\|(u_n^0)'\|_{L^\infty} \leq C\lambda_n^{-\frac{1}{2}}$ and $\|(v_n^0)'\|_{L^\infty} \leq C(\lambda_n^{-\frac{1}{2}} + \varepsilon_n)$.

      We fix n_0 so large that $T \|(u_n^0)'\|_{L^\infty} \leq \frac{1}{2}$, $T \|(v_n^0)'\|_{L^\infty} \leq \frac{1}{2}$ for $n \geq n_0$.

   b) Deduce that for $n \geq n_0$ the solutions u_n, v_n of (8.50) with data u_n^0, v_n^0 exist on $[0, T]$.

3. Using question 2 Part 2 we set,

$$y + tu_n^0(y) = x \Leftrightarrow y = \kappa_1(t, x), \qquad y + tv_n^0(y) = x \Leftrightarrow y = \kappa_2(t, x).$$

Using question 6a) Part 2 prove that there exists $C > 0$ such that, for $j = 1, 2$, for $n \geq n_0$, and all $t \in [0, T]$ we have,

$$\frac{2}{3} \leq |\partial_x \kappa_j(t, x)| \leq 2, \qquad |\partial_x^2 \kappa_j(t, x)| \leq Ct(\lambda_n^{\frac{1}{2}} |\chi''(\lambda_n \kappa_j(t, x))| + 1). \qquad (8.54)$$

4. Prove that there exists $C > 0$ such that for all $n \geq n_0$ and all $t \in [0, T]$ we have,

$$\|(u_n^0)''(\kappa_1(t, \cdot))\|_{L^2} \geq C. \qquad (8.55)$$

5. Recall that from Part 1 the solutions u_n, v_n of (8.50) can be written as,

$$u_n(t, x) = u_n^0(\kappa_1(t, x)), \quad v_n(t, x) = v_n^0(\kappa_2(t, x)).$$

Then we write,

$$v_n(t, x) - u_n(t, x) = A_n(t, x) + B_n(t, x),$$

$$A_n(t, x) = (v_n^0 - u_n^0)(\kappa_2(t, x)) = \varepsilon_n \chi(\kappa_2(t, x)), \qquad (8.56)$$

$$B_n(t, x) = u_n^0(\kappa_2(t, x)) - u_n^0(\kappa_1(t, x)).$$

Prove that $\lim_{n \to +\infty} \|A_n(t, \cdot)\|_{H^2} = 0$.

6. Prove that for $j = 1, 2$ we have $\lim_{n \to +\infty} \|u_n^0(\kappa_j(t, \cdot))\|_{H^1} = 0$. Deduce that,

$$\|B_n(t, \cdot)\|_{H^2} = \|\partial_x^2 \left[u_n^0(\kappa_2(t, \cdot)) - u_n^0(\kappa_1(t, \cdot)) \right] \|_{L^2} + o(1), \quad n \to +\infty.$$

7. a) Using (8.54) prove that for $j = 1, 2$,

$$\lim_{n \to +\infty} \left\| (u_n^0)'(\kappa_j(t, \cdot)) \partial_x^2 \kappa_j(t, \cdot) \right\|_{L^2} = 0.$$

b) Deduce from (8.56) and questions 5, 6, 7a) that when $n \to +\infty$ we have,

$$\|v_n(t, \cdot) - u_n(t, \cdot)\|_{H^2} = \|f_n(t, \cdot) - g_n(t, \cdot)\|_{L^2} + o(1), \text{ where,}$$

$$f_n(t, x) = (u_n^0)''(\kappa_2(t, x)) (\partial_x \kappa_2(t, x))^2,$$

$$g_n(t, x) = (u_n^0)''(\kappa_1(t, \cdot)) (\partial_x \kappa_1(t, x))^2.$$

Notice that, by construction, we have supp $g_n(t, \cdot) \subset \{x : \lambda_n |\kappa_1(t, x)| \leq 1\}$.

c) Let $x \in$ supp $g(t, \cdot)$. Writing,

$$\kappa_2(t, y + tu_n^0(y)) = \kappa_2(t, y + tv_n^0(y)) + t(u_n^0(y) - v_n^0(y)) \partial_x \kappa_2(t, x^*)$$

prove that,

$$\kappa_2(t, y + tu_n^0(y)) = y + t(u_n^0(y) - v_n^0(y)) \partial_x \kappa_2(t, x^*) = y - t\varepsilon_n \partial_x \kappa_2(t, x^*).$$

Setting $y + t u_n^0(y) = x$ deduce that for $t > 0$ and n large enough we have,

$$\lambda_n |\kappa_2(t, x)| \geq 2$$

and therefore that $x \notin \operatorname{supp} f(t, \cdot)$.

d) Let $\alpha, \beta \in L^2$ be such that $(\operatorname{supp} \alpha) \cap (\operatorname{supp} \beta) = \emptyset$. Prove that $\|\alpha - \beta\|_{L^2} \geq \frac{1}{\sqrt{2}} (\|\alpha\|_{L^2} + \|\beta\|_{L^2})$. Deduce that when $n \to +\infty$ we have,

$$\|v_n(t, \cdot) - u_n(t, \cdot)\|_{H^2} \geq C \|(u_n^0)''(\kappa_1(t, \cdot))(\partial_x \kappa_1(t, \cdot))^2\|_{L^2} + o(1).$$

e) Using (8.54) and (8.55) prove that there exists $\varepsilon_0 > 0$ such that for n large enough we have $\sup_{t \in [0,T]} \|v_n(t, \cdot) - u_n(t, \cdot)\|_{H^2} \geq \varepsilon_0$.

Part II
Solutions of the Problems and Classical Results

Chapter 9
Solutions of the Problems

This chapter is devoted to the detailed solutions of the problems stated in the previous chapters.

Solution 1

1. a) Let $\chi \in C_0^\infty$, $\chi(x) = 1$ for $|x| \leq 1$ and $0 \leq \chi \leq 1$. Set for $k \geq 1$, $\chi_k(x) = \chi(\frac{x}{k})$. If $u \in V$ set $u_k = \chi_k u$. Then,

$$\| \langle x \rangle (u_k - u)\|_{L^2}^2 = \int_{\mathbf{R}^d} |1 - \chi\left(\frac{x}{k}\right)|^2 \langle x \rangle^2 |u(x)|^2 \, dx.$$

The right-hand side tends to zero by the Lebesgue dominated convergence; indeed, for fixed x the interior of the integral tends to zero when $k \to +\infty$ and $|1 - \chi\left(\frac{x}{k}\right)|^2 \langle x \rangle^2 |u(x)|^2 \leq \langle x \rangle^2 |u(x)|^2 \in L^1$. Eventually we have,

$$\|\nabla_x (u_k - u)\|_{L^2} \leq \frac{1}{k}\|\nabla_x \chi\|_{L^\infty}\|u\|_{L^2} + \|(1 - \chi_k)\nabla_x u\|_{L^2}.$$

When $k \to +\infty$ the first term in the right-hand side tends to zero obviously and the second one tends to zero as well by the Lebesgue dominated convergence as above.

b) Let $\theta \in C_0^\infty$, supp $\theta \subset \{x : |x| \leq 1\}$, $\theta \geq 0$, $\int_{\mathbf{R}^d} \theta(x) \, dx = 1$. For $\varepsilon > 0$ set $\theta_\varepsilon(x) = \varepsilon^{-d}\theta\left(\frac{x}{\varepsilon}\right)$. For $u \in V_c$ with supp $u \subset \{x : |x| \leq R\}$ set $u_\varepsilon = \theta_\varepsilon \star u$. Then $u_\varepsilon \in C^\infty(\mathbf{R}^d)$ and supp $u_\varepsilon \subset$ supp $u + B(0, \varepsilon) \subset B(0, R+1)$ if $\varepsilon \leq 1$. It follows that,

$$\| \langle x \rangle (u_\varepsilon - u)\|_{L^2} \leq \langle R + 1 \rangle \|u_\varepsilon - u\|_{L^2},$$

and the right-hand side tends to zero when $\varepsilon \to 0$ using the Hint.

© The Editor(s) (if applicable) and The Author(s), under exclusive license to Springer Nature Switzerland AG 2020
T. Alazard, C. Zuily, *Tools and Problems in Partial Differential Equations*, Universitext, https://doi.org/10.1007/978-3-030-50284-3_9

On the other hand, since $\nabla_x u_\varepsilon = \theta_\varepsilon \star \nabla_x u$ we have,

$$\|u_\varepsilon - u\|_{H^1}^2 = \|u_\varepsilon - u\|_{L^2}^2 + \|\theta_\varepsilon \star \nabla_x u - \nabla_x u\|_{L^2}^2,$$

and the right-hand side tends to zero when $\varepsilon \to 0$ using the Hint.

2. Let B be the unit ball in V. Let (u_n) be a sequence in B. We have to show that there is a subsequence $(u_{\sigma(n)})$ which converges in $L^2(\mathbf{R}^d)$. Let $\varepsilon > 0$. Let $R_0 \geq 1$ be such that $\frac{1}{R_0} \leq \frac{\varepsilon}{2}$. Let $\chi \in C_0^\infty$, $\chi(x) = 1$ if $|x| \leq 1$, $\chi(x) = 0$ if $|x| \geq 2$, $0 \leq \chi \leq 1$. Set $\chi_{R_0}(x) = \chi\left(\frac{x}{R_0}\right)$ and write $u_n = \chi_{R_0} u_n + (1 - \chi_{R_0})u_n$. We have,

$$\|\chi_{R_0} u_n\|_{H^1} \leq \|\chi_{R_0} u_n\|_{L^2} + \frac{1}{R_0}\|\nabla_x \chi\|_{L^\infty}\|u_n\|_{L^2} + \|\chi_{R_0} \nabla_x u_n\|_{L^2}$$

$$\leq (1 + \|\nabla_x \chi\|_{L^\infty})\|u_n\|_{L^2} + \|\nabla_x u_n\|_{L^2}$$

$$\leq (1 + \|\nabla_x \chi\|_{L^\infty})\|u_n\|_{H^1} \leq (1 + \|\nabla_x \chi\|_{L^\infty}).$$

Therefore, $(\chi_{R_0} u_n)$ belongs to a bounded set of H_K^1 where K is the compact $\{x \in \mathbf{R}^d : |x| \leq 2R_0\}$. Since the map $u \mapsto u$ from H_K^1 to L^2 is compact we deduce that one can find a subsequence $(\chi_{R_0} u_{\sigma(n)})$ which converges in L^2 to some $v \in L^2$, that is, there exists $N \in \mathbf{N}$ such that for $n \geq N$ we have,

$$\|\chi_{R_0} u_{\sigma(n)} - v\|_{L^2} \leq \frac{\varepsilon}{2}. \tag{9.1}$$

Now since $1 - \chi_{R_0} = 0$ for $|x| \leq R_0$ and $0 \leq 1 - \chi_{R_0} \leq 1$ we have,

$$\|(1 - \chi_{R_0})u_{\sigma(n)}\|_{L^2}^2 \leq \int_{|x| \geq R_0} |u_{\sigma(n)}(x)|^2 dx$$

$$\leq \int_{|x| \geq R_0} \frac{1}{|x|^2} \langle x \rangle^2 |u_{\sigma(n)}(x)|^2 dx$$

$$\leq \frac{1}{R_0^2} \|\langle x \rangle u_{\sigma(n)}\|_{L^2}^2 \leq \frac{1}{R_0^2} \leq \frac{\varepsilon^2}{4},$$

since $(u_n) \subset B$ and $R_0 \geq \frac{2}{\varepsilon}$. It follows from (9.1) that $\|u_{\sigma(n)} - v\|_{L^2} \leq \varepsilon$ for $n \geq N$.

3. a) The function $v(x - he_k)$ belongs obviously to H^1 and the function $\langle x \rangle v(x - he_k)$ belongs to L^2 since $\|\langle x \rangle v(x - he_k)\|_{L_x^2} = \|\langle y + he_k \rangle v\|_{L_y^2}$ and $\langle y + he_k \rangle \leq 2\langle y \rangle$. Therefore, $D_h v \in V$.

   b) Since $\mathcal{F}(w(x - he_k))(\xi) = e^{-ih\xi_k}\widehat{w}(\xi)$ we have,

$$\|D_h w + \partial_{x_k} w\|_{H^{s-1}}^2 = \int_{\mathbf{R}^d} \langle \xi \rangle^{2s-2} \left|\frac{e^{-ih\xi_k} - 1 - (-ih\xi_k)}{h}\right|^2 |\widehat{w}(\xi)|^2 d\xi.$$

For fixed ξ the function in the interior of the integral tends to zero with h since, using the Hint with $\theta = -ih\xi_k$ we have,

$$\left| \frac{e^{-ih\xi_k} - 1 - (-ih\xi_k)}{h} \right| \le \frac{1}{2} h\xi_k^2.$$

Moreover, we have, again by the Hint,

$$\langle \xi \rangle^{2s-2} \left| \frac{e^{-ih\xi_k} - 1}{h} + i\xi_k \right|^2 |\widehat{w}(\xi)|^2 \le 4\xi_k^2 \langle \xi \rangle^{2s-2} |\widehat{w}(\xi)|^2 \in L^1(\mathbf{R}^d)$$

since $w \in H^s$. By the Lebesgue dominated convergence theorem we have $\lim_{h \to 0} \| D_h w + \partial_{x_k} w \|_{H^{s-1}}^2 = 0$.

c) Since for $\theta \in \mathbf{R}$, $|e^{i\theta} - 1| \le |\theta|$ we have,

$$\| D_h w \|_{L^2}^2 = \int_{\mathbf{R}^d} \left| \frac{e^{-ih\xi_k} - 1}{h} \right|^2 |\widehat{w}(\xi)|^2 \frac{d\xi}{(2\pi)^d}$$

$$\le \int_{\mathbf{R}^d} \xi_k^2 |\widehat{w}(\xi)|^2 \frac{d\xi}{(2\pi)^d} = \| \partial_{x_k} w \|_{L^2}^2.$$

By the same way,

$$\| D_h g \|_{H^{-1}}^2 = \int_{\mathbf{R}^d} \left| \frac{e^{-ih\xi_k} - 1}{h} \right|^2 \langle \xi \rangle^{-2} |\widehat{g}(\xi)|^2 d\xi$$

$$\le \int_{\mathbf{R}^d} \frac{\xi_k^2}{\langle \xi \rangle^2} |\widehat{g}(\xi)|^2 d\xi \le (2\pi)^d \| g \|_{L^2}^2.$$

4. The left-hand side of (2.1) is the scalar product on V. Since $f \in V'$ the Riesz theorem gives the answer. If we take $v \in C_0^\infty$ we see that the unique solution $u \in V$ of (2.1) solves the equation $-\Delta u + \langle x \rangle^2 u = \overline{f}$ in the space of distributions. Moreover, taking $u = v \in V$ and using the fact that $L \in V'$ we obtain,

$$\| u \|_V^2 \le C \| L \|_{V'} \| u \|_V,$$

from which we deduce the desired inequality.

5. a) This follows immediately from the Cauchy–Schwarz inequality applied to the two integrals.

b) If $v = \varphi \in S$ then,

$$(2\pi)^{-\frac{d}{2}} \int_{\mathbf{R}^d} \langle \xi \rangle^{-1} \widehat{f_1}(\xi) \langle \xi \rangle \widehat{\varphi}(-\xi) d\xi = (2\pi)^{-\frac{d}{2}} \langle \widehat{f_1}, \widehat{\varphi}(-\xi) \rangle_{S' \times S}$$

$$= \langle f_1, \varphi \rangle_{S' \times S},$$

and

$$\int_{\mathbf{R}^d} \langle x \rangle^{-1} \, f_2(x) \, \langle x \rangle \, \varphi(x) \, dx = \langle f_2, \varphi \rangle_{S' \times S},$$

so $L_{(f_1, f_2)}(\varphi) = \langle f, \varphi \rangle_{S' \times S}$.

c) It follows from question b) that $L_{(f_1, f_2)}(\varphi) = L_{(g_1, g_2)}(\varphi)$ for all $\varphi \in S(\mathbf{R}^d)$. By question a) these are two linear and continuous form on V which coincide on $S(\mathbf{R}^d)$ which is dense in V by question 1. Therefore, they coincide on V.

d) This follows from the inequality (2.2) proved in question a), taking the infimum on all decompositions $f = f_1 + f_2$ and using the definition of the norm in E.

e) Φ is obviously linear. Its continuity follows from the previous question and the fact that $\|L_f\|_{V'} = \sup_{\|v\|_E = 1} |L_f(v)|$. Eventually it is injective since, if $L_f = 0$ then by question b) we have, $0 = L_f(\varphi) = \langle f, \varphi \rangle_{S' \times S}$ for all $\varphi \in S$, so $f = 0$ in S'.

f) This is the distributional formulation of the equality (2.1). Now, since $u \in V$ we have, $\langle \xi \rangle^{-1} \widehat{f_1} = \langle \xi \rangle^{-1} |\xi|^2 \widehat{u} \in L^2$ and $\langle x \rangle^{-1} f_2 = \langle x \rangle^{-1} \langle x \rangle^2 u \in L^2$. Moreover, $L(\varphi) = \langle f_1 + f_2, \varphi \rangle = L_f(\varphi)$ for all $\varphi \in S$. Then L and L_f are two linear and continuous forms on V which coincide on the dense subspace S; therefore, they coincide on V.

g) E and V' are Banach spaces. Φ is linear, bijective, and continuous. Therefore, Φ^{-1} is continuous.

Solution 2

Part 1

1. Let A be an orthogonal matrix and denote by B its transposed matrix. Set $z = By$ in the integral in the middle below; since $|y| = |z|$ and $dy = dz$ we can write,

$$\int_{|y| \le r} e^{i\lambda \langle A\theta, y \rangle} dy = \int_{|y| \le r} e^{i\lambda \langle \theta, By \rangle} dy = \int_{|z| \le r} e^{i\lambda \theta \cdot z} dz.$$

Using the polar coordinates, $y = \rho\omega$, the equality between the first and the last integrals can be written as,

$$\int_0^r \int_{S^{d-1}} e^{i\lambda \rho \langle A\theta, \omega \rangle} \, d\omega \, dr = \int_0^r \int_{S^{d-1}} e^{i\lambda \rho \langle \theta, \omega \rangle} \, d\omega \, dr.$$

Differentiating both members with respect to r and taking $r = 1$ we obtain,

$$F(\lambda, A\theta) = \int_{S^{d-1}} e^{i\lambda \langle A\theta, \omega \rangle} \, d\omega = \int_{S^{d-1}} e^{i\lambda \langle \theta, \omega \rangle} \, d\omega = F(\lambda, \theta).$$

2. There exists an orthogonal matrix A such that $A\theta = (0\ldots,0,1)$. It follows from the previous question that,

$$F(\lambda,\theta) = \int_{S^{d-1}} e^{i\lambda\,\omega_d}\,d\omega.$$

If $\omega = (\omega',\omega_d) \in S^{d-1}$ where $\omega' = (\omega_1,\ldots\omega_{d-1})$ we have $\omega_d^2 + |\omega'|^2 = 1$. Therefore,

$$S^{d-1} = E_+ \cup E_-,\quad E_\pm = \left\{\omega \in \mathbf{R}^d : |\omega'| \le 1,\ \omega_d = \pm\sqrt{1-|\omega'|^2}\right\}.$$

It can be parametrized by. $|\omega'| \le 1$, $\omega_d = \pm\sqrt{1-|\omega'|^2}$.

Now recall that the surface measure on a graph $\{\omega_d = \psi(\omega')\}$ is given by $\sqrt{1+|\nabla_{\omega'}\psi|^2}\,d\omega'$. Here $\psi(\omega') = \pm\sqrt{1-|\omega'|^2}$. It follows that we have,

$$\partial_{\omega_j}\psi(\omega') = \mp\frac{\omega_j}{\sqrt{1-|\omega'|^2}},\quad |\nabla_{\omega'}\psi|^2 = \frac{|\omega'|^2}{1-|\omega'|^2},\quad \sqrt{1+|\nabla_{\omega'}\psi|^2} = \frac{1}{\sqrt{1-|\omega'|^2}}.$$
$$(9.2)$$

Summing up we can write,

$$F(\lambda,\theta) = \int_{|\omega'|<1} e^{i\lambda\sqrt{1-|\omega'|^2}}\,\frac{d\omega'}{\sqrt{1-|\omega'|^2}} + \int_{|\omega'|<1} e^{-i\lambda\sqrt{1-|\omega'|^2}}\,\frac{d\omega'}{\sqrt{1-|\omega'|^2}}$$
$$=: F^+(\lambda,\theta) + F^-(\lambda,\theta).$$

3. It follows from (9.2) that $\nabla_\omega\psi(\omega') = 0$ is equivalent to $\omega' = 0$. Thus $\omega' = 0$ is the only critical point. Now again by (9.2) we have,

$$\partial_{\omega_k}\partial_{\omega_j}\psi = \mp\delta_{jk}(1-|\omega'|^2)^{-\frac{1}{2}} \mp \omega_j\omega_k(1-|\omega'|^2)^{-\frac{3}{2}},\quad 1 \le j,k \le d-1,$$

where δ_{jk} is the Kronecker symbol. Therefore, if we denote by $\mathcal{H}(\omega') = \left(\partial_{\omega_k}\partial_{\omega_j}\psi(\omega')\right)_{1\le j,k\le d-1}$ the Hessian matrix, $\mathcal{H}(0)$ is the $\mp$ the identity matrix. Therefore, $\omega' = 0$ is a nondegenerate critical point of ψ.

Part 2

1. Since $\omega' = 0$ is the unique nondegenerate critical point of ψ on the support of χ_1 we can apply the stationary phase formula (1.3) with d replaced by $d-1$. We have $\psi(0) = 1$, $|\det(\psi''(0))| = 1$, $\chi_1(0) = 1$ and the signature of $\psi''(0)$ is equal to $\mp(d-1)$. Therefore, we have, when $\lambda \to +\infty$,

$$G^\pm(\lambda,\theta) = c_\pm\lambda^{-\frac{d-1}{2}}(1+o(1)),\quad c_\pm = (2\pi)^{\frac{d-1}{2}}e^{\pm i\lambda}e^{\mp i(d-1)\frac{\pi}{4}}.$$

Part 3

1. The first claim is obvious. Now it follows from (9.2) that,

$$L = -\frac{1}{i\lambda} \frac{\sqrt{1 - |\omega'|^2}}{|\omega'|^2} \sum_{j=1}^{d-1} \omega_j \partial_{\omega_j}. \tag{9.3}$$

Notice that on the support of χ_2 we have $|\omega'| \geq \delta$.

2. The claim is true when $m = 0$ with $a_0(\omega') = \chi_2(\omega')$. Let us prove it when $m = 1$. We have $L(e^{i\lambda\psi} - e^{-i\lambda\psi}) = e^{i\lambda\psi} + e^{-i\lambda\psi}$. By (9.3) we have,

$$A := H^+(\lambda, \theta) + H^-(\lambda, \theta),$$

$$= -\frac{1}{i\lambda} \sum_{j=1}^{d-1} \int_{|\omega'|<1} \partial_{\omega_j} (e^{i\lambda\psi(\omega')} - e^{-i\lambda\psi(\omega')}) \frac{\omega_j}{|\omega'|^2} \chi_2(\omega') \, d\omega'.$$

By the Gauss–Green formula, since the unitary exterior normal n to the boundary $|\omega'| = 1$ is $n = \omega'$ we have, $A = A_1 + A_2$, where,

$$A_1 = \frac{1}{i\lambda} \sum_{j=1}^{d-1} \int_{|\omega'|<1} (e^{i\lambda\psi(\omega')} - e^{-i\lambda\psi(\omega')}) \partial_{\omega_j} \left(\frac{\omega_j}{|\omega'|^2} \chi_2(\omega') \right) d\omega',$$

$$A_2 = -\frac{1}{i\lambda} \sum_{j=1}^{d-1} \int_{|\omega'|=1} (e^{i\lambda\psi(\omega')} - e^{-i\lambda\psi(\omega')}) \frac{\omega_j^2}{|\omega'|^2} \chi_2(\omega') \, d\sigma.$$

For $|\omega'| = 1$ we have $\psi(\omega') = 0$, so $e^{\pm i\lambda\psi(\omega')} = 1$. Therefore, the term A_2 is equal to zero. It follows that,

$$A = \frac{1}{\lambda} \int_{|\omega'|<1} (e^{i\lambda\psi} - e^{-i\lambda\psi}) b_1(\omega') \, d\omega',$$

where $b_1(\omega') = \frac{1}{i} \sum_{j=1}^{d-1} \partial_{\omega_j} \left(\frac{\omega_j}{|\omega'|^2} \chi_2(\omega') \right)$. This proves our claim for $m = 1$.

Assume the claim true for m and lets us prove it for $m + 1$. If m is even so $m + 1$ is odd the same proof as above gives the result. Therefore, assume m odd so $m + 1$ is even. Then,

$$A := H^+(\lambda, \theta) + H^+(\lambda, \theta) = \frac{1}{\lambda^m} \int_{|\omega'|<1} \left(e^{i\lambda\psi(\omega')} - e^{-i\lambda\psi(\omega')} \right) b_m(\omega') \, d\omega'.$$

We use the fact that $e^{i\lambda\psi(\omega')} - e^{-i\lambda\psi(\omega')} = L\left(e^{i\lambda\psi(\omega')} + e^{-i\lambda\psi(\omega')}\right)$ and we integrate by parts as before. According to (9.3) we obtain $A = A_1 + A_2$ where,

$$A_1 = -\frac{1}{i\lambda^{m+1}} \sum_{j=1}^{d-1} \int_{|\omega'|<1} \left(e^{i\lambda\psi(\omega')} + e^{-i\lambda\psi(\omega')}\right).$$

$$\partial_{\omega_j}\left(\sqrt{1-|\omega'|^2}\frac{\omega_j}{|\omega'|^2}b_m(\omega')\right)d\omega',$$

$$A_2 = \frac{1}{i\lambda^{m+1}} \sum_{j=1}^{d-1} \int_{|\omega'|=1} \left(e^{i\lambda\psi(\omega')} + e^{-i\lambda\psi(\omega')}\right)\sqrt{1-|\omega'|^2}\frac{\omega_j^2}{|\omega'|^2}b_m(\omega')\,d\sigma.$$

Due to the factor $\sqrt{1-|\omega'|^2}$ we have $A_2 = 0$. Setting $c_j(\omega') = \frac{\omega_j}{|\omega'|^2}b_m(\omega')$ we have,

$$\partial_{\omega_j}\left(\sqrt{1-|\omega'|^2}c_j(\omega')\right) = \frac{1}{\sqrt{1-|\omega'|^2}}\left(-\omega_j c_j(\omega') + (1-|\omega'|^2)\partial_{\omega_j}c_j(\omega')\right),$$

$$:= \frac{d_j(\omega')}{\sqrt{1-|\omega'|^2}},$$

where d_j is a C^∞ function whose support is contained in $\{|\omega'| \geq \delta\}$. Therefore,

$$A = \frac{1}{\lambda^{m+1}} \int_{|\omega'|<1} \left(e^{i\lambda\psi(\omega')} + e^{-i\lambda\psi(\omega')}\right)\frac{a_{m+1}(\omega')}{\sqrt{1-|\omega'|^2}}\,d\omega'.$$

This proves our claim for $m+1$.

3. It follows from the previous question that,

$$|H^+(\lambda,\theta) + H^+(\lambda,\theta)| \leq \frac{2}{\lambda^m} \int_{|\omega'|<1} \frac{|a_m(\omega')|}{\sqrt{1-|\omega'|^2}}\,d\omega', \quad m \text{ even,}$$

$$|H^+(\lambda,\theta) + H^+(\lambda,\theta)| \leq \frac{2}{\lambda^m} \int_{|\omega'|<1} |b_m(\omega')|\,d\omega', \quad m \text{ odd.}$$

Since a_m, b_m are continuous on the set $\{\omega' : |\omega'| \leq 1\}$ they are bounded. Moreover, using the polar coordinates in $\mathbf{R}^{d-1}$ we see easily that the integral $\int_{|\omega'|<1} \frac{1}{\sqrt{1-|\omega'|^2}}\,d\omega'$ is finite. It follows that for every $m \in \mathbf{N}$ there exists $C_m > 0$ such that,

$$|H^+(\lambda,\theta) + H^+(\lambda,\theta)| \leq C_m \lambda^{-m}.$$

Using the result in Part 2. we conclude that,

$$F(\lambda, \theta) = c_0 \lambda^{-\frac{d-1}{2}} (1 + o(1)) \quad \lambda \to +\infty.$$

Solution 3

Using the Hint we write, $\widehat{\partial^\alpha a} = \widehat{\partial^\alpha a} \cdot \chi_R = \widehat{a} \cdot (i\xi)^\alpha \chi_R$, from which we deduce that $\partial^\alpha a = a \star \mathcal{F}^{-1}((i\xi)^\alpha \chi_R)$. Now, $\mathcal{F}^{-1}((i\xi)^\alpha \chi_R)(x) = R^{d+|\alpha|} \mathcal{F}^{-1}((i\eta)^\alpha \chi)(Rx)$. Define q by $\theta = 1 - \frac{1}{q}$ where $q = 1$ if $\theta = 0$ and $q = +\infty$ if $\theta = 1$. Then $1 \leq q \leq +\infty$ and by definition we have, $\frac{1}{r} = \frac{1}{p} + \frac{1}{q} - 1$. We can therefore apply the Young inequality and write,

$$\|\partial^\alpha a\|_{L^r(\mathbf{R}^d)} \leq C R^{d+|\alpha|} \|a\|_{L^p(\mathbf{R}^d)} \|\mathcal{F}^{-1}((i\eta)^\alpha \chi)(R\cdot)\|_{L^q(\mathbf{R}^d)},$$

$$\leq C R^{d+|\alpha|} R^{-\frac{d}{q}} \|a\|_{L^p(\mathbf{R}^d)} \|\mathcal{F}^{-1}((i\eta)^\alpha \chi)\|_{L^q(\mathbf{R}^d)}.$$

To conclude we have just to notice that $d - \frac{d}{q} = d\theta$.

Solution 4

1. Let $t_0 \in (a, b)$, $r > 0$ be such that $[t_0 - r, t_0 + r] \subset (a, b)$ and n_0 be so large that $t_n \in [t_0 - r, t_0 + r]$ for $n \geq n_0$. By hypothesis (H1) there exists $M > 0$ such that,

$$\|u(t_0, \cdot)\|_{H^s} + \|u(t_n, \cdot)\|_{H^s} \leq M, \quad n \geq n_0.$$

Let $\varepsilon > 0$; since $H^{-s+\delta}$ is dense in H^{-s} there exists $\varphi_0 \in H^{-s+\delta}$ such that, $\|\varphi - \varphi_0\|_{H^{-s}} \leq \frac{\varepsilon}{2M}$. We write for $n \geq n_0$,

$$A_n = \langle u(t_n, \cdot) - u(t_0, \cdot), \varphi - \varphi_0 \rangle + \langle u(t_n, \cdot) - u(t_0, \cdot), \varphi_0 \rangle := B_n + C_n.$$

We have

$$|B_n| \leq \|u(t_n, \cdot) - u(t_0, \cdot)\|_{H^s} \|\varphi - \varphi_0\|_{H^{-s}} \leq M \|\varphi - \varphi_0\|_{H^{-s}} \leq \frac{\varepsilon}{2},$$

$$|C_n| \leq \|u(t_n, \cdot) - u(t_0, \cdot)\|_{H^{s-\delta}} \|\varphi_0\|_{H^{-s+\delta}}.$$

By hypothesis (H2) there exists $n_1 \geq n_0$ such that $|C_n| \leq \frac{\varepsilon}{2}$ for $n \geq n_1$. Therefore, $|A_n| \leq \varepsilon$ for $n \geq n_1$.

2. Let us first prove the Hint. We have,

$$(u, v)_{H^s} = \int_{\mathbf{R}^d} \widehat{u}(\xi) \, \langle \xi \rangle^{2s} \, \overline{\widehat{v}(\xi)} \, d\xi.$$

Since, $\langle \xi \rangle^{2s} \, \overline{\widehat{v}(\xi)} = \mathcal{F}((I - \Delta)^s \overline{v})(-\xi)$, we have,

$$(u, v)_{H^s} = (2\pi)^d \left\langle u, (I - \Delta)^s \overline{v} \right\rangle_{H^s \times H^{-s}}.$$

Now we write,

$$\|u(t_n, \cdot) - u(t_0, \cdot)\|_{H^s}^2 = \|u(t_n, \cdot)\|_{H^s}^2 + \|u(t_0, \cdot)\|_{H^s}^2 - 2\mathrm{Re}\,(u(t_n, \cdot), u(t_0, \cdot))_{H^s}.$$

By hypothesis (H 1) we have,

$$\lim_{n \to +\infty} \|u(t_n, \cdot)\|_{H^s}^2 = \|u(t_0, \cdot)\|_{H^s}^2.$$

On the other hand, by the Hint we can write,

$$(u(t_n, \cdot), u(t_0, \cdot))_{H^s} = (2\pi)^d \left\langle u(t_n, \cdot), \Lambda^{2s} \overline{u(t_0, \cdot)} \right\rangle_{H^s \times H^{-s}}.$$

Since $\Lambda^{2s} \overline{u(t_0, \cdot)} \in H^{-s}$ we deduce from question 1 that,

$$\lim_{n \to +\infty} (u(t_n, \cdot), u(t_0, \cdot))_{H^s} = (2\pi)^d \left\langle u(t_0 \cdot), \Lambda^{2s} \overline{u(t_0, \cdot)} \right\rangle_{H^s \times H^{-s}} = \|u(t_0, \cdot)\|_{H^s}^2,$$

therefore,

$$\lim_{n \to +\infty} \|u(t_n, \cdot) - u(t_0, \cdot)\|_{H^s}^2 = 0.$$

Solution 5

Part 1

1. Let $v \in C_0^\infty(\mathbf{R} \times \mathbf{R}^d)$ and set $f(z) = \|v(z, \cdot)\|_{H^s}^2$ where $H^s = H^s(\mathbf{R}^d)$. Let $z_0 \in \mathbf{R}$. We have,

$$f'(z_0) = \lim_{h \to 0} (A(h) + B(h)),$$

$$A(h) = \left(\frac{v(z_0 + h, \cdot) - v(z_0, \cdot)}{h}, v(z_0 + h, \cdot) \right)_{H^s}, \tag{9.4}$$

$$B(h) = \left(v(z_0, \cdot), \frac{v(z_0 + h, \cdot) - v(z_0, \cdot)}{h} \right)_{H^s}.$$

The Taylor formula with integral reminder at the order two gives for all $x \in \mathbf{R}^d$,

$$\frac{v(z_0 + h, x) - v(z_0, x)}{h} = (\partial_z v)(z_0, x) + h \int_0^1 (1 - t)(\partial_z^2 v)(z_0 + th, x) \, dt,$$

so that we have $A(h) = A_1(h) + A_2(h) + A_3(h)$ where,

$$A_1 = ((\partial_z v)(z_0, \cdot), v(z_0, \cdot))_{H^s},$$

$$A_2(h) = ((\partial_z v)(z_0, \cdot), v(z_0 + h, \cdot) - v(z_0, \cdot))_{H^s},$$

$$= \left((\partial_z v)(z_0, \cdot), h \int_0^1 (\partial_z v)(z_0 + th, \cdot) \, dt \right)_{H^s},$$

$$A_3(h) = h \left(\int_0^1 (1 - t)(\partial_z^2 v)(z_0 + th, \cdot) \, dt, v(z_0 + h, \cdot) \right)_{H^s}.$$

We have,

$$|A_2(h)| \le h \|(\partial_z v)(z_0, \cdot)\|_{H^s} \sup_{z \in \mathbf{R}} \|(\partial_z v)(z, \cdot)\|_{H^s}.$$

Therefore, $\lim_{h \to 0} A_2(h) = 0$. By the same way,

$$|A_3(h)| \le \frac{h}{2} \sup_{z \in \mathbf{R}} \|v(z, \cdot)\|_{H^s} \sup_{z \in \mathbf{R}} \|(\partial_z^2 v)(z, \cdot)\|_{H^s},$$

so $\lim_{h \to 0} A_3(h) = 0$. If follows that,

$$\lim_{h \to 0} A(h) = ((\partial_z v)(z_0, \cdot), v(z_0, \cdot))_{H^s}.$$

The same proof shows that,

$$\lim_{h \to 0} B(h) = (v(z_0, \cdot), (\partial_z v)(z_0, \cdot))_{H^s}.$$

It follows from (9.4) that

$$f'(z_0) = 2\mathrm{Re}\,((\partial_z v)(z_0, \cdot), v(z_0, \cdot))_{H^s}.$$

2. Since v has compact support in z there exists $A > 0$ such that $v(z, \cdot)$ vanishes for $z \le -A$. Integrating the above inequality between $-A$ and $z \in \mathbf{R}$, using

the inequality $|(f, g)_{H^s}| \leq \|f\|_{H^{s+\frac{1}{2}}} \|g\|_{H^{s-\frac{1}{2}}}$ and then the Cauchy–Schwarz inequality we obtain,

$$\sup_{z \in \mathbf{R}} \|v(z, \cdot)\|_{H^s} \leq C \|v\|_{L_z^2(\mathbf{R}, H^{s+\frac{1}{2}})}^{\frac{1}{2}} \|\partial_z v\|_{L_z^2(\mathbf{R}, H^{s-\frac{1}{2}})}^{\frac{1}{2}} \leq C' \|v\|_{W^s(\mathbf{R})}. \quad (9.5)$$

3. Let now $u \in W^s(\mathbf{R})$ and let (u_j) be a sequence in $C_0^\infty(\mathbf{R} \times \mathbf{R}^d)$ which converges to u in $W^s(\mathbf{R})$. By inequality (9.5) applied to $u_j - u_k$ we see that (u_j) is Cauchy sequence in $L^\infty(\mathbf{R}, H^s(\mathbf{R}^d))$. This space being complete, this sequence converges to an element $\tilde{u} \in L^\infty(\mathbf{R}, H^s(\mathbf{R}^d))$. The $u_j's$ being continuous in z and the convergence being uniform on $\mathbf{R}$ with values in $H^s(\mathbf{R}^d)$ we have $\tilde{u} \in C^0(\mathbf{R}, H^s(\mathbf{R}^d))$. Eventually, the convergence in $W^s(\mathbf{R})$ and in $L^\infty(\mathbf{R}, H^s(\mathbf{R}^d))$ implying the convergence in the space of distributions on $\mathbf{R} \times \mathbf{R}^d$, we have $u = \tilde{u}$ in $\mathcal{D}'(\mathbf{R} \times \mathbf{R}^d)$. Then we write (9.5) with $v = u_j$ and we pass to the limit to obtain the same inequality for u.

Part 2

1. If $u \in W^s((0, +\infty))$ there exists $\tilde{u} \in W^s(\mathbf{R})$ such that $u = \tilde{u}$ for $z \in (0, +\infty)$ and $\|\tilde{u}\|_{W^s(\mathbf{R})} \sim \|u\|_{W^s((0,+\infty))}$. If u is smooth this is the even extension, that is,

$$\tilde{u}(z, \cdot) = u(z, \cdot) \text{ if } z > 0, \quad \tilde{u}(z, \cdot) = u(-z, \cdot) \text{ if } z < 0.$$

Otherwise we use the density of smooth functions. Now Part 1 implies that $\tilde{u} \in C^0(\mathbf{R}, H^s(\mathbf{R}^d))$; by definition we have $u \in C^0([0, +\infty), H^s(\mathbf{R}^d))$ and we have an estimate similar to (9.5) for u.

2. Obvious.

Part 3

1. Let $I = (-1, 0)$. There exist $\chi_j \in C^\infty(\mathbf{R})$, $j = 1, 2$ such that $\text{supp}\,\chi_1 \subset (-\infty, -\frac{1}{4}]$, $\text{supp}\,\chi_2 \subset [-\frac{3}{4}, +\infty)$ and $\chi_1(z) + \chi_2(z) = 1$ on $(-1, 0)$. Let $u \in W^s(I)$. We write $u = u_1 + u_2$ where $u_j(z, \cdot) = \chi_j(z)u(z, \cdot), j = 1, 2$. Then $u_1 \in W^s((-1, +\infty))$ and $u_2 \in W^s((-\infty, 0))$.

2. By Part 2 we have $u_1 \in C^0([-1, +\infty), H^s(\mathbf{R}^d)) \subset C^0([-1, 0], H^s(\mathbf{R}^d))$, $u_2 \in C^0((-\infty, 0], H^s(\mathbf{R}^d)) \subset C^0([-1, 0], H^s(\mathbf{R}^d))$. So $u \in C^0([-1, 0], H^s(\mathbf{R}^d))$. Eventually, for $z \in [-1, 0]$ can write

$$\|u(z, \cdot)\|_{H^s(\mathbf{R}^d)} \leq \|u_1(z, \cdot)\|_{H^s(\mathbf{R}^d)} + \|u_2(z, \cdot)\|_{H^s(\mathbf{R}^d)},$$

$$\leq C\left(\|u_1\|_{W^s((-1,+\infty))} + \|u_2\|_{W^s((-\infty,0))}\right) \leq C'\|u\|_{W^s((-1,0))}$$

since u_1 vanishes for $z \geq 0$, u_2 vanishes for $z \leq -1$, and $u_j = \chi_j u$.

3. Obvious.

Solution 6

Part 1
1. Setting $z = x + iy$ we can write,

$$|e^{z \cdot \xi} \widehat{u}(\xi)| = e^{-y \cdot \xi} |\widehat{u}(\xi)| = \left[\langle \xi \rangle^{-s} e^{-\sigma |\xi|} e^{-y \cdot \xi} \right] \times \left[\langle \xi \rangle^{s} e^{\sigma |\xi|} |\widehat{u}(\xi)| \right] := U_1 \cdot U_2. \tag{9.6}$$

By hypothesis we have $U_2 \in L^2(\mathbf{R})$. Now since $-y \cdot \xi \le |y||\xi|$ we have for $|y| < \sigma$, $U_1 \le \langle \xi \rangle^{-s} e^{-(\sigma - |y|)|\xi|} \in L^2(\mathbf{R})$. It follows that $U_1 \cdot U_2 \in L^1(\mathbf{R})$.
2. Set $f(z, \xi) = e^{z \cdot \xi} \widehat{u}(\xi)$. For fixed ξ the function $z \mapsto f(z, \xi)$ is holomorphic. Moreover, if K is a compact subset of S_σ and $z = x + iy \in K$ we have $|x| \le M$ and $|y| \le \sigma - \varepsilon$ where $\varepsilon > 0$. The inequality (9.6) implies that,

$$|f(z, \xi)| \le \langle \xi \rangle^{-s} e^{-\varepsilon |\xi|} \langle \xi \rangle^{s} e^{\sigma |\xi|} |\widehat{u}(\xi)| \in L^1(\mathbf{R}).$$

It follows that U is holomorphic in S_σ. Moreover, $U|_{y=0} = \mathcal{F}^{-1} \widehat{u} = u$.
3. Since $iz \cdot \xi = ix \cdot \xi - y \cdot \xi$ we have $U(x + iy) = \mathcal{F}_{\xi \to x}^{-1}(e^{-y \cdot \xi} \widehat{u})$, so that $\mathcal{F}_{x \to \xi}(U(x + iy)) = e^{-y \cdot \xi} \widehat{u}(\xi)$. Therefore, since $-y \cdot \xi \le |y||\xi| \le \sigma |\xi|$ we get,

$$\|U(x + iy)\|_{H_x^s(\mathbf{R})}^2 \le \int_{\mathbf{R}} \langle \xi \rangle^{2s} e^{2\sigma |\xi|} |\widehat{u}(\xi)|^2 \, d\xi = \|u\|_{G^{s,\sigma}}^2.$$

Part 2
1. a) By (4.2) we have $u \in H^s(\mathbf{R})$. The function inside the integral is, for fixed ξ, holomorphic in z. Then, since $-y \cdot \xi \le \sigma |\xi|$ and supp $\psi_\lambda \subset \{\xi : |\xi| \le 2\lambda\}$ we have,

$$|e^{iz \cdot \xi} \psi_\lambda(\xi) \widehat{u}(\xi)| = |e^{-y \cdot \xi} \psi_\lambda(\xi) \widehat{u}(\xi)| \le e^{\sigma |\xi|} |\psi_\lambda(\xi) \widehat{u}(\xi)| \le e^{2\sigma\lambda} |\psi_\lambda(\xi) \widehat{u}(\xi)|,$$

$$\le \left[e^{\sigma\lambda} \langle \xi \rangle^{-s} |\psi_\lambda(\xi)| \right] \left[\langle \xi \rangle^{s} |\widehat{u}(\xi)| \right] \in L^1(\mathbf{R}),$$

since it is the product of two $L^2(\mathbf{R})$ functions. So F is holomorphic in S_σ.
2. a) The holomorphy being a local property it is enough to show that V_j is holomorphic in any ball $B(z_0, R) \subset S_\sigma$. If $z \in B(z_0, R)$ and $t \in$ supp χ_j we have $z - t \in K$ which is a compact subset of S_σ. The function U being holomorphic in S_σ it belongs to $L_{loc}^1(S_\sigma)$. Therefore, we have $V_j(z) = \int_{\mathbf{R}} U(z - t) \chi_j(t) \, dt$. On the other hand, for $z \in B(z_0, R)$ and $t \in$ supp χ_j we have,

$$|U(z - t) \chi_j(t)| \le \sup_{K} |U||\chi_j(t)| \in L^1(\mathbf{R}).$$

This proves that V_j is holomorphic in S_σ.

b) We have for $z \in S_\sigma$,

$$|V_j(z) - V(z)| = |\langle U(z-t), \chi_j(t) - \varphi_\lambda(t)\rangle| \le \|U(z-t)\|_{H_t^s} \|\chi_j - \varphi_\lambda\|_{H^{-s}},$$

$$\le \sup_{|y|<\sigma} \|U(x+iy)\|_{H_x^s} \|\chi_j - \varphi_\lambda\|_{H^{-s}} \le M_0 \|\chi_j - \varphi_\lambda\|_{H^{-s}}.$$

Therefore, the sequence (V_j) converges uniformly to V on S_σ, which ensures that V is holomorphic on S_σ.

3. We have $V|_{y=0} = u \star \varphi_\lambda$ so that $\mathcal{F}(V|_{y=0}) = \psi_\lambda(\xi)\widehat{u}(\xi)$ and therefore,

$$V|_{y=0} = (2\pi)^{-1} \int_{\mathbf{R}^d} e^{ix\cdot\xi} \psi_\lambda(\xi)\widehat{u}(\xi)\,d\xi = F|_{y=0}.$$

By analytic continuation we have, $V(z) = F(z)$ for all $z \in S_\sigma$.

4. We deduce from the previous question that,

$$\mathcal{F}_{x\to\xi}(V(\cdot+iy))(\xi) = \mathcal{F}_{x\to\xi}(U(\cdot+iy))(\xi)\psi_\lambda(\xi),$$

$$= \mathcal{F}_{x\to\xi}(F(\cdot+iy))(\xi) = e^{-y\cdot\xi}\psi_\lambda\widehat{u}(\xi)$$

since $F(x+iy) = \mathcal{F}_{\xi\to x}^{-1}(e^{-y\cdot\xi}\psi_\lambda(\xi)\widehat{u})$. Now $\psi_\lambda(\xi) = 1$ for $|\xi| \le \lambda$ so letting λ go to $+\infty$ we obtain,

$$\mathcal{F}_{x\to\xi}(U(\cdot+iy))(\xi) = e^{-y\cdot\xi}\widehat{u}(\xi).$$

It follows that,

$$\int_{\mathbf{R}} e^{-2y\cdot\xi}\,\langle\xi\rangle^{2s}\,|\widehat{u}(\xi)|^2\,d\xi = \|U(x+iy)\|_{H_x^s}^2 \le M_0^2, \quad \forall z \in S_\sigma. \tag{9.7}$$

5. By (9.7) we have for all $|y| < \sigma$,

$$I_1 = \int_0^{+\infty} e^{-2y\cdot\xi}\,\langle\xi\rangle^{2s}\,|\widehat{u}(\xi)|^2\,d\xi \le M_0^2,$$

$$I_2 = \int_{-\infty}^0 e^{-2y\cdot\xi}\,\langle\xi\rangle^{2s}\,|\widehat{u}(\xi)|^2\,d\xi \le M_0^2.$$

Let $0 \le b < \sigma$. In I_1 let us take $y = -b$ and in I_2, $y = b$ and add. We get,

$$G(b) := \int_{\mathbf{R}} e^{2b|\xi|}\,\langle\xi\rangle^{2s}\,|\widehat{u}(\xi)|^2\,d\xi \le 2M_0^2.$$

The function G is bounded above and nondecreasing. Therefore, it has a limit $\ell \leq 2M_0^2$ when b goes to σ. By the Fatou lemma we can write,

$$\int_{\mathbf{R}} \liminf_{b \to \sigma} e^{2b|\xi|} \langle \xi \rangle^{2s} |\widehat{u}(\xi)|^2 \, d\xi = \int_{\mathbf{R}} e^{2\sigma|\xi|} \langle \xi \rangle^{2s} |\widehat{u}(\xi)|^2 \, d\xi,$$

$$\leq \liminf_{b \to \sigma} G(b) = \ell \leq 2M_0^2,$$

which proves (4.3).

Part 3

1. We have $\|au\|_{G^{\sigma,0}} = \|v\|_{L^2(\mathbf{R})}$ where $v(\xi) = e^{\sigma|\xi|}\widehat{au}(\xi)$. Since $e^{\sigma|\xi|} \leq e^{\sigma|\xi-\eta|}e^{\sigma|\eta|}$ we have,

$$|v(\xi)| \leq \int_{\mathbf{R}} e^{\sigma|\xi-\eta|}|\widehat{a}(\xi-\eta)|e^{\sigma|\eta|}|\widehat{u}(\eta)| \, d\eta = (e^{\sigma|\cdot|}|\widehat{a}|) \star (e^{\sigma|\cdot|}|\widehat{u}|).$$

It follows that,

$$\|v\|_{L^2(\mathbf{R})} \leq \|e^{\sigma|\cdot|}\widehat{a}\|_{L^1(\mathbf{R})} \|e^{\sigma|\cdot|}\widehat{u}\|_{L^2(\mathbf{R})}.$$

Using the Cauchy–Schwarz inequality and the fact that $2s_0 > 1$ we can write,

$$\|e^{\sigma|\cdot|}\widehat{a}\|_{L^1(\mathbf{R})} = \int_{\mathbf{R}} \langle \xi \rangle^{-s_0} e^{\sigma|\xi|} \langle \xi \rangle^{s_0} |\widehat{a}(\xi)| \, d\xi \leq C\|e^{\sigma|\xi|} \langle \xi \rangle^{s_0} \widehat{a}\|_{L^2(\mathbf{R})},$$

which proves our inequality.

2. a) We use the same method as above and the fact that for $s \geq 0$ there exists $C_s > 0$ such that $\langle \xi \rangle^s \leq C_s(\langle \xi - \eta \rangle^s + \langle \eta \rangle^s)$.
   b) If $s > \frac{1}{2}$ we can take $s = s_0$ in the previous question.

Solution 7

Part 1

1. Let $\varphi \in C^\infty(\mathbf{R}^d)$ with support in $[-1, 1]^d$ be such that $0 \leq \varphi \leq 1$ and $\varphi = 1$ on $[-\frac{1}{2}, \frac{1}{2}]^d$. Set,

$$\phi(x) = \sum_{q \in \mathbf{Z}^d} \varphi(x - q).$$

Then $\phi(x - q') = \phi(x)$ for all $q' \in \mathbf{Z}^d$. Moreover, since for all $x \in \mathbf{R}^d$ there exists $q \in \mathbf{Z}^d$ such that $|x - q| = \sup_{1 \leq j \leq d} |x_j - q_j| \leq \frac{1}{2}$ we have $\phi(x) \geq 1$

for all $x \in \mathbf{R}^d$. Set $\chi(x) = \frac{\varphi(x)}{\phi(x)}$. Then supp $\chi \subset [-1, 1]^d$ and $\chi(x - q) = \frac{\varphi(x-q)}{\phi(x-q)} = \frac{\varphi(x-q)}{\phi(x)}$, so we have, $\sum_{q \in \mathbf{Z}^d} \chi(x - q) = 1$ for every $x \in \mathbf{R}^d$.

2. Let $(u_j)_{j \in \mathbf{N}}$ be a Cauchy sequence in $H^s_{ul}(\mathbf{R}^d)$. For all $q \in \mathbf{Z}^d$ $(\chi_q u_j)$ is then a Cauchy sequence in $H^s(\mathbf{R}^d)$. Therefore, there exists $v_q \in H^s(\mathbf{R}^d)$ such that,

$$\chi_q u_j \to v_q \quad \text{in } H^s(\mathbf{R}^d). \tag{9.8}$$

Let us show that the sequence (u_j) converges in $\mathcal{D}'(\mathbf{R}^d)$ to a distribution u. For this it is enough to prove that for all $\varphi \in C^\infty_0(\mathbf{R}^d)$ the sequence $(\langle u_j, \varphi \rangle)$ is convergent in $\mathbf{C}$. We write,

$$\langle u_j, \varphi \rangle = \left\langle u_j, \sum_{q \in \mathbf{Z}^d} \chi_q \varphi \right\rangle.$$

Now $\sum_{q \in \mathbf{Z}^d} \chi_q \varphi = \sum_{|q| \leq N} \chi_q \varphi$ (where N depends only on φ), so that,

$$\langle u_j, \varphi \rangle = \sum_{|q| \leq N} \langle u_j, \chi_q \varphi \rangle = \sum_{|q| \leq N} \langle \chi_q u_j, \varphi \rangle \to \sum_{|q| \leq N} \langle v_q, \varphi \rangle$$

by (9.8).

This implies that $\chi_q u_j \to \chi_q u$ in $\mathcal{D}'(\mathbf{R}^d)$ for all $q \in \mathbf{Z}^d$. It follows from (9.8) that $\chi_q u = v_q \in H^s(\mathbf{R}^d)$ and that,

$$\chi_q u_j \to \chi_q u \text{ in } H^s(\mathbf{R}^d), \quad \forall q \in \mathbf{Z}^d. \tag{9.9}$$

The fact that (u_j) is a Cauchy sequences reads,

$$\forall \varepsilon > 0 \quad \exists j_0 : \forall j, k \geq j_0 \quad \|\chi_q u_j - \chi_q u_k\|_{H^s(\mathbf{R}^d)} \leq \varepsilon, \quad \forall q \in \mathbf{Z}^d. \tag{9.10}$$

Letting in (9.10), k go to $+\infty$ and using (9.9) we obtain,

$$\forall \varepsilon > 0 \quad \exists j_0 : \forall j \geq j_0 \quad \|\chi_q u_j - \chi_q u\|_{H^s(\mathbf{R}^d)} \leq \varepsilon, \quad \forall q \in \mathbf{Z}^d,$$

which proves that $u_j \to u$ in $H^s_{ul}(\mathbf{R}^d)$.

3. The constant functions belong to $L^2_{ul}(\mathbf{R}^d)$ but of course not to $L^2(\mathbf{R}^d)$. Indeed let $u(x) = 1$ for all $x \in \mathbf{R}^d$. Then $\|\chi_q u\|_{L^2(\mathbf{R}^d)} = \|\chi(\cdot - q)\|_{L^2(\mathbf{R}^d)} = \|\chi\|_{L^2(\mathbf{R}^d)}$, so $\sup_{q \in \mathbf{Z}^d} \|\chi_q u\|_{L^2(\mathbf{R}^d)} = \|\chi\|_{L^2(\mathbf{R}^d)} < +\infty$.

Part 2

1. a) We use an induction on $|\alpha| = k$. The claim is true for $k = 0$. Assume it is
true for $k \geq 0$ and let $|\gamma| = k + 1$. Then $\partial_x^\gamma = \partial_{x_j}\partial_x^\alpha$ where $|\alpha| = k$. Set
$u(x) = 1 + |x|^2$. We have,

$$\partial_x^\gamma u(x)^\rho = \partial_{x_j}[P_\alpha(x)u(x)^{\rho-|\alpha|}],$$

$$= u(x)^{\rho-|\alpha|-1}[(\partial_{x_j}P_\alpha)(x)u(x) + 2(\rho - |\alpha|)x_j P_\alpha(x)],$$

$$= u(x)^{\rho-|\gamma|}P_\gamma(x),$$

where P_γ is a polynomial of order $|\gamma|$.

b) Use a) and the fact that $|P_\alpha(x)| \leq C_\alpha(1 + |x|)^{|\alpha|} \leq C'_\alpha(1 + |x|^2)^{\frac{1}{2}|\alpha|}$.

c) By induction on $|\alpha| = k$. The claim is true for $k = 0$. Assume it is true for
$k \geq 0$ and let $|\gamma| = k+1$. Differentiating the inequality given in the Hint and
using the Leibniz formula we obtain,

$$u(x)^\rho\partial_x^\gamma[u(x)^{-\rho}] = - \sum_{\substack{\gamma_1+\gamma_2=\gamma \\ \gamma_1\neq 0}} \binom{\gamma}{\gamma_1}\partial_x^{\gamma_1}[u(x)^\rho]\partial_x^{\gamma_2}[u(x)^{-\rho}].$$

Since $|\gamma_2| \leq k$, by induction and by question a), each term of the sum in
the right-hand side is bounded by $Cu(x)^{\rho-\frac{1}{2}|\gamma_1|}u(x)^{-\rho-\frac{1}{2}|\gamma_2|} = u(x)^{-\frac{1}{2}|\gamma|}$.
Dividing both members by $u(x)^\rho$ we obtain our claim at the rank $k + 1$.

d) This is obvious since $\partial_x^\alpha[f(x-a)] = (\partial_x^\alpha f)(x-a)$.

2. By the Leibniz formula for $|\alpha| \leq M$ we have,

$$\partial_x^\alpha\Phi_{k,q}(x) = \langle k - q\rangle^N \sum_{\alpha_1+\alpha_1=\alpha} \binom{\alpha}{\alpha_1}\partial_x^{\alpha_1}\left(\frac{1}{\langle x-q\rangle^N}\right)\partial_x^{\alpha_2}\tilde\chi(x-k).$$

Using the result in question 1d) with $\rho = \frac{N}{2}$ we get,

$$|\partial_x^\alpha\Phi_{k,q}(x)| \leq C_\alpha\langle k-q\rangle^N \sum_{\alpha_1+\alpha_1=\alpha}\frac{1}{\langle x-q\rangle^{N+|\alpha_1|}}|\partial_x^{\alpha_2}\tilde\chi(x-k)|.$$

On the support of $\tilde\chi_k$ we have $|x-k| \leq A$. Assume $|k-q| \geq 2A$. Then $|x-q| = |x-k+k-q| \geq |k-q| - |x-k| \geq \frac{1}{2}|k-q|$. Since $|\alpha_1| \geq 0$ we get,

$$|\partial_x^\alpha\Phi_{k,q}(x)| \leq C_{\alpha,N}\sum_{|\gamma|\leq M}|\partial_x^\gamma\tilde\chi(x-k)|.$$

If now $|k - q| < 2A$ we have,

$$|\partial_x^\alpha \Phi_{k,q}(x)| \leq C_{\alpha,N,A} \sum_{|\gamma| \leq M} |\partial_x^\gamma \widetilde{\chi}(x - k)|.$$

It follows that,

$$\|\partial_x^\alpha \Phi_{k,q}\|_{H^M(\mathbf{R}^d)} \leq C_{M,N,A} \|\widetilde{\chi}\|_{H^M(\mathbf{R}^d)}, \quad \forall k, q \in \mathbf{Z}^d.$$

3. By the definition of $\Phi_{k,q}$, and since $\psi_q \chi_q = \chi_q$ we can write,

$$\widetilde{\chi}_k \chi_q u(x) = \langle k - q \rangle^{-N} \Phi_{k,q}(x) \left\{ \langle x - q \rangle^N \psi_q(x) \right\} \chi_q u(x).$$

Fix $N > d + 1$. Since $u = \sum_{q \in \mathbf{Z}^d} \chi_q u$, using the result in the previous question and the fact that the function $\langle x - q \rangle^N \psi_q$ is $C^\infty-$ bounded independently of q we can write,

$$\|\widetilde{\chi}_k u\|_{H^s(\mathbf{R}^d)} \leq \sum_{q \in \mathbf{Z}^d} \|\widetilde{\chi}_k \chi_q u\|_{H^s(\mathbf{R}^d)} \leq C \sum_{q \in \mathbf{Z}^d} \langle k - q \rangle^{-N} \sup_{q \in \mathbf{Z}^d} \|\chi_q u\|_{H^s(\mathbf{R}^d)}.$$

4. Since the sum $\sum_{q \in \mathbf{Z}^d} \langle k - q \rangle^{-N}$ is finite, we have,

$$\sup_{k \in \mathbf{Z}^d} \|\widetilde{\chi}_k u\|_{H^s(\mathbf{R}^d)} \leq \sup_{q \in \mathbf{Z}^d} \|\chi_q u\|_{H^s(\mathbf{R}^d)}$$

for all $\widetilde{\chi} \in C_0^\infty(\mathbf{R}^d)$. Exchanging the roles of $\widetilde{\chi}$ and χ we see that the fact that $\sup_{q \in \mathbf{Z}^d} \|\chi_q u\|_{H^s(\mathbf{R}^d)}$ is finite is independent of the function χ.

Solution 8

1. Let $u \in W^{m,\infty}(\mathbf{R}^d)$. We have $\|\chi_q u\|_{H^m(\mathbf{R}^d)} = \left(\sum_{|\alpha| \leq m} \|\partial^\alpha (\chi_q u)\|_{L^2(\mathbf{R}^d)}^2 \right)^{\frac{1}{2}}$. Now, by the Leibniz formula, we have for $|\alpha| \leq m$,

$$\|\partial^\alpha (\chi_q u)\|_{L^2} \leq \sum_{\alpha_1 + \alpha_2 = \alpha} \binom{\alpha}{\alpha_1} \|(\partial^{\alpha_1} \chi_q)(\partial^{\alpha_2} u)\|_{L^2}$$

$$\leq C_\alpha \sum_{\alpha_1 + \alpha_2 = \alpha} \|\partial^{\alpha_1} \chi_q\|_{L^2} \|\partial^{\alpha_2} u\|_{L^\infty}$$

$$\leq C \|u\|_{W^{m,\infty}} \sup_{|\alpha| \leq m} \|\partial^\alpha \chi\|_{L^2},$$

since $(\partial^\alpha \chi_q)(x) = (\partial^\alpha \chi)(x - q)$. So

$$\sup_{q \in \mathbf{Z}^d} \|\chi_q u\|_{H^m(\mathbf{R}^d)} \leq C_m \|u\|_{W^{m,\infty}(\mathbf{R}^d)}.$$

2. If $\varepsilon \in (0, 1)$ then we have by question 1,

$$C_*^{m+\varepsilon}(\mathbf{R}^d) = W^{m+\varepsilon,\infty}(\mathbf{R}^d) \subset W^{m,\infty}(\mathbf{R}^d) \subset H_{ul}^m(\mathbf{R}^d).$$

If $\varepsilon \geq 1$ we have $C_*^{m+\varepsilon}(\mathbf{R}^d) \subset C_*^{m+\varepsilon'}(\mathbf{R}^d)$ where $\varepsilon' < 1$.

3. a) Since $s = m+\sigma$, if $u \in W^{s+\varepsilon,\infty}(\mathbf{R}^d)$ we have, $\partial^\alpha u \in W^{\sigma+\varepsilon,\infty}(\mathbf{R}^d)$, $|\alpha| \leq m$ so $\partial^\alpha u \in H_{ul}^\sigma(\mathbf{R}^d)$, $|\alpha| \leq m$ and therefore $u \in H_{ul}^s(\mathbf{R}^d)$.

b) We write, $\|\chi_q u\|_{L^2(\mathbf{R}^d)} \leq \|\chi_q\|_{L^2(\mathbf{R}^d)} \|u\|_{L^\infty(\mathbf{R}^d)} = \|\chi\|_{L^2(\mathbf{R}^d)} \|u\|_{L^\infty(\mathbf{R}^d)}$.

c) Since $|\chi_q(x)u(x) - \chi_q(y)u(y)|^2 \leq 2(|\chi_q(x)u(x)|^2 + |\chi_q(y)u(y)|^2)$ we have,

$$(3) \leq C \int |\chi_q(x)u(x)|^2 \left(\int_{|x-y|>1} \frac{dy}{|x-y|^{d+2\sigma}} \right) dx \leq C' \|u\|_{L^\infty(\mathbf{R}^d)}^2.$$

d) In the set $\{|x - y| < 1\}$ if $|y - q| \geq 2$ we have $|x - q| \geq 1$ so $f_q(x, y) = 0$.

e) This follows from the equality,

$$\chi_q(x)u(x) - \chi_q(y)u(y) = \chi_q(x)\left(u(x) - u(y)\right) + \left(\chi_q(x) - \chi_q(y)\right)u(y).$$

f) Since $|u(x) - u(y)| \leq |x - y|^{\sigma+\varepsilon} \|u\|_{W^{\sigma+\varepsilon}(\mathbf{R}^d)}$ we have,

$$\iint_\Omega g_q(x, y) \, dx \, dy \leq \int_{\mathbf{R}^d} |\chi_q(x)|^2 \left(\int_{|x-y|<1} \frac{dy}{|x-y|^{d-2\varepsilon}} \right) dx \cdot \|u\|_{W^{\sigma+\varepsilon,\infty}(\mathbf{R}^d)}^2,$$

$$\leq C \|u\|_{W^{\sigma+\varepsilon}(\mathbf{R}^d)}^2.$$

Moreover, since we have $|\chi_q(x) - \chi_q(y)| \leq C_0|x - y|$ where $C_0 = \|\chi'\|_{L^\infty(\mathbf{R}^d)}$, we can write,

$$\iint_\Omega h_q(x, y) \, dx \, dy \leq C_0 \int_{|y-q|<2} |u(y)|^2 \left(\int_{|x-y|<1} \frac{dx}{|x-y|^{d+2\sigma-2}} \right) dy\cdot,$$

$$\leq C \|u\|_{L^\infty(\mathbf{R}^d)}^2,$$

since $d + 2\sigma - 2 < d$.

g) It follows from the previous questions that, $\|\chi_q u\|_{H^\sigma(\mathbf{R}^d)}^2 \leq C \|u\|_{W^{\sigma+\varepsilon,\infty}(\mathbf{R}^d)}^2$, where C is independent of q. This proves that $W^{\sigma+\varepsilon,\infty}(\mathbf{R}^d) \subset H_{ul}^\sigma(\mathbf{R}^d)$ and by questions 2 and 3a) that $C_*^{s+\varepsilon}(\mathbf{R}^d) \subset H_{ul}^s(\mathbf{R}^d)$.

Solution 9

Part 1

1. We have,

$$\|u_\varepsilon - u\|^2_{L^2(\mathbf{R}^d)} = \int_{\mathbf{R}^d} |1 - \chi\left(\frac{x}{\varepsilon}\right)|^2 |u(x)|^2 \, dx.$$

The right-hand side tends to zero when $\varepsilon \to 0$, by the Lebesgue dominated convergence. Indeed for fixed x, we have, $\frac{|x|}{\varepsilon} \to +\infty$ so $|1 - \chi\left(\frac{x}{\varepsilon}\right)|^2 \to 0$ and $|1 - \chi\left(\frac{x}{\varepsilon}\right)|^2 |u(x)|^2 \leq |u(x)|^2 \in L^1(\mathbf{R}^d)$. Then we have,

$$\left\|\partial_{x_j}(u_\varepsilon - u)\right\|_{L^2(\mathbf{R}^d)} \leq \frac{1}{\varepsilon} \left\|(\partial_{x_j}\chi)\left(\frac{x}{\varepsilon}\right) u\right\|_{L^2(\mathbf{R}^d)} + \left\|\left(1 - \chi\left(\frac{x}{\varepsilon}\right)\right)\partial_{x_j}u\right\|_{L^2(\mathbf{R}^d)}.$$

The second term in the right-hand side tends to zero with ε by the same argument as that used above (the Lebesgue dominated convergence). For the first term we write,

$$\frac{1}{\varepsilon}\left\|(\partial_{x_j}\chi)\left(\frac{x}{\varepsilon}\right) u\right\|_{L^2(\mathbf{R}^d)} \leq \frac{1}{\varepsilon}\left\|(\partial_{x_j}\chi)\left(\frac{x}{\varepsilon}\right)\right\|_{L^2(\mathbf{R}^d)} \|u\|_{L^\infty(\mathbf{R}^d)},$$

$$\leq \frac{1}{\varepsilon}\varepsilon^{\frac{d}{2}}\|\partial_{x_j}\chi\|_{L^2(\mathbf{R}^d)}\|u\|_{L^\infty(\mathbf{R}^d)},$$

if we set $x = \varepsilon y$ in the integral involving $\partial_{x_j}\chi$. Since $\frac{d}{2} - 1 > 0$ this term converges also to zero with ε.

Notice that this argument does not work when $d = 2$ since we would have $\frac{d}{2} - 1 = 0$.

2. Let $v \in H^1(\mathbf{R}^d)$ and $\delta > 0$. Since $C_0^\infty(\mathbf{R}^d)$ is dense in $H^1(\mathbf{R}^d)$ one can find $u \in C_0^\infty(\mathbf{R}^d)$ such that $\|u - v\|_{H^1(\mathbf{R}^d)} \leq \frac{\delta}{2}$. By the question 1 one can find u_ε such that $\|u_\varepsilon - u\|_{H^1(\mathbf{R}^d)} \leq \frac{\delta}{2}$. Therefore, $\|u_\varepsilon - v\|_{H^1(\mathbf{R}^d)} \leq \delta$ and $u_\varepsilon = \chi_\varepsilon u \in C_0^\infty(\mathbf{R}^d \setminus \{0\})$.

Part 2

1. By continuity, there exists a neighborhood V_{ω_0} of ω_0 such that $|P_k(\omega)| \geq \frac{c_0}{2}$ for $\omega \in V_{\omega_0}$. Our claim is trivial if $k = 0$. Assume $k \geq 1$. For $0 \leq j \leq k - 1$ there exists $C_j > 0$ such that $|P_j(r\omega)| = r^j|P_j(\omega)| \leq C_j r^j, \forall \omega \in V_{\omega_0}$. Then, for $\omega \in V_{\omega_0}$ there exists $r_0 > 0$ such that,

$$|P(r\omega)| \geq |P_k(r\omega)| - \sum_{j=0}^{k-1} |P_j(r\omega)| \geq \frac{c_0}{2}r^k - \sum_{j=0}^{k-1} C_j r^j \geq \frac{c_0}{4}r^k, \quad r \geq r_0.$$

2. If supp $U \subset \{0\}$ we have $U = \sum_{|\alpha|\le k} c_\alpha \delta_0^{(\alpha)}$. Then $\widehat{U}(\xi) = \sum_{|\alpha|\le k} c_\alpha \xi^\alpha =:$
$P(\xi)$ where P is exactly of order $k \in \mathbf{N}$. By hypothesis we should have,

$$(1 + |\xi|^2)^{-\frac{1}{2}} \widehat{U}(\xi) = (1 + |\xi|^2)^{-\frac{1}{2}} P(\xi) \in L^2(\mathbf{R}^d).$$

But this is impossible if P does not vanish identically. Indeed, by the previous question we have for $R \gg r_0 > 0$,

$$\int_{r_0}^R \int_{V_{\omega_0}} (1 + r^2)^{-1} |P(r\omega)|^2 r^{d-1}\, dr\, d\omega \ge \frac{c_0^2}{16} |V_{\omega_0}| \int_{r_0}^R (1 + r^2)^{-1} r^{d-1+2k}\, dr$$

and the right-hand side tends to $+\infty$ if $R \to +\infty$, since $d \ge 2$ and $k \ge 0$ imply
$d - 3 + 2k \ge -1$.
3. If $u \in E^\perp$ we have,

$$(u, v)_{H^1} = (u, v)_{L^2} + \sum_{j=1}^d (\partial_j u, \partial_j v)_{L^2} = 0 \quad \forall v \in E.$$

Taking $v \in C_0^\infty(\mathbf{R}^d \setminus \{0\}) \subset E$ we deduce that $U = -\Delta u + u = 0$ in $\mathcal{D}'(\mathbf{R}^d \setminus \{0\})$.
4. Therefore, we have supp $U \subset \{0\}$ and $U \in H^{-1}(\mathbf{R}^d)$ since $u \in H^1(\mathbf{R}^d)$. We
deduce from Part 1 that $U = 0$ in $\mathbf{R}^d$ and by Fourier transform that $u = 0$ since,
$\widehat{U} = 0 = (1 + |\xi|^2)\widehat{u}$.
5. It follows that we have $E^\perp = \{0\}$, so $H^1(\mathbf{R}^d) = E^{\perp\perp} = \overline{E} = E$ since, by
definition, E is closed in $H^1(\mathbf{R}^d)$.
6. Assume this is true. Let $u \in C_0^\infty(\mathbf{R}^d) \subset H^1(\mathbf{R}^d)$ be such that $u(0) = 1$ and
let $(\varphi_j) \subset C_0^\infty(\mathbf{R}^d \setminus \{0\})$ which converges to u in $H^1(\mathbf{R}^d)$. Then (φ_j) would
converge to u in $C^0(\overline{B}(0; \varepsilon))$, therefore at each point. Since for all $j \in \mathbf{N}$ we
have $\varphi_j(0) = 0$ we would necessarily have $u(0) = 0$ which is false.

Part 3
1. There exists a continuous map $P_1 : H^1(\Omega) \to H^1(\mathbf{R}^d)$ such that $P_1 u = u$ on Ω.
Moreover, the restriction map $R : v \to Rv = v|_\Omega$ is continuous from $H^1(\mathbf{R}^d)$ to
$H^1(\Omega)$.

Let $u \in H^1(\Omega)$. Then $P_1 u \in H^1(\mathbf{R}^d)$. By Part 2 there exists a sequence
$(\psi_k) \subset C_0^\infty(\mathbf{R}^d \setminus \{0\})$ such that $(\psi_k) \to P_1 u$ in $H^1(\mathbf{R}^d)$. Set $\varphi_k = R\psi_k$. Then
by definition $(\varphi_k) \subset C_0^\infty(\overline{\Omega} \setminus \{0\})$. Now since $R(\psi_k - P_1 u) = \varphi_k - u$ we can
write,

$$\|\varphi_k - u\|_{H^1(\Omega)} = \|R(\psi_k - P_1 u)\|_{H^1(\Omega)} \le C\|\psi_k - P_1 u\|_{H^1(\mathbf{R}^d)} \to 0.$$

Solution 10

Part 1

1. If this is not true there exists $\alpha_0 > 0$ such that for every $C > 0$ one can find $u \in H^1(\Omega)$ such that $|\{x \in \Omega : u(x) = 0\}| \geq \alpha_0$ but $\int_\Omega |u(x)|^2 dx > C \int_\Omega |\nabla u(x)|^2 dx$. Taking $C = k \geq 1$ one can find $u_k \in H^1(\Omega)$ such that $|\{x \in \Omega : u_k(x) = 0\}| \geq \alpha_0$ with,

$$\int_\Omega |u_k(x)|^2 dx = 1, \quad \int_\Omega |\nabla u_k(x)|^2 dx < \frac{1}{k}. \tag{9.11}$$

Then the sequence $(u_k)_k$ is bounded in $H^1(\Omega)$. Therefore, there exists a subsequence $(u_{\sigma(k)})_k$ which converges weakly in $H^1(\Omega)$ to $u_0 \in H^1(\Omega)$. By the compacity of the injection $H^1(\Omega) \to L^2(\Omega)$ this sequence converges strongly to u_0 in $L^2(\Omega)$. Therefore, $\int_\Omega |u_0(x)|^2 dx = 1$. Moreover, $(\nabla u_{\sigma(k)})_k$ converges to ∇u_0 in $\mathcal{D}'(\Omega)$ and to zero, by (9.11). Therefore, $\nabla u_0 = 0$ so that u_0 is a nonzero constant. Now we can write,

$$\int_\Omega |u_{\sigma(k)}(x) - u_0|^2 dx \geq \int_{\{x \in \Omega : u_{\sigma(k)}(x)=0\}} |u_{\sigma(k)}(x) - u_0|^2 dx$$

$$\geq |u_0|^2 |\{x \in \Omega : u_{\sigma(k)}(x) = 0\}| \geq \alpha_0 |u_0|^2.$$

Now the right-hand side is strictly positive while the left-hand side tends to zero with k. This is a contradiction.

Part 2

1. The function $u = 1$ belongs to $H^1(\Omega)$, and it satisfies $(\star\star)$ but it does not satisfy the inequality (4.6).
2. Set $I = (a, b)$. We know that $H^1(I) \subset C^0(\overline{I})$ and that $C^\infty(\overline{I})$ is dense in $H^1(I)$. Let $u \in H^1(I)$ satisfying $(\star\star\star)$ and let x_0 be such that $u(x_0) = 0$. Let $(u_k)_k \subset C^\infty(\overline{I})$ be such that $(u_k) \to u$ in $H^1(I)$. Then $(u_k) \to u$ in $C^0(\overline{I})$. It follows that $u_k(x_0) \to u(x_0) = 0$. Now we can write, for $x \in I$,

$$u_k(x) = u_k(x_0) + \int_{x_0}^x u_k'(t)\, dt.$$

Then using the Hölder inequality we obtain,

$$|u_k(x)|^2 \leq 2\left(|u_k(x_0)|^2 + (b-a)\int_a^b |u_k'(t)|^2 dt\right),$$

which implies that,

$$\int_a^b |u_k(x)|^2\, dx \le 2(b-a)\left(|u_k(x_0)|^2 + (b-a)\int_a^b |u_k'(t)|^2\, dt\right).$$

To conclude we have just to notice that $u_k \to u$ and $u_k' \to u'$ in $L^2(I)$.

3. By Part 3 of the Problem 9 we know that $C_0^\infty(\overline{\Omega} \setminus \{0\})$ is dense in $H^1(\Omega)$. Let $u \in H^1(\Omega)$ and $(\varphi_k) \subset C_0^\infty(\overline{\Omega} \setminus \{0\})$ converging to u in $H^1(\Omega)$. Since φ_k vanishes in a neighborhood of 0 it satisfies the condition $(\star\star\star\star)$. If (4.6) was true with an absolute constant C (independent of the function) we would have for every k,

$$\int_\Omega |\varphi_k(x)|^2\, dx \le C \int_\Omega |\nabla\varphi_k((x)|^2\, dx$$

and by density we would have the same inequality for any $u \in H^1(\Omega)$. But this inequality is trivially false for instance if $u = 1 \in H^1(\Omega)$.

Solution 11

Part 1

1. a) If $m = 1$ we have by hypothesis, $|G(y)| \le M|y|$ and $|\partial_{y_k} G(y)| \le M$, for all $y \in \mathbf{R}^d$. It follows that,

$$\|G(u)\|_{H^1} \le \|G(u)\|_{L^2} + \sum_{k=1}^d \|(\nabla_y G)(u) \cdot \partial_{x_k} u\|_{L^2},$$

$$\le M\|u\|_{L^2} + \sum_{k=1}^d \|\nabla_y G\|_{L^\infty}\|\partial_{x_k} u\|_{L^2} \le C\|u\|_{H^1}.$$

b) Let $m = 2$. By hypothesis we have, $|G(y)| \le M|y|^2$, $|(\partial_{y_k} G)(y)| \le M|y|$ and $|(\partial_{y_j}\partial_{y_k} G)(y)| \le M$ for all $y \in \mathbf{R}^d$. On the other hand, the Sobolev embedding gives $H^2(\mathbf{R}^d) \subset L^\infty(\mathbf{R}^d)$. First of all we have, $\|G(u)\|_{L^2} \le M\|u^2\|_{L^2} \le M\|u\|_{L^\infty}\|u\|_{L^2} \le K\|u\|_{H^2}^2$.
   Then we have,

$$\left\|\partial_{x_k}(G(u))\right\|_{L^2} = \left\|(\nabla_y G)(u) \cdot \partial_{x_k} u\right\|_{L^2} \le \left\|(\nabla_y G)(u)\right\|_{L^\infty}\|\partial_{x_k} u\|_{L^2},$$

$$\le M\|u\|_{L^\infty}\|\nabla u\|_{L^2} \le K\|u\|_{H^2}^2.$$

Eventually denoting by ∂ any partial derivative in x we have,

$$\partial^2(G(u)) = (\nabla_y^2 G)(u)(\partial u)^2 + (\nabla_y G)(u)\partial^2 u.$$

It follows that,

$$\|\partial^2(G(u))\|_{L^2} \leq M\|(\partial u)^2\|_{L^2} + M\|u\|_{L^\infty}\|\partial^2 u\|_{L^2}.$$

The second term in the right-hand side is bounded by $K\|u\|_{H^2}^2$. Let us consider the first one. If $d = 1$ then $H^1 \subset L^\infty$. Therefore, $\|(\partial u)^2\|_{L^2} \leq \|\partial u\|_{L^\infty}\|\partial u\|_{L^2} \leq K\|u\|_{H^2}^2$. If $d = 2$ then $H^1 \subset L^p$ for any $p \geq 2$. Then $\|(\partial u)^2\|_{L^2} = \|\partial u\|_{L^4}^2 \leq K\|\partial u\|_{H^1}^2 \leq K\|u\|_{H^2}^2$. Eventually if $d = 3$ then $H^1 \subset L^p$ for $2 \leq p \leq 6$ and the same estimate holds.

c) By question b) the inequality is true for $m = 2$. Assume it is true for $m \geq 2$ and let us prove it for $m + 1$. Since $m > 3/2 \geq \frac{d}{2}$, $H^m(\mathbf{R}^d)$ is an algebra and $H^m(\mathbf{R}^d) \subset L^\infty(\mathbf{R}^d)$. We have, $\|G(u)\|_{H^{m+1}} \leq C(\|G(u)\|_{L^2} + \|\nabla_x(G(u))\|_{H^m})$. First of all we have,

$$\|G(u)\|_{L^2} \leq M\|u^{m+1}\|_{L^2} \leq K\|u\|_{L^\infty}^m \|u\|_{L^2} \leq \|u\|_{H^{m+1}}^{m+1}.$$

Then,

$$\|\nabla_x(G(u))\|_{H^m} = \left\|(\nabla_y G)(u)\nabla_x u\right\|_{H^m} \leq K\left\|(\nabla_y G)(u)\right\|_{H^m} \|\nabla_x u\|_{H^m}.$$

By assumption $G \in C^\infty(\mathbf{R}^p)$ and $|\partial^\alpha \nabla_y G(y)| \leq M|y|^{m+1-|\alpha|-1} = M|y|^{m-|\alpha|}$ for all $y \in \mathbf{R}^p$ and $|\alpha| \leq m$. Hence the function $\nabla_y G$ satisfies the estimates (4.7) at the order m. Therefore, we can apply the induction assumption and we obtain,

$$\left\|(\nabla_y G)(u)\right\|_{H^m} \leq K\|u\|_{H^m}^m.$$

Therefore, $\|G(u)\|_{H^{m+1}} \leq K\|u\|_{H^{m+1}}^{m+1}$.

Part 2

1. By assumption, $F \in C^\infty(\mathbf{R}^p \setminus \{0\})$. Differentiating the equality (4.8) $|\alpha|$ times, for $|\alpha| \leq m$, we find $(\partial^\alpha F)(\lambda y) = \lambda^{m-|\alpha|}(\partial^\alpha F)(y)$ for $y \in \mathbf{R}^p \setminus \{0\}$. Let $y \in \mathbf{R}^p \setminus \{0\}$. Using this equality with $\lambda = \frac{1}{|y|}$ we obtain,

$$|(\partial^\alpha F)(y)| = |y|^{m-|\alpha|}\left|(\partial^\alpha F)\left(\frac{y}{|y|}\right)\right| \leq \|\partial^\alpha F\|_{L^\infty(\mathbf{S}^{p-1})}|y|^{m-|\alpha|}.$$

2. Using the Leibniz formula we can write,

$$\partial^\alpha F_\ell(y) = (1 - \chi(\ell y))(\partial^\alpha F)(y) - \sum_{\substack{\beta \leq \alpha \\ \beta \neq 0}} \binom{\alpha}{\beta} \ell^{|\beta|} (\partial^\beta \chi)(\ell y)(\partial^{\alpha - \beta} F)(y).$$

Using question 1 we obtain for all $y \in \mathbf{R}^p$,

$$|\partial^\alpha F_\ell(y) \leq M|y|^{m-|\alpha|} + \sum_{\substack{\beta \leq \alpha \\ \beta \neq 0}} \binom{\alpha}{\beta} \ell^{|\beta|} |(\partial^\beta \chi)(\ell y)| M |y|^{m-|\alpha|+|\beta|},$$

$$\leq M|y|^{m-|\alpha|} + \sum_{\substack{\beta \leq \alpha \\ \beta \neq 0}} \binom{\alpha}{\beta} \sup_{z \in \mathbf{R}^d} (|z|^{|\beta|} |(\partial^\beta \chi)(z)|) M |y|^{m-|\alpha|}$$

$$\leq C(\alpha, \chi) M |y|^{m-|\alpha|}.$$

3. This follows immediately from question 1a) Part 1 since F_ℓ satisfies the uniform estimates (4.7).
4. a) It follows from the definition (4.8) that $F(0) = 0$. Therefore, if $u(x) = 0$ we have $F(u(x)) = 0$ and if $u(x) \neq 0$ we have by question 1, $|F(u(x))| \leq M|u(x)|^m$.

   If $m = 1, d \geq 1$ and $u \in H^1(\mathbf{R}^d)$ we have $|F(u(x))\varphi(x)| \leq M|u(x)\varphi(x)| \in L^1(\mathbf{R}^d)$ by the Cauchy–Schwarz inequality.

   If $m \geq 2, 1 \leq d \leq 3$ and $u \in H^m(\mathbf{R}^d) \subset L^\infty(\mathbf{R}^d)$ we have,

$$|F(u(x))\varphi(x)| \leq M|u(x)|^m |\varphi(x)| \leq CM\|u\|_{L^\infty}^{m-2} |u(x)|^2 |\varphi(x)| \in L^1(\mathbf{R}^d).$$

   b) By the Lebesgue dominated convergence theorem we have,

$$\int_{\mathbf{R}^d} F(u(x))\varphi(x)\, dx = \lim_{\ell \to +\infty} \int_{\mathbf{R}^d} F_\ell(u(x))\varphi(x)\, dx.$$

On the other hand, by question 3 we have,

$$\left| \int_{\mathbf{R}^d} F_\ell(u(x))\varphi(x)\, dx \right| \leq \|F_\ell(u)\|_{H^m} \|\varphi\|_{H^{-m}} \leq K \|u\|_{H^m}^m \|\varphi\|_{H^{-m}}.$$

It follows that,

$$\left| \int_{\mathbf{R}^d} F(u(x))\varphi(x)\, dx \right| \leq K \|u\|_{H^m}^m \|\varphi\|_{H^{-m}},$$

which implies $F(u) \in H^m(\mathbf{R}^d)$ together with $\|F(u)\|_{H^m} \leq K \|u\|_{H^m}^m$.

Solution 12

We have trivially $\|Hu\|_{L^2(\mathbf{R})} \leq \|u\|_{L^2(\mathbf{R})}$, so the map $u \mapsto Hu$ is continuous from $L^2(\mathbf{R})$ to itself. Now if $u \in H^1(\mathbf{R})$ then u is continuous and we have, $\partial_x(Hu) = u\delta_0 + H\partial_x u = u(0)\delta_0 + H\partial_x u = H\partial_x u$ so,

$$\|\partial_x(Hu)\|_{L^2(\mathbf{R})} = \|H\partial_x u\|_{L^2(\mathbf{R})} \leq \|\partial_x u\|_{L^2(\mathbf{R})} \leq \|u\|_{H^1(\mathbf{R})}.$$

This shows that the map $u \mapsto Hu$ is continuous from $H^1(\mathbf{R})$ to itself. By interpolation, for $0 < s < 1$, the map $u \mapsto Hu$ is continuous from $H^s(\mathbf{R})$ to itself and there exists $C > 0$ such that $\|Hu\|_{H^s(\mathbf{R})} \leq C\|u\|_{H^s(\mathbf{R})}$.

Solution 13

1. Set $J = \int_0^{+\infty} \frac{|u(r\omega)|^2}{r^2} r^{d-1}\, dr$. Integrating by parts we can write,

$$J = \frac{1}{d-2}\int_0^{+\infty} |u(r\omega)|^2 \frac{d}{dr} r^{d-2}\, dr = -\frac{1}{n-2}\int_0^{+\infty} \frac{d}{dr}\left(|u(r\omega)|^2\right) r^{d-2}\, dr,$$

since the function $|u(r\omega)|^2 r^{d-2}$ vanishes at $r = 0$ and $r = +\infty$. Now,

$$\frac{d}{dr}\left(|u(r\omega)|^2\right) = 2\mathrm{Re}\left\{u(r\omega)\frac{d}{dr}u(r\omega)\right\} = 2\mathrm{Re}\left\{u(r\omega)\sum_{i=1}^{d}\omega_i(\partial_{x_i}u)(r\omega)\right\},$$

which proves the equality (4.10).
2. Integrating both members of the equality (4.10) on $\mathbf{S}^{d-1}$ we can write,

$$\iint_{I\times S^{d-1}} \frac{|u(r\omega)|^2}{r^2} r^{d-1}\, dr\, d\omega$$

$$= -\frac{2}{d-2}\mathrm{Re}\iint_{I\times S^{d-1}} \frac{u(r\omega)}{r} r^{\frac{d-1}{2}}(\omega\cdot\nabla_x u)(r\omega)r^{\frac{d-1}{2}}\, dr\, d\omega,$$

where $I = (0, +\infty)$. By the Cauchy–Schwarz inequality we obtain,

$$\iint_{I\times S^{d-1}} \frac{|u(r\omega)|^2}{r^2} r^{d-1}\, dr\, d\omega \leq \frac{2}{d-2}\left(\iint_{I\times S^{d-1}} \frac{|u(r\omega)|^2}{r^2} r^{d-1}\, dr\, d\omega\right)^{\frac{1}{2}}$$

$$\cdot\left(\iint_{I\times S^{d-1}} |\omega\cdot\nabla_x u(r\omega)|^2 r^{d-1}\, dr\, d\omega\right)^{\frac{1}{2}}.$$

Using the fact that $|\omega \cdot \nabla_x u| \leq |\nabla_x u|$ (since $|\omega| = 1$), then setting $x = r\omega$, which implies that $dx = r^{d-1} dr\, d\omega$, we obtain,

$$\int_{\mathbf{R}^d} \frac{|u(x)|^2}{|x|^2}\, dx \leq \frac{2}{d-2} \left(\int_{\mathbf{R}^d} \frac{|u(x)|^2}{|x|^2}\, dx \right)^{\frac{1}{2}} \left(\int_{\mathbf{R}^d} |\nabla_x u(x)|^2\, dx \right)^{\frac{1}{2}}.$$

We may of course assume that u does not vanish identically, which allows us to divide both members by $\left(\int_{\mathbf{R}^d} \frac{|u(x)|^2}{|x|^2}\, dx \right)^{\frac{1}{2}}$. The desired result follows.

Solution 14

1. Take $A = \left\{ x \in \mathbf{R}^d : |f(x)| > \lambda \right\}$ which is a measurable set of finite measure. We have for all $0 < \lambda < +\infty$,

$$\|f\|_{L_w^q(\mathbf{R}^d)} \geq \left(\mu \left\{ x \in \mathbf{R}^d : |f(x)| > \lambda \right\} \right)^{\frac{1}{q}-1} \int_{\{x \in \mathbf{R}^d : |f(x)| > \lambda\}} |f(x)|\, dx,$$

$$\geq \lambda \left(\mu \left\{ x \in \mathbf{R}^d : |f(x)| > \lambda \right\} \right)^{\frac{1}{q}},$$

so $S(f) \leq \|f\|_{L_w^q(\mathbf{R}^d)}$.

2. a) Set $g(x) = 1_A(x) |f(x)|$. Using the Fubini theorem we can write,

$$\int_A |f(x)|\, dx = \int_{\mathbf{R}^3} g(x)\, dx = \int_{\mathbf{R}^3} \left(\int_0^{g(x)} d\lambda \right) dx,$$

$$= \int_0^{+\infty} \left(\int_{\{x : g(x) > \lambda\}} dx \right) d\lambda = \int_0^{+\infty} \mu \left\{ x \in \mathbf{R}^d : |g(x)| > \lambda \right\} d\lambda,$$

$$= \int_0^{+\infty} \mu \{ x \in A : |f(x)| > \lambda \}\, d\lambda,$$

$$= \int_0^{+\infty} \mu \left(\left\{ x \in \mathbf{R}^d : |f(x)| > \lambda \right\} \cap A \right) d\lambda,$$

$$\leq \int_0^{+\infty} \min \left(\mu \left\{ x \in \mathbf{R}^d : |f(x)| > \lambda \right\}, \mu(A) \right) d\lambda.$$

b) By hypothesis we have, $\mu \left\{ x \in \mathbf{R}^d : |f(x)| > \lambda \right\} \leq \lambda^{-q} S(f)^q$. Therefore,

$$\int_A |f(x)|\, dx \leq \int_0^{+\infty} \min \left(\lambda^{-q} S(f)^q, \mu(A) \right) d\lambda.$$

Set $\lambda_0 = S(f)\mu(A)^{-\frac{1}{q}}$ and $h(\lambda) = \min\left(\lambda^{-q}S(f)^q, \mu(A)\right)$. We have,

$$\int_A |f(x)|\, dx \le \int_0^{\lambda_0} h(\lambda)\, d\lambda + \int_{\lambda_0}^{+\infty} h(\lambda)\, d\lambda = I_1 + I_2.$$

In I_1 we have $\mu(A) \le \lambda^{-q}S(f)^q$ so that,

$$I_1 = \int_0^{\lambda_0} \mu(A)\, d\lambda = \mu(A)^{1-\frac{1}{q}}S(f).$$

In I_2 we have $\mu(A) \ge \lambda^{-q}S(f)^q$ so that, since $q > 1$,

$$I_2 = S(f)^q \int_{\lambda_0}^{+\infty} \lambda^{-q}\, d\lambda = S(f)^q \frac{\lambda_0^{1-q}}{q-1} = \frac{1}{q-1}\mu(A)^{1-\frac{1}{q}}S(f).$$

It follows that,

$$\mu(A)^{\frac{1}{q}-1}\int_A |f(x)|\, dx \le \frac{q}{q-1}S(f).$$

c) The fact that $\|f\|_{L_w^q(\mathbf{R}^d)}$ is a norm on $L_w^q(\mathbf{R}^d)$ is easy.

Solution 15

1. If $f \in L^\infty(\mathbf{R}^d)$ we have $|f_B| \le \frac{1}{|B|}\int_B |f(x)|dx \le \frac{1}{|B|}|B|\|f\|_{L^\infty} \le \|f\|_{L^\infty}$ and

$$|f(x) - f_B| \le |f(x)| + |f_B| \le 2\|f\|_{L^\infty}.$$

It follows that,

$$\frac{1}{|B|}\int_B |f(x) - f_B|dx \le 2\|f\|_{L^\infty}\frac{1}{|B|}|B| \le 2\|f\|_{L^\infty}.$$

2. We have,

$$(f_\delta)_{B(x_0,r)} = \frac{1}{|B(x_0,r)|}\int_{|x-x_0|<r} f(\delta x)dx = \frac{\delta^{-d}}{|B(x_0,r)|}\int_{|y-\delta x_0|<\delta r} f(y)dy.$$

On the other hand, in $\mathbf{R}^d$ we have, $|B(\delta x_0, \delta r)| = \delta^d|B(\delta x_0, r)| = \delta^d|B(x_0, r)|$
so that,

$$(f_\delta)_{B(x_0,r)} = \frac{1}{|B(\delta x_0, \delta r)|} \int_{B(\delta x_0, \delta r)} f(y)dy = f_{B(\delta x_0, \delta r)}.$$

Setting $y = \delta x$ we have $dy = \delta^d\, dx$ so,

$$\frac{1}{|B(x_0, r)|} \int_{B(x_0,r)} |f(\delta x) - (f_\delta)_{B(x_0,r)}|dx$$

$$= \frac{1}{|B(\delta x_0, \delta r)|} \int_{B(\delta x_0, \delta r)} |f(y) - f_{B(\delta x_0, \delta r)}|dx \le A.$$

3. a) We have, $c_B - f_B = c_B - \frac{1}{|B|}\int_B f(x)dx = \frac{1}{|B|}\int_B(c_B - f(x))dx$ so that,

$$|c_B - f_B| \le \frac{1}{|B|} \int_B |c_B - f(x)|dx \le A'.$$

b) If $f \in \tilde{E}$ we can write,

$$\frac{1}{|B|} \int_B |f(x) - f_B|\, dx \le \frac{1}{|B|} \int_B |f(x) - c_B|\, dx + \frac{1}{|B|} \int_B |c_B - f_B|\, dx \le 2A'.$$

4. a) Since $f \in L^1_{loc}(\mathbf{R}^d)$ we have,

$$\int_{|x|<1} \frac{|f(x)|}{(1+|x|)^{d+\varepsilon}} dx \le \int_{|x|<1} |f(x)|\, dx < +\infty.$$

Now if $|x| > 1$ we can write,

$$\frac{|f(x)|}{(1+|x|)^{d+\varepsilon}} \le \frac{|f(x) - f_{B_0}|}{|x|^{d+\varepsilon}} + \frac{|f_{B_0}|}{|x|^{d+\varepsilon}}.$$

Since $\int_{|x|>1} \frac{dx}{|x|^{d+\varepsilon}} dx < +\infty$ it is enough to prove (1).
b) We have just to use the definition of BMO and to notice that $|B_k| = c_d 2^{dk}$.
c) We write,

$$|f_{B_{k+1}} - f_{B_k}| \le |f_{B_{k+1}} - f(x)| + |f(x) - f_{B_k}|$$

so,

$$\int_{B_k} |f_{B_{k+1}} - f_{B_k}| \, dx \le \int_{B_k} |f_{B_{k+1}} - f(x)| \, dx + \int_{B_k} |f(x) - f_{B_k}| \, dx$$

$$\le \int_{B_{k+1}} |f_{B_{k+1}} - f(x)| \, dx + \int_{B_k} |f(x) - f_{B_k}| \, dx.$$

Since $|B_k| = c_d 2^{dk}$ we deduce from question b),

$$c_d 2^{dk} |f_{B_{k+1}} - f_{B_k}| \le C_1 (2^{d(k+1)} A + 2^{dk} A).$$

Dividing both members of this inequality by $c_d 2^{dk}$ we obtain the inequality (3) in the statement. The estimate (4) follows by iteration.

d) It is an immediate consequence of (4) since $|B_k| = c_d 2^{dk}$.

e) Since we have a disjoint union we can write,

$$I := \int_{|x|>1} \frac{|f(x) - f_{B_0}|}{|x|^{d+\varepsilon}} \, dx = \sum_{k=1}^{+\infty} \int_{2^{k-1} \le |x| < 2^k} \frac{|f(x) - f_{B_0}|}{|x|^{d+\varepsilon}} \, dx.$$

Using in the integral the lower bound $|x| \ge 2^{(k-1)}$ we obtain,

$$I \le \sum_{k=1}^{+\infty} 2^{-(k-1)(d+\varepsilon)} \int_{B_k} |f(x) - f_{B_0}| \, dx$$

which, using question d), implies that,

$$I \le \sum_{k=1}^{+\infty} 2^{-(k-1)(d+\varepsilon)} C_4 k A 2^{kd} \le C_4 A 2^{d+\varepsilon} \sum_{k=1}^{+\infty} k 2^{-k\varepsilon} \le C_5 A < +\infty.$$

5. a) Let $x_0 \in \mathbf{R}^d$. If $|x_0 - a| < 1$ there exists $\varepsilon_0 > 0$ such that $|x_0 - a| \le 1 - \varepsilon_0$ and there exists k_0 such that $|a_k - a| \le \frac{\varepsilon_0}{2}$ for $k \ge k_0$. Then, $|x_0 - a_k| \le 1 - \frac{\varepsilon_0}{2}$ so $1_{B(a_k,1)}(x_0) = 1$ for all $k \ge k_0$ and $1_{B(a,1)}(x_0) = 1$. If $|x_0 - a| > 1$ there exist $\varepsilon_0 > 0$ such that $|x_0 - a| > 1 + \varepsilon_0$ and k_0 such that $|a_k - a| \le \frac{\varepsilon_0}{2}$ for $k \ge k_0$. Then, $|x_0 - a_k| \ge |x_0 - a| - |a_k - a| \ge 1 + \frac{\varepsilon_0}{2}$. Therefore, $1_{B(a_k,1)}(x_0) = 0$ for all $k \ge k_0$ and $1_{B(a1)}(x_0) = 0$. Eventually the set $\{x \in \mathbf{R}^d : |x - a| = 1\}$ has Lebesgue measure zero.

b) Let $a \in \mathbf{R}^d$ and let (a_k) be a sequence which converges to a. Set $f_k(x) = 1_{B(a_k,1)} f(x)$. From question a) the sequence (f_k) converges to the function $1_{B(a,1)} f(x)$ almost everywhere. On the other hand, if k is large enough we have, $B(a_k, 1) \subset B(a, 2)$ so that $|f_k(x)| \le 1_{B(a,2)}(x)|f(x)| \in L^1(\mathbf{R}^d)$ since $f \in L^1_{loc}(\mathbf{R}^d)$. We conclude by the Lebesgue dominated convergence.

c) There exists $\varepsilon_0 > 0$ such that for $|x| \le \varepsilon_0$ we have $|x|^{\frac{1}{2}}|\ln|x|| \le 1$. It follows that,

$$\int_{|x|\le\varepsilon_0} |\ln|x||\,dx \le \int_{|x|\le\varepsilon_0} \frac{dx}{|x|^{\frac{1}{2}}} < +\infty.$$

On the other hand, the function $\ln|x|$ is continuous so locally integrable on the set $\{x : |x| > \varepsilon_0\}$.

d) Using the notation in question 5b) we have, $\int_{|y-y_0|<1} |\ln|y||\,dy = \Phi(y_0)$. This function being continuous it is bounded on the compact set $\{y : |y| \le 2\}$.

e) Since $\frac{d}{dx}\ln|x| = \frac{1}{x}$, the Taylor formula with integral reminder gives,

$$\ln|y| - \ln|y_0| = \int_0^1 \frac{|y| - |y_0|}{t|y| + (1-t)|y_0|}\,dt.$$

Since $|y_0| > 2$ and $|y - y_0| < 1$ we have, $|y| > |y_0| - |y - y_0| > 1$ so that $t|y| + (1-t)|y_0| > 1$. Moreover, $||y| - |y_0|| \le |y - y_0|$. Summing up we obtain, $|\ln|y| - \ln|y_0|| \le |y - y_0.|$. It follows that

$$\int_{|y-y_0|<1} |\ln|y| - \ln|y_0||\,dy \le \int_{|y-y_0|<1} |y - y_0|\,dy = 1.$$

f) It follows from questions 5d) and e) that there exists $A > 0$ such that for all $y_0 \in \mathbf{R}^d$ there exists a constant c_{y_0} such that $\int_{|y-y_0|<1} |\ln|y| - c_{y_0}|\,dy \le A$. Let $r > 0$. Setting in the integral $y = \frac{x}{r}$ et $y_0 = \frac{x_0}{r}$ we obtain,

$$\frac{1}{r}\int_{|x-x_0|<r} |\ln|x| - \ln r - c_{\frac{x_0}{r}}|\,dx \le A.$$

Set $B = \{x \in \mathbf{R} : |x - x_0| < r\}$ and $c_B = \ln r + c_{\frac{x_0}{r}}$. We have $|B| = 2r$ so,

$$\frac{1}{|B|}\int_B |\ln|x| - c_B|\,dx \le \frac{1}{2}A.$$

Therefore, $f \in \tilde{E} = BMO$.

Solution 16

1. Since m is increasing its derivative is positive and we deduce that the latter is integrable on $\mathbf{R}$ by writing that

$$\int_a^b m'(x)\,dx = m(b) - m(a) \le \lim_{x\to+\infty} m - \lim_{x\to-\infty} m = 1.$$

By letting a go to $-\infty$ in the previous inequality, we find that for any $b \in \mathbf{R}$ we have,

$$m(b) = \int_{-\infty}^{b} m'(x) \, dx \leq \|m'\|_{L^1},$$

which proves that $\|m\|_{L^\infty} \leq \|\partial_x m\|_{L^1}$. This immediately implies the desired inequality for $\|mu\|_{L^2}$.

2. The Fourier transform of $S_{j-2}(m)\Delta_j u$ is supported in $\{A^{-1}2^j \leq |\xi| \leq A2^j\}$ with $A = 1/4$. Therefore, by almost orthogonality we have that,

$$\|T_m u\|_{\dot{H}^s}^2 \lesssim \sum_{j \geq 2} 2^{2js} \|S_{j-2}(m)\Delta_j u\|_{L^2}^2.$$

Now we recall that, $\|S_{j-2}(m)\|_{L^\infty} \lesssim \|m\|_{L^\infty}$ and that $\sum_{j \geq -1} 2^{2js}\|\Delta_j u\|_{L^2}^2 \lesssim \|u\|_{\dot{H}^s}^2$. These estimates imply that $\|T_m u\|_{\dot{H}^s}^2 \lesssim \|m\|_{L^\infty}^2 \|u\|_{\dot{H}^s}^2$.

3. The estimate for u follows directly from the Bernstein inequality (use Problem 3 with $d = 1, r = +\infty, \theta = \frac{1}{2}, R = 2^p$),

$$\|\Delta_p u\|_{L^\infty} \lesssim 2^{p/2}\|\Delta_p u\|_{L^2}.$$

The Bernstein inequality (use Problem 3 with $d = 1, r = 2, p = 1, \theta = \frac{1}{2}, R = 2^j$) also implies that,

$$\|\Delta_j m\|_{L^2} \lesssim 2^{j/2}\|\Delta_j m\|_{L^1}$$

and

$$\|\Delta_j m\|_{L^1} \lesssim 2^{-j}\|\Delta_j m'\|_{L^1}.$$

This implies the wanted result since $\|\Delta_j m'\|_{L^1} \leq \|m'\|_{L^1}$.

4. Since $k \geq 3$, the Fourier transform of $(\Delta_{j-k}u)\Delta_j m$ is included in a dyadic ring of size 2^j. By almost orthogonality, this implies that

$$\left\|\sum_{j \geq k-1}(\Delta_{j-k}u)\Delta_j m\right\|_{\dot{H}^s}^2 \lesssim \sum_{j \geq k-1} 2^{2sj}\|(\Delta_{j-k}u)\Delta_j m\|_{L^2}^2.$$

Then we write,

$$\left\|(\Delta_{j-k}u)\Delta_j m\right\|_{L^2} \leq \|\Delta_{j-k}u\|_{L^\infty}\|\Delta_j m\|_{L^2},$$

and next use the previous question to bound these two factors. We thus find that,

$$
\left\| \sum_{j \geq k-1} (\Delta_{j-k} u) \Delta_j m \right\|_{\dot{H}^s}^2 \lesssim \sum_{j \geq k-1} 2^{2sj} \left(2^{(j-k)/2} \|\Delta_{j-k} u\|_{L^2} \right)^2 \left(2^{-j/2} \|m'\|_{L^1} \right)^2
$$

$$
\lesssim 2^{(2s-1)k} \|m'\|_{L^1}^2 \sum_{j \geq k-1} 2^{(j-k)2s} \|\Delta_{j-k} u\|_{L^2}^2
$$

$$
\lesssim 2^{(2s-1)k} \|m'\|_{L^1}^2 \sum_{p \geq -1} 2^{2ps} \|\Delta_p u\|_{L^2}^2
$$

$$
\lesssim 2^{(2s-1)k} \|m'\|_{L^1}^2 \|u\|_{\dot{H}^s}^2 .
$$

By taking the square root, this shows that,

$$
\left\| \sum_{j \geq k-1} (\Delta_{j-k} u) \Delta_j m \right\|_{\dot{H}^s} \leq 2^{\frac{2s-1}{2}k} \|m'\|_{L^1} \|u\|_{\dot{H}^s} .
$$

5. It follows from the definition of the paraproduct $T_u m$ that

$$
T_u m = \sum_{j=2}^{\infty} S_{j-2}(u) \Delta_j m.
$$

On the other hand, by definition of $S_{j-2}(u)$ and using the convention $\Delta_p u = 0$ for $p \leq -2$, we have

$$
S_{j-2}(u) = \sum_{\ell=-1}^{j-3} \Delta_\ell u = \sum_{k=3}^{+\infty} \Delta_{j-k} u.
$$

This implies that

$$
T_u m = \sum_{j=2}^{\infty} \sum_{k=3}^{+\infty} (\Delta_{j-k} u) \Delta_j m,
$$

which gives the wanted identity

$$
T_u m = \sum_{k=3}^{+\infty} \sum_{j=2}^{\infty} (\Delta_{j-k} u) \Delta_j m.
$$

Now, recall from the previous question that

$$
\left\| \sum_{j \geq k-1} (\Delta_{j-k} u) \Delta_j m \right\|_{\dot{H}^s} \leq 2^{\frac{2s-1}{2}k} \|m'\|_{L^1} \|u\|_{\dot{H}^s} .
$$

Since $2s - 1 < 0$, the series $\sum_k 2^{\frac{2s-1}{2}k}$ is convergent and hence we deduce that

$$\|T_u m\|_{\dot{H}^s} \lesssim \|m'\|_{L^1} \|u\|_{\dot{H}^s}.$$

6. We now have to estimate the reminder term,

$$R(u, m) = \sum_{p \geq -1} \sum_{\substack{q \geq -1 \\ |p-q| \leq 2}} (\Delta_p u)(\Delta_q m).$$

When $|p - q| \leq 2$, the support of the Fourier transform of $(\Delta_p u)(\Delta_q m)$ is included in the ball of center 0 and radius 2^{p+4}. Consequently we have,

$$(\Delta_p u)(\Delta_q m) = \sum_{\ell=0}^{p+6} \Delta_{p+5-\ell} \left((\Delta_p u)(\Delta_q m)\right),$$

and hence

$$R(u, m) = \sum_{\ell=0}^{\infty} \sum_{p \geq \max\{-1, \ell-6\}} \sum_{\substack{q \geq -1 \\ |p-q| \leq 2}} \Delta_{p+5-\ell} \left((\Delta_p u)(\Delta_q m)\right).$$

Now, by proceeding as above, we find that, for each fixed q and ℓ,

$$\left\| \sum_p \sum_q \Delta_{p+5-\ell} \left((\Delta_p u)(\Delta_q m)\right) \right\|_{\dot{H}^s}^2 \lesssim \sum_p \sum_q 2^{2(p-\ell)s} \|(\Delta_p u)(\Delta_q m)\|_{L^2}^2$$

$$\lesssim \sum_j \sum_q 2^{2(p-\ell)s} \|\Delta_p u\|_{L^2}^2 \|\Delta_q m\|_{L^\infty}^2$$

$$\lesssim 2^{-2s\ell} \|m\|_{L^\infty}^2 \sum_{k \geq -1} 2^{2ks} \|\Delta_k u\|_{L^2}^2$$

$$\lesssim 2^{-2s\ell} \|m\|_{L^\infty}^2 \|u\|_{\dot{H}^s}^2,$$

so, by taking the square root,

$$\left\| \sum_{p \geq \max\{-1, \ell-6\}} \Delta_{p+5-\ell} \left((\Delta_p u)(\Delta_q m)\right) \right\|_{\dot{H}^s} \lesssim 2^{-s\ell} \|m\|_{L^\infty} \|u\|_{\dot{H}^s},$$

where the implicit constant does not depend on ℓ. Since $s > 0$, the sum on ℓ converges and we obtain that,

$$\|R(m, u)\|_{\dot{H}^s} \lesssim \|m\|_{L^\infty} \|u\|_{\dot{H}^s},$$

which completes the proof.

7. a) This follows from the fact that the function arctan(x) is nonnegative on
    $[0, +\infty)$, strictly positive on $(0, +\infty)$, bounded by $\frac{\pi}{2}$ on $[0, +\infty)$ and is a
    C^∞ function on $\mathbf{R}$. Moreover, we have,

$$f'(x) = \frac{1}{\sqrt{2\pi}} \frac{1}{(1 + x^2)(\arctan(x))^{\frac{1}{2}}}.$$

This shows that f' is nonincreasing. Now, for x close to zero $f'(x)$ is
equivalent to $\frac{C_1}{\sqrt{x}}$ and for x large $f'(x)$ is bounded by $\frac{C_2}{x^2}$, so f' belongs to
$L^1((0, +\infty))$.

b) Since ρ_ε is in $C_0^\infty(\mathbf{R})$ and $\tilde{f}$ is in $L^\infty(\mathbf{R})$ then m_ε is a C^∞ function on $\mathbf{R}$.
    Moreover, setting $x - y = \varepsilon z$ in the integral we have,

$$m_\varepsilon(x) = \frac{1}{\varepsilon} \int_0^{+\infty} \rho\left(\frac{x - y}{\varepsilon}\right) f(y)\, dy = \int_{-\infty}^{\frac{x}{\varepsilon}} \rho(z) f(x - \varepsilon z)\, dz.$$

Then m_ε is nondecreasing since f is nondecreasing. Now, if $x < 0$ in the last
integral we have $z < 0$ so $\rho(z) = 0$. Therefore, $m_\varepsilon(x) = 0$. If $x \geq 1$ we have
$\frac{x}{\varepsilon} \geq 1$ so $m_\varepsilon(x) = \int_{-\infty}^1 \rho(z) f(x - \varepsilon z)\, dz$. The fact that $\lim_{x \to +\infty} m_\varepsilon(x) =$
1 follows from the Lebesgue dominated convergence since for fixed z we
have, $\lim_{x \to +\infty} f(x - \varepsilon z) = 1$ and $|\rho(z) f(x - \varepsilon z)| \leq \frac{\pi}{2} \rho(z) \in L^1(0, 1)$.
Eventually, by the Young inequality we have,

$$\|m_\varepsilon\|_{L^\infty(\mathbf{R})} \leq \|\rho_\varepsilon\|_{L^1(\mathbf{R})} \|\tilde{f}\|_{L^\infty(\mathbf{R})} \leq \|\rho\|_{L^1(\mathbf{R})} \|f\|_{L^\infty((0,+\infty))} \leq 1.$$

c) Since ρ_ε is compactly supported, it follows that, for any $x \in \mathbf{R}$, we have

$$m_\varepsilon'(x) = \rho\left(\frac{x}{\varepsilon}\right) f(0) + \int_{-\infty}^{\frac{x}{\varepsilon}} \rho(z) f'(x - \varepsilon z)\, dy = \int_{-\infty}^{\frac{x}{\varepsilon}} \rho(z) f'(x - \varepsilon z)\, dy,$$

$$= \int_0^{+\infty} \rho_\varepsilon(x - y) f'(y)\, dy = \rho_\varepsilon \star g,$$

where $g = 0$ for $x < 0$ and $g = f'$ for $x > 0$.

Now $\rho_\varepsilon \in L^p(\mathbf{R})$ for all $p \in [1, +\infty)$ and by question a) $g \in L^1(\mathbf{R})$.
Therefore, by the Young inequality we have $\rho_\varepsilon \star g \in L^p(\mathbf{R})$. In particular,

$$\|m_\varepsilon'\|_{L^1(\mathbf{R})} \leq \|\rho_\varepsilon\|_{L^1(\mathbf{R})} \|g\|_{L^1(\mathbf{R})} \leq \|g\|_{L^1(\mathbf{R})}. \tag{9.12}$$

d) We now want to prove that the functions m_ε' are not bounded in L^2. To do so,
    we are going to prove the pointwise lower bound,

$$\forall x \in [\varepsilon, 1], \quad m_\varepsilon'(x) \geq \frac{C}{\sqrt{x}},$$

for some positive constant C. Consider $x \in [\varepsilon, 1]$. Then $\frac{x}{\varepsilon} \geq 1$. Since supp $\rho \subset [0, 1]$ we have from the formula proved in question c) that,

$$m_\varepsilon'(x) = \int_0^1 \rho(z) f'(x - \varepsilon z) \, dz.$$

By question a) we know that f' is nonincreasing; therefore, $f(x - \varepsilon z) \geq f'(x)$ so,

$$m_\varepsilon'(x) \geq f'(x) \int_0^1 \rho(z) \, dz = f'(x) \geq \frac{C}{\sqrt{x}}$$

for $x \in [\varepsilon, 1]$. This implies that,

$$\int_0^1 m_\varepsilon'(x)^2 \, dx \geq C \, |\log(\varepsilon)| . \tag{9.13}$$

e) We are now in a position to conclude the proof. By contradiction, let us assume that there exists a constant $C > 0$ such that, for all $\varepsilon \in (0, 1]$ and all u in $H^1(\mathbf{R})$,

$$\|m_\varepsilon u\|_{H^1} \leq C \|m_\varepsilon'\|_{L^1} \|u\|_{H^1}. \tag{9.14}$$

We fix a C^∞ function u supported in $[-2, 2]$ and satisfying, $u(x) = 1$ for $x \in [-1, 1]$. Then the right-hand side of (9.14) is bounded by $C\|u\|_{H^1}$.

Since by question b) we have, $\|m_\varepsilon\|_{L^\infty(\mathbf{R})} \leq 1$ we obtain,

$$\|m_\varepsilon u\|_{L^2} \leq \|m_\varepsilon\|_{L^\infty} \|u\|_{L^2} \leq \|u\|_{L^2}. \tag{9.15}$$

Similarly,

$$\|m_\varepsilon u'\|_{L^2} \leq \|m_\varepsilon\|_{L^\infty} \|u'\|_{L^2} \leq \|u'\|_{L^2}. \tag{9.16}$$

Therefore, if (9.14) holds we would have,

$$\|m_\varepsilon' u\|_{L^2} \leq C \|u\|_{H^1}.$$

This is impossible. Indeed since u equals 1 on $[0, 1]$, we have, using (9.13), $\|m_\varepsilon' u\|_{L^2(\mathbf{R})} \geq \|m_\varepsilon'\|_{L^2([0,1])} \geq C' |\log(\varepsilon)|^{\frac{1}{2}} \to +\infty$ when $\varepsilon \to 0$. .

Solution 17

1. Since $\beta > \frac{d}{2}$ we have $H^\beta(\mathbf{R}^d) \subset L^\infty(\mathbf{R}^d)$ with continuous embedding. So we can write,

$$2^{j\alpha}\|v_j\|_{L^2} \leq 2^{j\alpha}\|S_{j-2}(a)\|_{L^\infty}\|\Delta_j u\|_{L^2} \leq C\|a\|_{L^\infty}2^{j\alpha}\|\Delta_j u\|_{L^2},$$

$$\leq C\|a\|_{H^\beta}2^{j(\alpha-\gamma)}c_j\|u\|_{H^\gamma} \leq Cc_j\|a\|_{H^\beta}\|u\|_{H^\gamma},$$

since $\alpha \leq \gamma$, where $\sum_j c_j^2 < +\infty$. It follows that $T_a u$ belongs to $H^\alpha(\mathbf{R}^d)$ and that we have the inequality (4.14).

2. Since $\gamma - \alpha > 0$ we have $\gamma - \alpha + \frac{d}{2} > \frac{d}{2}$. It follows that,

$$2^{j\alpha}\|v_j\|_{L^2} \leq 2^{j\alpha}\|S_{j-2}(a)\|_{L^\infty}\|\Delta_j u\|_{L^2},$$

$$\leq 2^{j\alpha}\|S_{j-2}(a)\|_{H^{\gamma-\alpha+\frac{d}{2}}}\|\Delta_j u\|_{L^2}.$$

We know (see (3.13)) that $S_{j-2}(a) = \psi(2^{-j+4}D)a$. It follows that,

$$\|S_{j-2}(a)\|_{H^{\gamma-\alpha+\frac{d}{2}}} \leq C2^{j(\gamma-\alpha)}\|a\|_{H^{\frac{d}{2}}}.$$

Therefore, we have,

$$2^{j\alpha}\|v_j\|_{L^2} \leq C2^{j\alpha}2^{j(\gamma-\alpha)}c_j 2^{-j\gamma}\|a\|_{H^{\frac{d}{2}}}\|u\|_{H^\gamma},$$

$$\leq Cc_j\|a\|_{H^{\frac{d}{2}}}\|u\|_{H^\gamma},$$

where $\sum_j c_j^2 < +\infty$. This proves (4.14) in this case.

3. In this case we have $\alpha \leq \beta + \gamma - \frac{d}{2}$. We write as before,

$$\|v_j\|_{L^2} \leq \|S_{j-2}(a)\|_{L^\infty}\|\Delta_j u\|_{L^2}.$$

Now we have

$$\|S_{j-2}(a)\|_{L^\infty} \leq \sum_{k=-1}^{j-3} \|\Delta_k a\|_{L^\infty}.$$

Let $\varphi_1 \in C_0^\infty(\mathbf{R}^d)$ be supported in a ring $\widetilde{C}_j$ and equal to one on the support of φ. Then,

$$\widehat{\Delta_k a}(\xi) = \varphi(2^{-k}\xi)a = \varphi_1(2^{-k}\xi)\varphi(2^{-k}\xi)a,$$

from which we deduce, taking the inverse Fourier transform of both members,

$$\Delta_k a = 2^{kd}\widehat{\varphi_1}(2^k\cdot) \star \Delta_k a.$$

Therefore, we have,

$$\|\Delta_k a\|_{L^\infty} \leq \|2^{kd}\widehat{\varphi_1}(2^k\cdot)\|_{L^2}\|\Delta_k a\|_{L^2},$$

so using the above inequality and the Cauchy–Schwarz inequality we get,

$$\|v_j\|_{L^2} \leq \left(\sum_{k=-1}^{j-3} 2^{\frac{kd}{2}} \|\widehat{\varphi_1}\|_{L^2}\|\Delta_k a\|_{L^2} \right) \|\Delta_j u\|_{L^2},$$

$$\leq \|\widehat{\varphi_1}\|_{L^2} \left(\sum_{k=-1}^{j-3} 2^{kd}2^{-2k\beta} \right)^{\frac{1}{2}} \left(\sum_{k=-1}^{j-3} 2^{2k\beta}\|\Delta_k a\|_{L^2}^2 \right)^{\frac{1}{2}} \|\Delta_j u\|_{L^2},$$

$$\leq C\|\widehat{\varphi_1}\|_{L^2} \left(\sum_{k=-1}^{j-3} 2^{-2k(\beta-\frac{d}{2})} \right)^{\frac{1}{2}} \|a\|_{H^\beta}\|\Delta_j u\|_{L^2}.$$

Since $\beta - \frac{d}{2} < 0$ we have $2^{-2k(\beta-\frac{d}{2})} \leq C2^{-2j(\beta-\frac{d}{2})}$ for $k \leq j-3$ so that,

$$\|v_j\|_{L^2} \leq C2^{-j(\beta-\frac{d}{2})}\|a\|_{H^\beta}\|\Delta_j u\|_{L^2},$$

therefore,

$$2^{j\alpha}\|v_j\|_{L^2} \leq C2^{-j(\beta-\alpha+\gamma-\frac{d}{2})}\|a\|_{H^\beta}2^{j\gamma}\|\Delta_j u\|_{L^2},$$

$$\leq C2^{-j(\beta-\alpha+\gamma-\frac{d}{2})}c_j\|a\|_{H^\beta}\|u\|_{H^\gamma},$$

where $\sum_j c_j^2 < +\infty$. Since $\beta - \alpha + \gamma - \frac{d}{2} \geq 0$ this proves (4.14).

4. Let us interpretate (4.14). If $\beta > \frac{d}{2}$ we can take $\gamma = \alpha$ in (4.13). Thus T_a sends $H^\alpha(\mathbf{R}^d)$ to $H^\alpha(\mathbf{R}^d)$. It is therefore of order zero. If $\beta = \frac{d}{2}$ we can take, by (4.13), (iii) $\gamma = \alpha + \varepsilon$ for every $\varepsilon > 0$. Thus T_a is of order ε. Eventually if $\beta < \frac{d}{2}$ then we can take $\gamma = \alpha - \beta + \frac{d}{2}$, thus T_a sends $H^\alpha(\mathbf{R}^d)$ to $H^{\alpha-\beta+\frac{d}{2}}(\mathbf{R}^d)$. It is therefore of order $\frac{d}{2} - \beta < 0$.

Solution 18

Part 1

1. To estimate the first term in the left-hand side of (4.16) we want to use (4.14) with $\alpha = s_0, \beta = s_1, \gamma = s_2, a = u_1, u = u_2$. Let us check (4.13). Using the hypotheses (i), (ii) we have, $\alpha \le \gamma, \alpha \le \beta + \gamma - \frac{d}{2}$ and if $\beta = s_1 = \frac{d}{2}$ then $\alpha = s_0 < s_2 = \gamma$. Therefore, (4.14) can be applied and gives the result. The estimate of the second term in (4.16) is symmetric.

Part 2

1. a) We have $\widehat{\Delta_p R_j} = \varphi(2^{-p}\xi)\widehat{R_j}$. The support of $\widehat{R_j}$ is contained in a ball of radius $C2^j$ and that of $\varphi(2^{-p}\xi)$ is contained in $\left\{\xi : 2^{p-1} \le |\xi| \le 2^{p+1}\right\}$. Therefore, if $C2^j < 2^{p-1}$ that is if $p - j > \frac{\mathrm{Ln}(2C)}{\mathrm{Ln}2}$, we have $\Delta_p R_j = 0$. Therefore, we have $\Delta_p R = \sum_{j \ge p - N_0} \Delta_p R_j$ where $N_0 = \left[\frac{\mathrm{Ln}(2C)}{\mathrm{Ln}2}\right] + 1$.

   b) We have $\Delta_p R_j = \mathcal{F}^{-1}(\varphi(2^{-p}\xi)) \star R_j = 2^{pd}(\mathcal{F}^{-1}\varphi)(2^p\cdot) \star R_j$. By the Young inequality on convolutions we have,

   $$\|\Delta_p R_j\|_{L^2(\mathbf{R}^d)} \le 2^{pd}\|(\mathcal{F}^{-1}\varphi)(2^p\cdot)\|_{L^2(\mathbf{R}^d)}\|R_j\|_{L^1(\mathbf{R}^d)},$$

   $$\le 2^{\frac{pd}{2}}\|\mathcal{F}^{-1}\varphi\|_{L^2(\mathbf{R}^d)}\|R_j\|_{L^1(\mathbf{R}^d)}.$$

   c) The first inequality follows from the fact that,

   $$\|R_j\|_{L^1(\mathbf{R}^d)} \le \sum_{k=j-2}^{j+2} \|\Delta_k u_1\|_{L^2(\mathbf{R}^d)}\|\Delta_j u_2\|_{L^2(\mathbf{R}^d)}.$$

   Moreover, we have, with $(c_k), (d_j) \in \ell^2$,

   $$\alpha_j \le C \sum_{k=j-2}^{j+2} c_k\|u_1\|_{H^{s_1}(\mathbf{R}^d)}d_j\|u_2\|_{H^{s_2}(\mathbf{R}^d)},$$

   and $|j - k| \le 2$ in the sum. Therefore, by Hölder inequality we get,

   $$\alpha_j \le C' \left(\sum_{k \ge -1} c_k^2\right) d_j\|u_1\|_{H^{s_1}(\mathbf{R}^d)}\|u_2\|_{H^{s_2}(\mathbf{R}^d)}.$$

2. a) From the questions 1a) and c) we have,

   $$\beta_p \le C_2\, 2^{p(s_1+s_2-\frac{d}{2})} \sum_{j \ge p - N_0} 2^{\frac{pd}{2}} 2^{-j(s_1+s_2)}\alpha_j = C_2 \sum_{j \ge p - N_0} 2^{(p-j)(s_1+s_2)}\alpha_j.$$

b) Using the Hölder inequality we can write,

$$\beta_p^2 \le C_2 \left(\sum_{j \ge p-N_0} 2^{(p-j)(s_1+s_2)} \right) \left(\sum_{j \ge p-N_0} 2^{(p-j)(s_1+s_2)} \alpha_j^2 \right).$$

Now we have,

$$\sum_{j \ge p-N_0} 2^{(p-j)(s_1+s_2)} = \sum_{k=-N_0}^{+\infty} 2^{-k(s_1+s_2)} < +\infty$$

since $s_1 + s_2 > 0$. Therefore, we get,

$$\sum_{p \ge -1} \beta_p^2 \le C_2 \sum_{p \ge -1} \sum_{j \ge p-N_0} 2^{(p-j)(s_1+s_2)} \alpha_j^2 = \sum_{j \ge -1} \left(\sum_{p=-1}^{j+N_0} 2^{(p-j)(s_1+s_2)} \right) \alpha_j^2,$$

$$\le C_3 \sum_{j \ge -1} \left(\sum_{k=N_0}^{j+1} 2^{-k(s_1+s_2)} \right) \alpha_j^2 \le C_4 \sum_{j \ge -1} \alpha_j^2.$$

c) Using the definition of β_p it follows from question 1c) that,

$$\sum_{p \ge -1} 2^{2p(s_1+s_2-\frac{d}{2})} \|\Delta_p R\|_{L^2(\mathbf{R}^d)}^2 \le C_5 \left(\sum_{j \ge -1} d_j^2 \right) \|u_1\|_{H^{s_1}(\mathbf{R}^d)}^2 \|u_2\|_{H^{s_2}(\mathbf{R}^d)}.$$

This proves (4.17).

d) Using (4.16) and (4.17) and the fact that $s_0 \le s_1 + s_2 - \frac{d}{2}$ we deduce that $u_1 u_2 \in H^{s_0}(\mathbf{R}^d)$ and that we have the estimate (4.15).

Solution 19

1. a) Set $a_{\alpha\beta} = \partial_x^\alpha \partial_\xi^\beta a$. (H_0) is true by hypothesis. Assume (H_k) satisfied and let us prove (H_{k+1}). We have then $|a_{\alpha\beta}(x,\xi)| \le C(1+|\xi|)^{m-\rho|\beta|-k(\rho-1)}$. Since $m - \rho|\beta| - k(\rho-1) < 0$ we have $\lim_{\xi_1 \to \pm\infty} a_{\alpha\beta}(x,\xi_1,\xi') = 0$. Therefore, for $\xi_1 > 0$,

$$a_{\alpha\beta}(x,\xi) = -\int_{\xi_1}^{+\infty} \partial_{\xi_1} a_{\alpha\beta}(x,\theta,\xi')\,d\theta.$$

It follows from (H_k), since inside the integral a is differentiated $|\beta| + 1$ times with respect to ξ that,

$$|a_{\alpha\beta}(x,\xi)| \leq C \int_{\xi_1}^{+\infty} (1 + \theta + |\xi'|)^{m - \rho(|\beta|+1) - k(\rho - 1)} \, d\theta. \qquad (9.17)$$

Let $M = m - \rho(|\beta| + 1) - k(\rho - 1)$. Then for $|\beta| \geq 0$ and $k \geq 0$ we have $M + 1 \leq m - \rho + 1 < 0$. Therefore, we can integrate the right-hand side of (9.17) and we obtain,

$$|a_{\alpha\beta}(x,\xi)| \leq C(1 + \xi_1 + |\xi'|)^{M+1}.$$

We have just to notice that $M + 1 = m - \rho|\beta| - (k + 1)(\rho - 1)$.

For $\xi_1 < 0$ we write,

$$a_{\alpha\beta}(x,\xi) = \int_{-\infty}^{\xi_1} \partial_{\xi_1} a_{\alpha\beta}(x, \theta, \xi') \, d\theta$$

and we argue as above. This proves (H_{k+1}).

b) If $N \in \mathbf{N}$ let us take $k \in \mathbf{N}$ such that $m - \rho|\beta| - k(\rho - 1) < -N$ which is possible since $\rho - 1 > 0$. Then $|a_{\alpha\beta}(x,\xi)| \leq C(1 + |\xi|)^{-N}$. So $a \in S^{-\infty}$. Conversely it is obvious that $S^{-\infty} \subset S_\rho^m$.

2. a) Since p is of order $m - 1$ we have $\partial_\xi^\beta p = 0$ if $|\beta| \geq m$. Now for $|\beta| \leq m - 1$ we have $|\partial_\xi^\beta p(\xi)| \leq C(1 + |\xi|)^{m-1-|\beta|}$. We have just to notice that $-(1 + |\beta|) \leq -\rho|\beta|$, which is equivalent to $(\rho - 1)|\beta| \leq 1$ or to $\varepsilon|\beta| \leq 1$, which is implied by our assumption $\varepsilon(m - 1) \leq 1$. However, $p \notin S^{-\infty}$.

b) Take $a(x,\xi) = b(x)$ where $b \in C_0^\infty(\mathbf{R}^d)$. Then $a \in S_\rho^0$ since for $\beta = 0$ we have $|\partial_x^\alpha b(x)| \leq C_\alpha$ and for $\beta \neq 0$, $\partial_x^\alpha \partial_\xi^\beta b(x) = 0$. However, $a \notin S^{-\infty}$.

Solution 20

We make an induction on $|\alpha| + |\beta| = \ell$. The claim is true for $\ell = 0$. Assume it is true for ℓ and let $|\tilde{\alpha}| + |\tilde{\beta}| = \ell + 1$. We have the $\partial_x^{\tilde{\alpha}} \partial_\xi^{\tilde{\beta}} = \partial_{x_i} \partial_x^\alpha \partial_\xi^\beta$ or $\partial_{\xi_i} \partial_x^\alpha \partial_\xi^\beta$ where $|\alpha| + |\beta| = \ell$. Let $\partial = \partial_{x_i}$ or ∂_{ξ_i}. By the induction we have,

$$\partial_x^{\tilde{\alpha}} \partial_\xi^{\tilde{\beta}} e^{-\sigma a} = \left(\partial d_{0,\alpha\beta} + \sum_{j=1}^{\ell} \sigma^j (\partial d_{j,\alpha\beta} - (\partial a) d_{j-1,\alpha\beta}) + \sigma^{\ell+1} (\partial a) d_{\ell,\alpha\beta} \right) e^{-\sigma a}.$$

Therefore, $d_{0,\tilde{\alpha}\tilde{\beta}} = \partial d_{0,\alpha\beta}$, $d_{j,\tilde{\alpha}\tilde{\beta}} = \partial d_{j,\alpha\beta} - (\partial a) d_{j-1,\alpha\beta}$, for $1 \leq j \leq \ell$, $d_{\ell+1,\tilde{\alpha}\tilde{\beta}} = (\partial a) d_{\ell,\alpha\beta}$.

Case 1. $\partial = \partial_{x_i}$. Then $\widetilde{\beta} = \beta$. We have, $d_{0,\widetilde{\alpha}\widetilde{\beta}} \in S^{-|\beta|}$, $d_{j,\widetilde{\alpha}\widetilde{\beta}} \in S^{j-|\beta|} + S^1 \times S^{j-1-|\beta|} \subset S^{j-|\beta|}$, eventually, $d_{\ell+1,\widetilde{\alpha}\widetilde{\beta}} \in S^1 \times S^{\ell-|\beta|} \subset S^{\ell+1-|\beta|}$.

Case 2. $\partial = \partial_{\xi_i}$. Then $|\widetilde{\beta}| = |\beta| + 1$. We have, $d_{0,\widetilde{\alpha}\widetilde{\beta}} \in S^{-(|\beta|+1)}$, $d_{j,\widetilde{\alpha}\widetilde{\beta}} \in S^{j-|\beta|-1} + S^0 \times S^{j-1-|\beta|} \subset S^{j-(|\beta|+1)}$, eventually, $d_{\ell+1,\widetilde{\alpha}\widetilde{\beta}} \in S^0 \times S^{\ell-|\beta|} \subset S^{\ell+1-(|\beta|+1)}$.

Solution 21

Part 1

1. The Fourier transform of v is given by,

$$\widehat{v}(\xi) = \int \big(p(\xi) - p(\eta) \big) \widehat{f}(\xi - \eta) \widehat{u}(\eta)\, d\eta,$$

which immediately implies the desired result by splitting the integral into two parts.

2. We can write, $p(\xi) - p(\eta) = \int_0^1 (\nabla_\xi p)(t\xi + (1-t)\eta)(\xi - \eta)\, dt$, so using the definition of S^m we obtain, $|p(\xi) - p(\eta)| \leq C \langle t\xi + (1-t)\eta \rangle^{m-1} |\xi - \eta|$. If $|\xi - \eta| \leq \frac{1}{2}|\eta|$ then $\langle t\xi + (1-t)\eta \rangle^{m-1} \leq C \langle \eta \rangle^{m-1}$ which gives the result.

3. a) If $|\xi - \eta| \leq \frac{1}{2}|\eta|$, then $|\xi| \leq |\xi - \eta| + |\eta| \leq \frac{3}{2}|\eta|$. Therefore, since $\sigma - m + 1 \geq 0$ we have, $\langle \xi \rangle^{\sigma - m + 1} \leq C \langle \eta \rangle^{\sigma - m + 1}$. So using question 2 we can write,

$$\langle \xi \rangle^{\sigma - m + 1} |\widehat{v_L}(\xi)| \leq K \int |\xi - \eta| \big| \widehat{f}(\xi - \eta) \big| \langle \eta \rangle^\sigma |\widehat{u}(\eta)|\, d\eta.$$

b) By the Young inequality we have, $\|h \star g\|_{L^2} \leq \|h\|_{L^1} \|g\|_{L^2}$. We deduce from question a), taking the L^2 norm of both members that, $\|v_L\|_{H^{\sigma - m + 1}} \leq \left(\int \langle \zeta \rangle |\widehat{f}(\zeta)|\, d\zeta \right) \|u\|_{H^\sigma}$. Now using the Cauchy–Schwarz inequality and the fact that $2(\sigma_0 - 1) > d$ we can write,

$$\int \langle \zeta \rangle |\widehat{f}(\zeta)|\, d\zeta = \int \langle \zeta \rangle^{-(\sigma_0 - 1)} \langle \zeta \rangle^{(\sigma_0 - 1)} \langle \zeta \rangle |\widehat{f}(\zeta)|\, d\zeta \leq C \|f\|_{H^{\sigma_0}}.$$

4. If $|\xi - \eta| \geq \frac{1}{2}|\eta|$ we obtain $|\eta| \leq 2|\xi - \eta|$ and $|\xi| \leq 3|\xi - \eta|$. It follows that for $t \in [0,1]$ we have $\langle t\xi + (1-t)\eta \rangle \leq C \langle \xi - \eta \rangle$. Since $m \geq 1$ we deduce that $|p(\xi) - p(\eta)| \leq C \langle \xi - \eta \rangle^m$. On the other hand, since $\sigma \geq m - 1$ we also have $\langle \xi \rangle^{\sigma - m + 1} \leq C \langle \xi - \eta \rangle^{\sigma - m + 1}$. Therefore,

$$\langle \xi \rangle^{\sigma - m + 1} |\widehat{v_H}(\xi)| \leq C \int \langle \xi - \eta \rangle^{\sigma + 1 - \sigma_0} \langle \xi - \eta \rangle^{\sigma_0} |\widehat{f}(\xi - \eta)| |\widehat{u}(\eta)|\, d\eta.$$

Since $\sigma+1-\sigma_0 \le 0$ and $|\xi-\eta| \ge \frac{1}{2}|\eta|$ we have, $\langle \xi - \eta \rangle^{\sigma+1-\sigma_0} \le C \langle \eta \rangle^{\sigma+1-\sigma_0}$.
It follows that,

$$\langle \xi \rangle^{\sigma-m+1} |\widehat{v_H}(\xi)| \le C' \int \langle \xi - \eta \rangle^{\sigma_0} |\widehat{f}(\xi - \eta)| \langle \eta \rangle^{\sigma+1-\sigma_0} |\widehat{u}(\eta)| \, d\eta.$$

Using the Young inequality as in question 3b) we deduce that,

$$\|v_H\|_{H^{\sigma-m+1}} \le C\|f\|_{H^{\sigma_0}} \int \frac{1}{\langle \eta \rangle^{\sigma_0-1}} \langle \eta \rangle^{\sigma} |\widehat{u}(\eta)| \, d\eta.$$

Using the Cauchy–Schwarz inequality and the fact that $2(\sigma_0 - 1) > d$ we obtain the desired inequality.

Part 2

1. At this stage, we have proved (6.1) for $\sigma \in [m - 1, \sigma_0 - 1]$. We now consider the case $\sigma \in [-\sigma_0 + m, 0]$. The two key observations are that

$$\mu := -\sigma + m - 1 \in [m - 1, \sigma_0 - 1], \quad \mu - m + 1 = -\sigma,$$

and that the adjoint of the operator C given by $Cu = P(fu) - fPu$ is a commutator of the same form. Consequently, the result of the previous question implies that,

$$\|C^*\varphi\|_{H^{-\sigma}} = \|C^*\varphi\|_{H^{\mu-m+1}} \le K \|f\|_{H^{\sigma_0}} \|\varphi\|_{H^\mu} = K \|f\|_{H^{\sigma_0}} \|\varphi\|_{H^{-\sigma+m-1}}.$$

This proves that $C^* \in \mathcal{L}(H^{-\sigma+m-1}, H^{-\sigma})$ with operator norm bounded by $K\|f\|_{H^{\sigma_0}}$, which in turn implies the desired result by a duality argument.

Solution 22

In this proof, we shall denote by $\|T\|_{X \to Y}$ the operator norm of a bounded linear operator $T: X \to Y$.

1. Observe that,

$$fg - T_f g = \sum_{j,p \ge -1} (\Delta_j f)(\Delta_p g) - \sum_{\substack{p \ge -1 \\ -1 \le j \le p-3}} (\Delta_j f)(\Delta_p g) = \sum_{\substack{p \ge -1 \\ j \ge p-2}} (\Delta_j f)(\Delta_p g)$$

$$= \sum_{j \ge -1} \sum_{-1 \le p \le j+2} (\Delta_p g)(\Delta_j f) = \sum_{j \ge -1} S_{j+3}(g)\Delta_j f.$$

Therefore, one can write,

$$\left\| fg - T_f g \right\|_{L^2} \leq \sum_{j \geq -1} \left\| S_{j+2}(g) \right\|_{L^2} \left\| \Delta_j f \right\|_{L^\infty}.$$

Now recall that, according to (3.12), the support of the Fourier transform of $S_{j+3}(g)$ is contained in a ball of radius 2^{j+3}. It follows that,

$$\left\| S_{j+3}(g) \right\|_{L^2} \leq C' 2^{j\theta} \|g\|_{H^{-\theta}}.$$

Then, using the characterization of Hölder spaces in terms of Littlewood–Paley decomposition and the assumption $\theta < \nu$ we can write,

$$\left\| fg - T_f g \right\|_{L^2} \leq C \sum_{j \geq -1} 2^{\theta j} \|g\|_{H^{-\theta}} 2^{-j\nu} \|f\|_{W^{\nu,\infty}} \leq C \|g\|_{H^{-\theta}} \|f\|_{W^{\nu,\infty}}.$$

2. Denote by $f^\flat$ the multiplication operator $u \mapsto fu$. We have,

$$[Q, f^\flat] = [Q, T_f] + Q(f^\flat - T_f) - (f^\flat - T_f)Q. \tag{9.18}$$

Notice that $Q = T_q$ since the symbol q does not depend on x. Consequently, the composition rule for paradifferential operators implies that,

$$\forall \sigma \in \mathbb{R}, \quad \left\| [Q, T_f] \right\|_{H^\sigma \to H^{\sigma+\nu}} \leq c(\nu, \sigma) \|f\|_{W^{\nu,\infty}}. \tag{9.19}$$

In particular,

$$\left\| [Q, T_f] \right\|_{H^{-\nu} \to L^2} \leq C \|f\|_{W^{\nu,\infty}}. \tag{9.20}$$

With regards to the two other terms in the right-hand side of (9.18) we claim that,

$$\left\| Q(f^\flat - T_f) \right\|_{H^{-\theta} \to L^2} + \left\| (f^\flat - T_f)Q \right\|_{H^{-\theta} \to L^2} \lesssim \|f\|_{W^{\nu,\infty}}. \tag{9.21}$$

This follows from the estimate proved in the first question and the fact that Q is bounded from H^σ to itself for any $\sigma \in \mathbf{R}$.

3. Denote by $\mathcal{H}_0$ the Fourier multiplier with symbol $-i(1 - \Phi(\xi))\xi/|\xi|$ where $\Phi \in C_0^\infty(\mathbf{R})$ is such that $\Phi(\xi) = 1$ on a neighborhood of the origin. With these notations, one can rewrite the commutator $[\mathcal{H}, f^\flat]$ as,

$$\begin{aligned}
[\mathcal{H}, f^\flat] &= [\mathcal{H}, T_f] + \mathcal{H}(f^\flat - T_f) - (f^\flat - T_f)\mathcal{H}, \\
&= [\mathcal{H}_0, T_f] + (\mathcal{H} - \mathcal{H}_0)T_f - T_f(\mathcal{H} - \mathcal{H}_0), \\
&\quad + \mathcal{H}(f^\flat - T_f) - (f^\flat - T_f)\mathcal{H}. \tag{9.22}
\end{aligned}$$

We want to prove that the five operators in the right-hand side of (9.22) are bounded from $H^{-\theta}$ to L^2 with operator norm controlled by $\|f\|_{W^{\nu,\infty}}$. Notice that the first three terms have been estimated in the second question (one can use the estimates (9.20) and (9.21) since $-i(1 - \Phi(\xi))\xi/|\xi|$ is a smooth symbol of order 0). So it remains only to study the last two terms. To do so, we notice that $\mathcal{H} - \mathcal{H}_0$ is a smoothing operator (that is an operator bounded from H^{σ} to $H^{\sigma+t}$ for any real numbers $\sigma, t \in \mathbb{R}$). On the other hand, recall the following classical estimate for paraproducts:

$$\forall \sigma \in \mathbb{R}, \quad \|T_f\|_{H^{\sigma} \to H^{\sigma}} \le c(\sigma) \|f\|_{L^{\infty}}.$$

We infer that,

$$\left\|(\mathcal{H} - \mathcal{H}_0)T_f\right\|_{H^{-\theta} \to L^2} \le \|\mathcal{H} - \mathcal{H}_0\|_{H^{-\theta} \to L^2} \left\|T_f\right\|_{H^{-\theta} \to H^{-\theta}} \le C \|f\|_{L^{\infty}},$$

$$\left\|T_f(\mathcal{H} - \mathcal{H}_0)\right\|_{H^{-\theta} \to L^2} \le \left\|T_f\right\|_{L^2 \to L^2} \|\mathcal{H} - \mathcal{H}_0\|_{H^{-\theta} \to L^2} \le C \|f\|_{L^{\infty}}.$$

This completes the proof.

Solution 23

Part 1
In what follows we shall denote $E = E(\mathbb{R}^d)$ if $E = L^p, H^s, C_0^{\infty}, S, S' \dots$

1. This follows from the fact that $\mathrm{supp}(\partial_{x_j} u) \subset \mathrm{supp}\, u$ and $\mathrm{supp}(au) \subset \mathrm{supp}\, a \cap \mathrm{supp}\, u$.
2. By definition we have $\mathrm{supp}(PQu) \subset \mathrm{supp}(Qu) \subset \mathrm{supp}\, u$. The same is true for the bracket.
3. We show that $(\mathrm{supp}\, u)^c \subset (\mathrm{supp}\, P^*u)^c$. Let $x_0 \notin \mathrm{supp}\, u$. There exists a neighborhood V_{x_0} such that for all $\varphi \in C_0^{\infty}(V_{x_0})$ we have $(u, \varphi)_{L^2(\mathbb{R}^d)} = 0$. Let $\theta \in C_0^{\infty}(V_{x_0})$. Since P is local we have $\mathrm{supp}(P\theta) \subset \mathrm{supp}\,\theta \subset V_{x_0}$. So $\varphi = P\theta \in C_0^{\infty}(V_{x_0})$. Therefore, $(u, P\theta)_{L^2} = 0$, so $(P^*u, \theta)_{L^2} = 0$ which proves that $x_0 \notin \mathrm{supp}\, P^*u$.
4. a) If $P^*P = 0$, for all $\varphi \in S$ we have $0 = (P^*P\varphi, \varphi)_{L^2} = \|P\varphi\|_{L^2}^2$. So $P\varphi = 0$ for all $\varphi \in S$, therefore $P = 0$.
   b) If $(P^*P)^{\ell} = 0$ for all $\varphi \in S$ we have,

$$0 = ((P^*P)^{\ell}\varphi, (P^*P)^{\ell-1}\varphi)_{L^2} = \|P(P^*P)^{\ell-1}\varphi\|_{L^2}^2,$$

so $P(P^*P)^{\ell-1} \equiv 0$. Then,

$$0 = (P(P^*P)^{\ell-1}\varphi, P(P^*P)^{\ell-2}\varphi)_{L^2} = \|(P^*P)^{\ell-1}\varphi\|_{L^2}^2.$$

By induction we deduce that $P^*P = 0$ and using question 4 that $P = 0$.

Part 2

1. Let $\varphi \in C_0^\infty$ be such that $\varphi(x) = 1$ if $|x| \leq 1, \varphi(x) = 0$ if $|x| \geq 2$. Set, for $k \geq 1, u_k(x) = (1 - \varphi(k(x - x_0)))u(x)$. This is an element of S which vanishes for $|x - x_0| \leq \frac{1}{k}$. Now,

$$\|u_k - u\|_{L^2}^2 = \int |\varphi(k(x - x_0))u(x)|^2 \, dx.$$

The right-hand side goes to zero, when $k \to +\infty$, by the dominated convergence theorem. Indeed if $x \neq x_0$ is fixed we have $k|x - x_0| \to +\infty$, so $|\varphi(k(x - x_0))u(x)|^2 \to 0$. Moreover, $|\varphi(k(x - x_0))u(x)|^2 \leq \|\varphi\|_{L^\infty}^2 |u(x)|^2 \in L^1$.

2. Since $-m > \frac{d}{2}$ the embedding from H^{-m} to L^∞ is continuous. Therefore, there exists $C_1 > 0$ such that $\sup_{\mathbf{R}^d} |Au(x)| \leq C_1 \|Au\|_{H^{-m}}$ for all $u \in S$. Now A being of order m it is continuous from L^2 to H^{-m}. So there exists $C_2 > 0$ such that $\|Au\|_{H^{-m}} \leq C_2\|u\|_{L^2}$ for all $u \in S$ and the result follows.

3. By the previous question we have,

$$|Au_k(x_0) - Au(x_0)| \leq C\|u_k - u\|_{L^2(\mathbf{R}^d)}. \tag{9.23}$$

If A is local we have $(\operatorname{supp}u_k)^c \subset (\operatorname{supp}Au_k)^c$. Since $u_k \equiv 0$ near x_0 we have $x_0 \in (\operatorname{supp}u_k)^c$ so that $x_0 \in (\operatorname{supp}Au_k)^c$. Therefore, $Au_k(x_0) = 0$ for $k \geq 1$. Since the right-hand side of (9.23) goes to zero, we have $Au(x_0) = 0$. This being true for all $x_0 \in \mathbf{R}^d$ and all $u \in S$, therefore for all $u \in S'$, we deduce that $A \equiv 0$.

4. The operator $(A^*A)^\ell$ is local by Part 1 and it is of order $2\ell m < -\frac{d}{2}$; by the previous question we have $(A^*A)^\ell \equiv 0$. It follows from Part 1 that $A \equiv 0$.

5. First of all if f is C^∞-bounded we have $f \in S^0$. Now if $P \in \operatorname{Op}(S^r)$ is local then the bracket $(ad\,f)(P)$ belongs to $\operatorname{Op}(S^{r-1})$ and it is local by Part 1. Applying this to $P = A, f = f_k$ then to $P = (ad\,f_k)(A), f = f_{k-1}$ and so on, eventually to $P = (ad\,f_2)\ldots(ad\,f_k)(A), f = f_1$ we see that $(ad\,f_1)(ad\,f_2)\ldots(ad\,f_k)(A)$ belongs to $\operatorname{Op}(S^{m-k})$ and it is local. Since $m-k < 0$ it follows from the previous question that this bracket vanishes.

6. For $\ell = 1$ we have $A(gw) = [A, g]w + gAw = -(ad\,g)(A)w + gAw$. Since $a \in S^m$ and $w \in C_0^\infty$ we have $Aw \in S$ and $|g(x)| \leq C|x - x_0|$. Assume the formula true at the order $\ell \geq 1$. Using the induction with w replaced by gw and the fact $(ad\,P)(Q) = -(ad\,Q)(P)$ we obtain,

$$(-1)^\ell A(g^\ell gw) = (ad\,g)^\ell(A)gw+O(|x-x_0|) = -(ad\,g)^{\ell+1}(A)w+O(|x-x_0|).$$

7. a) This follows from the Taylor formula with integral reminder. We have,

$$v_\beta(x) = \frac{k}{\beta!}\int_0^1 (1-t)^{k-1}(\partial_x^\beta u)(tx + (1-t)x_0)\,dt.$$

It is obviously a C^∞ function on $\mathbf{R}^d$.

b) We have $\theta^{k+1}u = u$ so that using the previous question we get,

$$Au(x) = \sum_{|\beta|=k} A\left((\theta(x)(x-x_0))^\beta(\theta v_\beta)(x)\right).$$

Setting for $1 \leq j \leq d$, $f_j(x) = \theta(x)(x_j - x_{0,j})$ we have,

$$Au(x) = \sum_{|\beta|=k} A\left(f_1^{\beta_1}f_2^{\beta_2}\ldots f_k^{\beta_k}(\theta v_\beta)\right).$$

Using question 6 with successively $g = f_1, \ldots, g = f_d$ and $w = \theta v_\beta$ we obtain,

$$Au(x) = \sum_{|\beta|=k} (-1)^{|\beta|}(ad\ f_1)^{\beta_1}\ldots(ad\ f_d)^{\beta_d}(A)(\theta v_\beta) + O(|x-x_0|).$$

c) Since A is local and $m < k$ we can use question 5 and deduce that, $Au(x) = O(|x-x_0|)$. So $Au(x_0) = 0$.

8. Let $u \in C_0^\infty$ and $w(x) = u(x) - \sum_{|\beta|\leq k-1} \frac{1}{\beta!}\chi(x)(x-x_0)^\beta \partial_x^\beta u(x_0)$ where $\chi \in C_0^\infty$, $\chi(x) = 1$ near x_0. Then $w \in C_0^\infty$ and $\partial_x^\alpha w(x_0) = 0$ for $|\alpha| \leq k-1$. By the previous question we have $Aw(x_0) = 0$, so,

$$Au(x_0) = \sum_{|\beta|\leq k-1} \frac{1}{\beta!}A\left(\chi(x)(x-x_0)^\beta\right)\partial_x^\beta u(x_0).$$

This being true for all $x_0 \in \mathbf{R}^d$ we have $Au(x) = \sum_{|\beta|\leq k-1} a_\beta(x)\partial_x^\beta u(x)$. Therefore, it is a differential operator of order $\leq k-1$.

Solution 24

1. We have,

$$\|A_k\|_{H^s} \leq C \sum_{|q-k|\leq 2} \|\chi_k P \chi_q u\|_{H^s}.$$

From (3.2) and the fact that $|\widehat{\chi_k}(\xi)| = |\widehat{\chi}(\xi)|$ we have,

$$\|\chi_k v\|_{H^t} \leq C_{t,d}\left(\int \langle\xi\rangle^{|t|}|\widehat{\chi}(\xi)|\,d\xi\right)\|v\|_{H^t}.$$

Using the continuity of the pseudo-differential operators in the usual Sobolev space we deduce that,

$$\|A_k\|_{H^s} \leq C_1(s, d, \chi) \sum_{|q-k|\leq 2} \|P\chi_q u\|_{H^s}$$

$$\leq C_2(s, d, \chi, p, m) \sum_{|q-k|\leq 2} \|\chi_q u\|_{H^{s+m}}$$

$$\leq C_3(s, d, \chi, p, m) \left(\sum_{|q-k|\leq 2} 1 \right) \|u\|_{H_{ul}^{s+m}}.$$

For fixed k, the cardinal of the set $\{q \in \mathbf{Z}^d : |k - q| \leq 2\}$ is finite and depends only on the dimension d. Therefore, we have,

$$\|A_k\|_{H^s} \leq C\|u\|_{H_{ul}^{s+m}},$$

where C is independent of k.

2. The support of $\partial^\alpha B_{k,q}$ is included in the support of χ_k. Let $\widetilde{\chi} \in C_0^\infty(\mathbf{R}^d)$ be equal to 1 on the support of χ. Then we have $\widetilde{\chi}_k \partial^\alpha B_{k,q} = \partial^\alpha B_{k,q}$ and therefore,

$$\|\partial^\alpha B_{k,q}\|_{L^2} = \|\widetilde{\chi}_k \partial^\alpha B_{k,q}\|_{L^2} \leq \|\widetilde{\chi}_k\|_{L^2} \|\partial^\alpha B_{k,q}\|_{L^\infty} = \|\widetilde{\chi}\|_{L^2} \|\partial^\alpha B_{k,q}\|_{L^\infty}.$$

3. This follows from the expression of the ΨDO and from the Leibniz formula. Indeed we have, with some constants $c_{\alpha_1,\alpha_2,\alpha_3}$,

$$\partial_x^\alpha B_{kq}(x) = \sum_{\alpha=\alpha_1+\alpha_2+\alpha_3} c_{\alpha_1,\alpha_2,\alpha_3} \partial_x^{\alpha_1} \chi_k(x) \int e^{ix\cdot\xi} \xi^{\alpha_2} \partial_x^{\alpha_3} p(x,\xi) \widehat{\chi_q u}(\xi) \, d\xi,$$

and $a_{\alpha_2}(x, \xi) =: \xi^{\alpha_2} \partial_x^{\alpha_3} p(x, \xi) \in S^{m+|\alpha_2|}$.

4. From the Lebesgue dominated convergence theorem and the Fubini theorem we can write,

$$C_{kq\alpha}(x) = \lim_{\varepsilon\to 0} \partial_x^{\alpha_1} \chi_k(x) \int e^{ix\cdot\xi} a_{\alpha_2}(x,\xi) \theta(\varepsilon \langle\xi\rangle) \widehat{\chi_q u}(\xi) \, d\xi,$$

$$= \lim_{\varepsilon\to 0} \partial_x^{\alpha_1} \chi_k(x) \int (\chi_q u)(y) \left(\int e^{i(x-y)\cdot\xi} a_{\alpha_2}(x,\xi) \theta(\varepsilon \langle\xi\rangle) \, d\xi \right) dy.$$

5. We have,

$$|C_{kq\alpha}(x)| \leq \|K_{kq\alpha}^\varepsilon(x, \cdot)\|_{H^{-(s+m)}} \|\chi_q u\|_{H^{s+m}(\mathbf{R}^d)} \leq \|K_{kq\alpha}^\varepsilon(x, \cdot)\|_{H^{n_1}} \|u\|_{H_{ul}^{s+m}}.$$

6. This follows again from the Leibniz formula. We have,

$$\partial_y^\beta K_{kq\alpha}^\varepsilon(x,y) = \sum_{\beta_1+\beta_2=\beta} \binom{\beta}{\beta_1} \partial_x^{\alpha_1}\chi_k(x)\partial_y^{\beta_1}\tilde\chi_q(y)$$

$$\cdot \left(\int e^{i(x-y)\cdot\xi}(-i\xi)^{\beta_2} a_{\alpha_2}(x,\xi)\theta(\varepsilon\langle\xi\rangle)\,d\xi \right),$$

and $b_{\alpha_2,\beta_2}(x,\xi) = (-i\xi)^{\beta_2} a_{\alpha_2}(x,\xi) \in S^{m+|\alpha_2|+|\beta_2|}$.

7. On the support of $K_{kq\alpha}^\varepsilon(x,y)$ we have $|x-k|\leq 1, |y-q|\leq \frac{3}{2}$ so that,

$$|x-y| = |x-k+k-q+q-y| \geq |k-q|-|x-k|-|q-y| \geq |k-q|-\frac{5}{2} \geq \frac{1}{6}|k-q|.$$

8. Obvious.
9. a) Easy by induction on N.
   b) Integrating by parts and using the Leibniz formula we find,

$$|K_{kq\alpha\beta}^\varepsilon(x,y)| \leq |\partial_x^{\alpha_1}\chi_k(x)\partial_y^{\beta_1}\tilde\chi_q(y)|\cdot$$

$$\sum_{|\gamma|=N}\sum_{\gamma_1+\gamma_2=\gamma} d_{\gamma,\gamma_1} M^{|\gamma|} \int \varepsilon^{|\gamma_1|}|\theta_1(\varepsilon\langle\xi\rangle)||\partial_\xi^{\gamma_2} b_{\alpha_2,\beta_2}(x,\xi)|\,d\xi,$$

where $\theta_1 \in C_0^\infty(\mathbf{R}^d)$. Now we have,

$$M^{|\gamma|}(x,y) \leq \frac{C}{|x-y|^N}, \quad |x-y|\geq\frac{1}{6}|k-q|, \quad |\partial_x^{\alpha_1}\chi_k(x)\partial_y^{\beta_1}\tilde\chi_q(y)|\leq C,$$

where C is independent of k,q, so that we can write,

$$|K_{kq\alpha\beta}^\varepsilon(x,y)| \leq \frac{C_N}{|k-q|^N}\int (\varepsilon\langle\xi\rangle)^{|\gamma_1|}|\theta_1(\varepsilon\langle\xi\rangle)\langle\xi\rangle^{-|\gamma_1|}|\partial_\xi^{\gamma_2} b_{\alpha_2,\beta_2}(x,\xi)|\,d\xi,$$

and $\langle\xi\rangle^{-|\gamma_1|}\partial_\xi^{\gamma_2} b_{\alpha_2,\beta_2} \in S^{m+n_0+n_1-N}$ since $|\alpha_2|\leq n_0, |\beta_2|\leq n_1$ and $|\gamma_1|+|\gamma_2|=N$. Then it is sufficient to bound $\varepsilon\langle\xi\rangle)^{|\gamma_1|}|\theta_1(\varepsilon\langle\xi\rangle)$ by a constant independent of ε and to take $m+n_0+n_1-N\leq-(d+1)$, or equivalently $N\geq m+n_0+n_1+d+1$, to ensure that the integral is absolutely convergent.
10. It follows from the previous questions that,

$$\sum_{|k-q|\geq3}\|B_{kq}\|_{H^s} \leq C\left(\sum_{|k-q|\geq3}\frac{1}{|k-q|^N}\right)\|u\|_{H^{s+m}} \leq C'\|u\|_{H^{s+m}},$$

since, when $N > d + 1$ the series $\sum_{|p| \geq 3} \frac{1}{|p|^N}$ is convergent. Using question 1 this proves the continuity of the ΨDO in the uniformly local Sobolev spaces.

Solution 25

1. Using the equality given in the statement we have,

$$\partial_\xi^\alpha p(\xi) = z^\mu \sum_{\alpha_1+\alpha_2=\alpha} \binom{\alpha}{\alpha_1} \left(\partial_\xi^{\alpha_1} \langle\xi\rangle^\mu\right) \sum_{j=0}^{|\alpha_2|} z^j q_{j\alpha_2}(\xi) e^{-za(\xi)},$$

It follows that,

$$|\partial_\xi^\alpha p(\xi)| \leq C_{\alpha\mu} z^\mu \sum_{\alpha_1+\alpha_2=\alpha} \langle\xi\rangle^{\mu-|\alpha_1|} \sum_{j=0}^{|\alpha_2|} z^j \langle\xi\rangle^{j-|\alpha_2|} e^{-za(\xi)},$$

$$\leq C'_{\alpha\mu} z^\mu \langle\xi\rangle^{\mu-|\alpha|} e^{-\frac{1}{2}za(\xi)} \sum_{\alpha_1+\alpha_2=\alpha} \sum_{j=0}^{|\alpha_2|} (z\langle\xi\rangle)^j e^{-\frac{1}{2}za(\xi)}.$$

Eventually since $a(\xi) \geq c_0 \langle\xi\rangle$ we have $(z\langle\xi\rangle)^j e^{-\frac{1}{2}za(\xi)} \leq C_j$.

2. From question 1 we have,

$$|\partial_\xi^\alpha p(\xi)| \leq C_{\alpha\mu} \langle\xi\rangle^{-|\alpha|} (z\langle\xi\rangle)^\mu e^{-\frac{1}{2}za(\xi)} \leq C'_{\alpha\mu} \langle\xi\rangle^{-|\alpha|}.$$

Taking $|\alpha| = d + 1$, we obtain,

$$|x^\alpha \widehat{p}(x)| = |\widehat{\partial_\xi^\alpha p}(x)| \leq \|\partial_\xi^\alpha p\|_{L^1(\mathbf{R}^d)} \leq C'_{\alpha\mu} \int \frac{d\xi}{\langle\xi\rangle^{d+1}} \leq C''_{\alpha\mu}. \qquad (9.24)$$

To conclude we have just to notice that $|x|^{d+1} \leq C_d \sum_{k=0}^d |x_k|^{d+1}$ and to take $\alpha = (0, \ldots, d+1, \ldots, 0)$ in the above estimate.

3. a) We have $z\langle\xi\rangle = (z^2 + z^2|\xi|^2)^{\frac{1}{2}}$ so that using (9.24) and the estimate in question 1 we obtain,

$$|x^\alpha \widehat{p}(x)| \leq C_{\alpha\mu} z^{|\alpha|} \int (z^2 + z^2|\xi|^2)^{\frac{\mu-|\alpha|}{2}} e^{-\frac{c_0}{2}(z^2+z^2|\xi|^2)^{\frac{1}{2}}} d\xi.$$

Set $z\xi = \eta$ so $d\xi = z^{-d} d\eta$. Then,

$$|x^\alpha \widehat{p}(x)| \leq C_{\alpha\mu} z^{|\alpha|-d} \int \langle z, \eta\rangle^{\mu-|\alpha|} e^{-\frac{c_0}{2}\langle z,\eta\rangle} d\eta.$$

b) Since $z \in (0, 1)$ we have $|\eta| \le \langle z, \eta \rangle \le \langle \eta \rangle$. Then,

$$\int_{|\eta| \ge 1} \langle z, \eta \rangle^{\mu - |\alpha|} e^{-\frac{c_0}{2} \langle z, \eta \rangle} d\eta \le C \int_{|\eta| \ge 1} e^{-\frac{c_0}{3} |\eta|} d\eta := M_0 < +\infty.$$

Now if $|\alpha| \ge 1$ we have,

$$\int_{|\eta| \le 1} \langle z, \eta \rangle^{\mu - |\alpha|} e^{-\frac{c_0}{2} \langle z, \eta \rangle} d\eta \le \int_{|\eta| \le 1} \frac{d\eta}{|\eta|^{|\alpha| - \mu}} = M_1 < +\infty,$$

since $|\alpha| = d$ or $d - 1$. If $\alpha = 0$ we have,

$$\int_{|\eta| \le 1} \langle z, \eta \rangle^{\mu} e^{-\frac{c_0}{2} \langle z, \eta \rangle} d\eta \le 2^{\frac{\mu}{2}} \int_{|\eta| \le 1} d\eta = M_2 < +\infty.$$

c) It follows from the above results that,

$$|x|^d |\widehat{p}(x)| \le M_2, \quad |x|^{d-1} |\widehat{p}(x)| \le M_3 z^{-1}.$$

Using the equality $|x|^{d-\varepsilon} |\widehat{p}(x)| = \left(|x|^d |\widehat{p}(x)| \right)^{1-\varepsilon} \left(|x|^{d-1} |\widehat{p}(x)| \right)^{\varepsilon}$ and the above estimates we obtain the desired result.

d) This follows from the fact that using questions 2 and 3 we have,

$$|\widehat{p}(x)| \le \frac{C}{|x|^{d+1}}, \quad |x| \ge 1, \quad |\widehat{p}(x)| \le C \frac{z^{-\varepsilon}}{|x|^{d-\varepsilon}}, \quad |x| \le 1.$$

4. Set $A = \|\Delta_j (e^{-za(D_x)} u)\|_{L^\infty}$. We have,

$$A = z^{-\mu} \|z^{\mu} \langle D_x \rangle^{\mu} e^{-za(D_x)} \langle D_x \rangle^{-\mu} \Delta_j u\|_{L^\infty} = z^{-\mu} \|p(D_x) \langle D_x \rangle^{-\mu} \Delta_j u\|_{L^\infty},$$

$$= z^{-\mu} \|\widehat{p} \star \left(\langle D_x \rangle^{-\mu} \Delta_j u \right) \|_{L^\infty} \le C z^{-\mu} \|\widehat{p}\|_{L^1} \| \langle D_x \rangle^{-\mu} \Delta_j u\|_{L^\infty},$$

$$\le C_\varepsilon z^{-\mu-\varepsilon} 2^{-j(\rho+\mu)} \|u\|_{C_*^\rho}.$$

5. Choose $\varepsilon > 0$ such that $\mu + \varepsilon < \frac{1}{2}$. Then,

$$\int_0^1 \left(\sup_{j \ge -1} 2^{j(\rho+\mu)} \|\Delta_j (e^{-za(D_x)} u)\|_{L^\infty(\mathbf{R}^d)} \right)^2 dz \le C_\varepsilon^2 \left(\int_0^1 \frac{dz}{z^{2(\mu+\varepsilon)}} \right) \|u\|_{C_*^\rho}^2,$$

which proves the result.

Solution 26

Part 1
1. We have,

$$|Tu(x)| \le \left(\int_{\mathbf{R}^d} |K(z)|\,dz \right) \|u\|_{L^\infty(\mathbf{R}^d)}, \qquad \forall x \in \mathbf{R}^d.$$

So T is continuous from L^∞ to itself and $\|T\|_{L^\infty \to L^\infty} \le \int_{\mathbf{R}^d} |K(z)|\,dz$. Now set $A = \{ z \in \mathbf{R}^d : K(z) \ne 0 \}$. If $\mu(A) = 0$ we have $K \equiv 0$, $T \equiv 0$ (since K is continuous). If $\mu(A) \ne 0$ let $x_0 \in \mathbf{R}^d$ be fixed. Then $Tu(x_0) = \int_{\{x_0 - y \in A\}} K(x_0 - y)u(y)\,dy$. Let u_0 be such that $u_0(y) = \frac{K(x_0-y)}{|K(x_0-y)|}$ if $x_0 - y \in A$ and $u_0(y) = 0$ otherwise. Then $u_0 \in L^\infty$, $\|u_0\|_{L^\infty(\mathbf{R}^d)} = 1$ and,

$$Tu_0(x_0) = \int_{\{x_0 - y \in A\}} |K(x_0 - y)|\,dy = \int_{\mathbf{R}^d} |K(z)|\,dz.$$

Therefore, $\frac{\|Tu_0\|_{L^\infty(\mathbf{R}^d)}}{\|u_0\|_{L^\infty(\mathbf{R}^d)}} \ge \int_{\mathbf{R}^d} |K(z)|\,dz$, so that $\|T\|_{L^\infty \to L^\infty} \ge \int_{\mathbf{R}^d} |K(z)|\,dz$ and we have the equality.

2. a) If $i \ne k$ we have $\ell_i(0) = (0, \ldots, 1, \ldots, 0)$ and $\ell_k'(\lambda) = (0, \ldots, \omega_k^2, \ldots, 0)$, which shows that $\det(\ell_1(0), \ldots, \ell_k'(0), \ldots, \ell_d(0)) = \omega_k^2$. Therefore, $F'(0) = \sum_{k=1}^d \omega_k^2 = 1$.
   b) Set $G_k(\lambda) = \det(\ell_1(\lambda) \ldots, \ell_k'(\lambda) \ldots \ell_d(\lambda))$. Since $\ell_k''(\lambda) \equiv 0$ we have

$$G_k'(\lambda) = \sum_{j \ne k} \det(\ell_1(\lambda), \ldots, \ell_j'(\lambda), \ldots, \ell_k'(\lambda), \ldots, \ell_d(\lambda)).$$

Now $\ell_j'(\lambda) = \omega_j(\omega_1, \ldots, \omega_d)$ and $\ell_k'(\lambda) = \omega_k(\omega_1, \ldots, \omega_d)$. Therefore, each determinant in the sum vanishes; then $G_k'(\lambda) \equiv 0$ and $F''(\lambda) = \sum_{k=1}^d G_k'(\lambda) \equiv 0$. It follows that we have $F(\lambda) = 1 + \lambda$.

Part 2
1. Indeed we have $\|\Delta_j u\|_{C_*^\sigma} = \sup_{k \ge -1} (2^{k\sigma} \|\Delta_k \Delta_j u\|_{L^\infty})$. On the other hand, due to the support of φ, we have $\Delta_k \Delta_j = 0$ if $|k - j| > 2$, so,

$$\|\Delta_j u\|_{C_*^\sigma} = \max \left(2^{(j-1)\sigma} \|\Delta_{j-1} \Delta_j u\|_{L^\infty}, 2^{j\sigma} \|\Delta_j \Delta_j u\|_{L^\infty}, \right.$$

$$\left. 2^{(j+1)\sigma} \|\Delta_{j+1} \Delta_j u\|_{L^\infty} \right).$$

We just have to notice that Δ_j is uniformly bounded on L^∞ to see that there exists $C > 0$ such that,

$$\|\Delta_j u\|_{C^\sigma_*} \leq C 2^{j\sigma} \|\Delta_j u\|_{L^\infty} \leq C' 2^{j\sigma} \|u\|_{L^\infty}.$$

2. Since Δ_j commutes with $S(-1)$, using the definition of C^s_*, the assumption, and the previous question we obtain,

$$2^s \|\Delta_j S(-1) \Delta_j u\|_{L^\infty} \leq \|S(-1) \Delta_j u\|_{C^s_*} \leq C \|\Delta_j u\|_{C^\sigma_*,} \leq C' 2^{j\sigma} \|u\|_{L^\infty},$$

for all $u \in L^\infty(\mathbf{R}^d)$.
3. a) We have

$$T_j u(x) = (2\pi)^{-d} \iint e^{i[(x-y)\cdot\xi + |\xi|^\alpha]} \varphi^2(2^{-j}\xi) u(y) \, dy \, d\xi,$$

$$= \int \left((2\pi)^{-d} \int e^{i[(x-y)\cdot\xi + |\xi|^\alpha]} \varphi^2(2^{-j}\xi) \, d\xi \right) u(y) \, dy.$$

Set $K_h(z) = (2\pi)^{-d} \int e^{i[z\cdot\xi + |\xi|^\alpha]} \varphi^2(2^{-j}\xi) \, d\xi$ then $\eta = h\xi$ in the integral. We obtain,

$$K_h(z) = (2\pi h)^{-d} \int e^{\frac{i}{h}[z\cdot\eta + h^{1-\alpha}|\eta|^\alpha]} \varphi^2(\eta) \, d\eta.$$

b) We see that

$$K_h = h^{-d} \mathcal{F}^{-1}\left(e^{ih^{-\alpha}|\eta|^\alpha} \varphi^2(\eta) \right) \left(\frac{z}{h} \right).$$

Since the support of φ is contained in $\left\{ \frac{1}{2} \leq |\eta| \leq 2 \right\}$ the function $e^{ih^{-\alpha}|\eta|^\alpha} \varphi^2(\eta)$ belongs to $C_0^\infty(\mathbf{R}^d)$. Its inverse Fourier transform belongs to $C^0 \cap L^1$.
c) It follows from Part 1 that T_j is continuous from $L^\infty(\mathbf{R}^d)$ to itself and setting $z = h^{1-\alpha}s$ that,

$$\|T_j\|_{L^\infty \to L^\infty} = \int |K_h(z)| \, dz = h^{d(1-\alpha)} \int |K_h(h^{1-\alpha}s)| \, ds,$$

$$= h^{d(1-\alpha)} \int |\tilde{K}_h(s)| \, ds,$$

where,

$$\tilde{K}_h(s) = (2\pi h)^{-d} \int e^{ih^{-\alpha}\phi(s,\eta)} \varphi^2(\eta) \, d\eta, \quad \phi(s, \eta) = s \cdot \eta + |\eta|^\alpha.$$

4. Recall that $\text{supp}\varphi \subset \left\{ \frac{1}{2} \leq |\eta| \leq 2 \right\}$. We have $\frac{\partial \phi}{\partial \eta}(s, \eta) = s + \alpha |\eta|^{\alpha-2}\eta$, so in the integral we have,

$$\frac{\partial \phi}{\partial \eta}(s, \eta) \geq \alpha |\eta|^{\alpha-1} - |s| \geq \frac{\alpha}{2^{|1-\alpha|}} - |s| \geq \frac{1}{2}\frac{\alpha}{2^{|1-\alpha|}}.$$

We consider then the operator,

$$L = \frac{h^\alpha}{i} \frac{1}{|\partial_\eta \phi|^2} \sum_{k=1}^{d} \frac{\partial \phi}{\partial \eta_k} \frac{\partial}{\partial \eta_k} := h^\alpha L_0 \left(s, \eta, \partial_\eta\right)$$

It satisfies $Le^{ih^{-\alpha}\phi(s,\eta)} = e^{ih^{-\alpha}\phi(s,\eta)}$ so $h^{\alpha N} L_0^N e^{ih^{-\alpha}\phi(s,\eta)} = e^{ih^{-\alpha}\phi(s,\eta)}$ for all $N \in \mathbf{N}$. Since $\varphi \in C_0^\infty(\mathbf{R}^d)$ we can integrate by parts and we obtain,

$$\widetilde{K}_h(s) = h^{\alpha N} \int e^{ih^{-\alpha}\phi(s,\eta)} ({}^t L_0)^N \varphi(\eta) \, d\eta,$$

where ${}^t L_0$ denotes the transpose operator of L_0. It follows that,

$$|\widetilde{K}_h(s)| \leq h^{\alpha N} \int |({}^t L_0)^N \varphi(\eta)| \, d\eta.$$

5. Since $|\eta|^{\alpha-1} \leq |\eta|^{|\alpha-1|} \leq 2^{|\alpha-1|}$ on the support of φ we have,

$$|\partial_\eta \phi(s, \eta)| \geq |s| - \alpha |\eta|^{\alpha-1} \geq |s| - \alpha 2^{|\alpha-1|} \geq \frac{1}{2}|s|. \tag{9.25}$$

Taking the same vector field as in question 4, we obtain by the same way,

$$|\widetilde{K}_h(s)| \leq h^{\alpha N} \int |({}^t L_0)^N \varphi(\eta)| \, d\eta.$$

Now in this region according to (9.25) the coefficients of L_0 are bounded by $\frac{C}{|s|}$. Then we have $|\widetilde{K}_h(s)| \leq C_N h^{\alpha N} |s|^{-N}$.

6. a) In this region, $\partial_\eta \phi(s, \eta) = s + \alpha \eta |\eta|^{\alpha-2} = 0$ implies that $|s| = \alpha |\eta|^{\alpha-1}$ so

$$|\eta| = \left(\frac{|s|}{\alpha}\right)^{\frac{1}{\alpha-1}} \text{ where } s = -\alpha \eta \left(\frac{|s|}{\alpha}\right)^{\frac{\alpha-2}{\alpha-1}} \text{ and eventually } \eta = c_\alpha s |s|^{\frac{2-\alpha}{\alpha-1}}.$$

On the other hand,

$$\frac{\partial^2 \phi}{\partial \eta_j \partial \eta_k} = \frac{\partial}{\partial \eta_k}\left(s_j + \alpha \eta_j |\eta|^{\alpha-2}\right) = \alpha \delta_{jk} |\eta|^{\alpha-2} + \alpha(\alpha - 2)\eta_j \eta_k |\eta|^{\alpha-4},$$

$$= \alpha |\eta|^{\alpha-2}(\delta_{jk} + (\alpha - 2)\omega_j \omega_k), \quad \omega = \frac{\eta}{|\eta|}.$$

b) Using question 2 of Part 1 we obtain,

$$\det\left(\frac{\partial^2\phi}{\partial\eta_j\partial\eta_k}\right) = (\alpha|\eta|^{\alpha-2})^d \det(\delta_{jk}+(\alpha-2)\omega_j\omega_k) = (\alpha-1)(\alpha|\eta|^{\alpha-2})^d \neq 0$$

since $\alpha \neq 1$.
c) Recall that,

$$\tilde{K}_h(s) = (2\pi h)^{-d} \int_{\mathbf{R}^d} e^{ih^{-\alpha}\phi(s,\eta)}\varphi^2(\eta)d\eta, \quad \phi(s,\eta) = s\cdot\eta + |\eta|^\alpha.$$

We apply the stationary phase formula. The large parameter is $\lambda = h^{-\alpha}$. Since $|\eta_c| = C_\alpha|s|^{\frac{1}{\alpha-1}}$ the square root of the determinant of the Hessian matrix at the critical point is equal to $c_{d\alpha}|s|^{\frac{(\alpha-2)d}{2(\alpha-1)}}$. The formula then gives,

$$\tilde{K}_h(s) = C_{\alpha,d}h^{-d}h^{\frac{\alpha d}{2}}\left\{\frac{e^{ih^{-\alpha}\phi(s,\eta_c)}}{|s|^{\frac{(\alpha-2)d}{2(\alpha-1)}}}\varphi^2(\eta_c) + O(h^\alpha)\right\}. \tag{9.26}$$

7. We have proved in question 3 that,

$$\|T_j\|_{L^\infty\to L^\infty} = h^{d(1-\alpha)}\int|\tilde{K}_h(s)|\,ds.$$

In the region where $|s| \leq \frac{1}{2}\frac{\alpha}{2^{|1-\alpha|}}$ we have proved in question 4 that $|\tilde{K}_h(s)| \leq C_N h^N$ for all $N \in \mathbf{N}$. The integral of $\tilde{K}_h$ on this domain gives a contribution $O(h^N)$. In the region where $|s| \geq 2^{1+|\alpha-1|}\alpha$ we have proved in question 5 that $|\tilde{K}_h(s)| \leq C_N h^N|s|^{-N}$. The integral of $\tilde{K}_h$ on this domain gives a contribution $O(h^N)$. Eventually in the intermediate region we have the equality (9.26) so the integral of $\tilde{K}_h$ on this domain gives a contribution which is bounded below by $Ch^{-d}h^{\frac{d\alpha}{2}}$. Therefore,

$$\|T_j\|_{L^\infty\to L^\infty} \geq Ch^{d(1-\alpha)}h^{-d}h^{\frac{d\alpha}{2}} - C_N h^N \geq C'h^{-\frac{d\alpha}{2}} = C2^{j\frac{d\alpha}{2}}.$$

8. From the definition of the operator norm, for all $\varepsilon > 0$ there exists $u_0 \in L^\infty(\mathbf{R}^d)$ nonvanishing identically such that,

$$\|T_ju_0\|_{L^\infty} \geq (C2^{j\frac{d\alpha}{2}} - \varepsilon)\|u_0\|_{L^\infty}.$$

Taking ε small enough and using question 2 we obtain,

$$C2^{j\sigma}\|u_0\|_{L^\infty} \geq 2^{js}\|T_ju_0\|_{L^\infty} \geq C2^{js}2^{j\frac{d\alpha}{2}}\|u_0\|_{L^\infty}.$$

This proves that we must have $\sigma \geq s + \frac{d\alpha}{2}$.

9. a) We have, setting $\xi = \lambda\eta$, $\lambda > 0$ in the integral,

$$\Delta_j S(-1)u(x) = (2\pi)^{-d} \int e^{ix\cdot\xi} e^{i|\xi|^\alpha} \varphi(2^{-j}\xi)\widehat{u}(\xi)\, d\xi,$$

$$= \lambda^d (2\pi)^{-d} \int e^{i\lambda x\cdot\eta} e^{i\lambda^\alpha |\eta|^\alpha} \varphi(2^{-j}\lambda\eta)\widehat{u}(\lambda\eta)\, d\eta$$

$$= \Delta_j^\lambda S(t_0)u_{\frac{1}{\lambda}}(\lambda x).$$

since $\widehat{u}(\lambda\eta) = \lambda^{-d}\widehat{u_{\frac{1}{\lambda}}}(\eta)$.

b)

$$\overline{\Delta_j S(-t_0)u(x)} = (2\pi)^{-d} \int e^{-ix\cdot\xi} e^{-it_0|\xi|^\alpha} \varphi(2^{-j}\xi)\overline{\widehat{u}(\xi)}\, d\xi,$$

$$= (2\pi)^{-d} \int e^{ix\cdot\eta} e^{-it_0|\eta|^\alpha} \varphi(2^{-j}\eta)\overline{\widehat{u}(-\eta)}\, d\xi = \Delta_j S(t_0)\overline{u}(x),$$

since $\overline{\widehat{u}(-\eta)} = \widehat{\overline{u}}(\eta)$.

c) The questions a) and b) show that if $S(t_0)$ is continuous from C_*^σ to C_*^s the same holds for $S(-1)$; so we have $s \le \sigma - \frac{d\alpha}{2}$.

Part 3

1. It is easy to see that u satisfies $(i\partial_t - |D_x|)u = 0$, $u|_{t=0} = u_0$. Then $\partial_t u|_{t=0} = -i|D_x|u_0$. It is sufficient to apply the operator $i\partial_t + |D_x|$ to the equation and to notice that $(i\partial_t + |D_x|)(i\partial_t - |D_x|) = -\square$.

2. Following the Hint we deduce that for $t_0 \ne 0$,

$$u(t_0, x) = -i\frac{t_0}{4\pi} \int_{S^2} (|D_x|u_0)(x - t_0\omega)\, d\omega + \frac{1}{4\pi}\int_{S^2} u_0(x - t_0\omega)\, d\omega$$

$$+ \frac{t_0}{4\pi}\sum_{i=1}^3 \int_{S^2} \frac{\partial u_0}{\partial x_i}(x - t_0\omega)\omega_i\, d\omega.$$

$$(9.27)$$

Let v_0 be one of the functions $|D_x|u_0, u_0, \frac{\partial u_0}{\partial x_i}$. Noticing that,

$$\Delta_j[v_0(x - t_0\omega)] = (\Delta_j v_0)(x - t_0\omega),$$

we have, $\|\Delta_j[v_0(\cdot - t_0\omega)]\|_{L^\infty(\mathbf{R}_x^3)} \le \|\Delta_j v_0\|_{L^\infty(\mathbf{R}^3)}$. Assuming that $u_0 \in C_*^\sigma(\mathbf{R}^3)$ and applying the operator Δ_j to both members of (9.27) we deduce that,

$$\|u(t_0, \cdot)\|_{C_*^{\sigma-1}(\mathbf{R}^3)} \le C\|u_0\|_{C_*^\sigma(\mathbf{R}^3)}.$$

Therefore, the operator $e^{-it_0|D_x|}$ is continuous from $C_*^\sigma(\mathbf{R}^3)$ to $C_*^\rho(\mathbf{R}^3)$ where $\rho = \sigma - 1$ and we do not have $\rho \leq \sigma - \frac{3}{2}$.

Solution 27

Part 1
In what follows we shall denote $L^2 = L^2(\mathbf{R}^d)$, $H^1 = H^1(\mathbf{R}^d)$ and by $v(t)$ the function $\mathbf{R}^d \to \mathbf{C} : x \mapsto v(t, x)$.

1. We have $u = e^{-\frac{k}{2}(t-T)^2} v$ so,

$$D_t u + A u = e^{-\frac{k}{2}(t-T)^2} \left(D_t v + A v - \frac{k}{i}(t-T)v\right).$$

2. a) This follows from the fact that modulo S^0 the symbol of A^* is $\bar{a} = a$ since a is real.
   b) Since $D_t = \frac{1}{i}\partial_t$ and $(f, g)_{L^2} = \int f(x)\overline{g(x)}\,dx$ we have,

$$(Xv(t), Yv(t))_{L^2} = -k(t-T)(\partial_t v(t), v(t))_{L^2} - ik(t-T)(Av(t), v(t))_{L^2},$$

$$=: f(t) + g(t).$$

We have $2\operatorname{Re} f(t) = -k(t-T)\partial_t(\|v(t)\|_{L^2}^2)$ so integrating by parts and using the fact that $v(0) = v(T) = 0$ we obtain,

$$2\operatorname{Re} \int_0^T f(t)\,dt = k\int_0^T \|v(t)\|_{L^2}^2\,dt.$$

Now $2\operatorname{Re} g(t) = 2k(t-T)\operatorname{Im}(Av(t), v(t))_{L^2}$. Since the symbol a of A is real we have $A^* = A - B_0$ where $B_0 \in \operatorname{Op}(S^0)$. Then $(Av(t), v(t))_{L^2} = (v(t), Av(t))_{L^2} + (B_0 v(t), v(t))_{L^2}$ which shows that $2i\operatorname{Im}(Av(t), v(t))_{L^2} = (B_0 v(t), v(t))_{L^2}$. The operators with symbols in S^0 being continuous on L^2 we have,

$$\left|2\operatorname{Re} \int_0^T g(t)\,dt\right| \leq Ck\int_0^T (T-t)\|v(t)\|_{L^2}^2 \leq CkT\int_0^T \|v(t)\|_{L^2}^2.$$

Let $T_0 > 0$ be such that $CT_0 \leq \frac{1}{2}$. For $T \leq T_0$ we obtain,

$$2\operatorname{Re} \int_0^T (Xv(t), Yv(t))_{L^2}\,dt \geq \frac{k}{2}\int_0^T \|v(t)\|_{L^2}^2\,dt.$$

3. According to question 1 and since $\widetilde{P} = X + Y$ we obtain,

$$\int_0^T e^{k(t-T)^2} \|Pu(t)\|_{L^2}^2 \, dt = \int_0^T \|\widetilde{P}v(t)\|_{L^2}^2 \, dt = \int_0^T \|Xv(t)\|_{L^2}^2 \, dt$$

$$+ \int_0^T \|Yv(t)\|_{L^2}^2 \, dt + 2\mathrm{Re}\,(Xv(t),\, Yv(t))_{L^2}.$$

It follows from the previous question that for all $v \in E_T$ we have,

$$k \int_0^T e^{k(t-T)^2} \|u(t)\|_{L^2}^2 \, dt \leq \int_0^T e^{k(t-T)^2} \|Pu(t)\|_{L^2}^2 \, dt.$$

4. a) For $u \in S'$ we have $\mathcal{F}(A^2 u) = \langle \xi \rangle \mathcal{F}(Au) = \langle \xi \rangle^2 \widehat{u} = (1 + |\xi|^2)\widehat{u} = \mathcal{F}((Id - \Delta_x)u)$.

   b) We have $Q = (D_t - A)(D_t + A) = D_t^2 - A^2 = -\partial_t^2 + \Delta_x - Id = -(\Box - Id)$.

   c) Using question 3 successively with $P = D_t \pm A$ we obtain,

$$k^2 \int_0^T e^{k(t-T)^2} \|u(t)\|_{L^2}^2 \, dt \leq C_1 k \int_0^T e^{k(t-T)^2} \|(D_t + A)u(t)\|_{L^2}^2 \, dt,$$

$$\leq C_2 \int_0^T e^{k(t-T)^2} \|(D_t - A)(D_t + A)u(t)\|_{L^2}^2 \, dt,$$

$$\leq C_3 \left(\int_0^T e^{k(t-T)^2} \|\Box u(t)\|_{L^2}^2 \, dt + \int_0^T e^{k(t-T)^2} \|u(t)\|_{L^2}^2 \, dt \right).$$

Taking $k \geq k_0$, where $k_0^2 \geq 2C_3$ we get, $k^2 - C_3 \geq \frac{1}{2}k^2$. Therefore, we can absorb the last term of the right-hand side by the left-hand side and we obtain the desired estimate.

Part 2

1. Same proof as in question 1 Part 1.

2. The term (1) corresponds to the integral $2\mathrm{Re} \int_0^T (D_t v(t), ik(t - T)v(t))_{L^2} \, dt$ which by question 2 of Part 1 is bounded below by $\frac{k}{2} \int_0^T \|v(t)\|_{L^2}^2 \, dt$.

3. a) This follows from the general theory ; $A^* \in \mathrm{Op}(S^1)$, $B \in \mathrm{Op}(S^1)$ so $A^* B \in \mathrm{Op}(S^2)$. Moreover, $c = a^* b + r_1, r_1 \in S^1$. Since $a^* = a + r_2, r_2 \in S^1$ we have $c = ab + r_3, r_3 \in S^1$ and ab is real. So, again by the general theory, we have $C - C^* \in \mathrm{Op}(S^1)$.

   b) We have $(2) = \mathrm{Im} \int_0^T (Av(t), Bv(t))_{L^2} \, dt = \mathrm{Im} \int_0^T (v(t), Cv(t))_{L^2} \, dt$. It follows from question a) that,

$$(v(t), Cv(t))_{L^2} = (Cv(t), v(t))_{L^2} + (Rv(t), v(t))_{L^2}, \quad R \in \mathrm{Op}(S^1).$$

Using the fact that $z - \bar{z} = 2i\,\mathrm{Im}\,z$, the Cauchy–Schwarz inequality and the continuity of R from H^1 to L^2 we obtain,

$$|\mathrm{Im}\,(v(t), Cv(t))_{L^2}| \le C_1 \|v(t)\|_{H^1} \|v(t)\|_{L^2}.$$

It follows that,

$$|(2)| \le C_2 \int_0^T \|v(t)\|_{H^1} \|v(t)\|_{L^2}\, dt.$$

4. a) We have $(3) = -2\mathrm{Re}\int_0^T (\partial_t v(t), Bv(t))_{L^2}\, dt$.

Since $v(0) = v(T) = 0$ we have $\int_0^T \partial_t\big[(v(t), Bv(t))_{L^2}\big]\, dt = 0$. Therefore,

$$\int_0^T (\partial_t v(t), Bv(t))_{L^2}\, dt + \int_0^T (v(t), B\,\partial_t v(t))_{L^2}\, dt = 0,$$

since ∂_t and B commute. Now since the symbol of B is real we have $B^* = B + R$ where $R \in \mathrm{Op}(S^0)$. Therefore,

$$\int_0^T (\partial_t v(t), Bv(t))_{L^2}\, dt + \int_0^T (Bv(t), \partial_t v(t))_{L^2}\, dt,$$

$$= -\int_0^T (Rv(t), \partial_t v(t))_{L^2}\, dt,$$

which implies that,

$$2\mathrm{Re}\int_0^T (\partial_t v(t), Bv(t))_{L^2}\, dt = -\int_0^T (Rv(t), \partial_t v(t))_{L^2}\, dt.$$

Since R is continuous from L^2 to L^2 we obtain,

$$|(3)| \le \int_0^T \|v(t)\|_{L^2} \|\partial_t v(t)\|_{L^2}\, dt.$$

b) We have $D_t = D_t + A - A = X - A$ so that,

$$|(3)| \le \int_0^T \|v(t)\|_{L^2} \|Xv(t)\|_{L^2}\, dt + \int_0^T \|v(t)\|_{L^2} \|Av(t)\|_{L^2}\, dt.$$

We use the fact that for all positive numbers a, b, ε, $ab \le \varepsilon a^2 + \frac{1}{4\varepsilon}b^2$ and the fact that A is continuous from H^1 to L^2.

5. This follows from the fact that,

$$\int_0^T \|\tilde{P}v(t)\|_{L^2}^2 = \int_0^T \left(\|Xv(t)\|_{L^2}^2 + \|Yv(t)\|_{L^2}^2 \right) dt$$

$$+ 2\text{Re} \int_0^T (Xv(t), Yv(t))_{L^2} \, dt$$

and from questions 2, 3, and 4 taking ε small enough.

6. a) Since $B \in \text{Op}(S^1)$ is elliptic there exists $C > 0$ such that,

$$\|v(t)\|_{H^1} \leq C(\|Bv(t)\|_{L^2} + \|v(t)\|_{L^2}).$$

The inequality to be proved follows then from the fact that $Y = iB + ik(t - T)$.

b) From question a) we have,

$$\int_0^T \|v(t)\|_{H^1} \|v(t)\|_{L^2} \, dt \leq C \int_0^T \|Yv(t)\|_{L^2} \|v(t)\|_{L^2} \, dt$$

$$+ C(1 + kT) \int_0^T \|v(t)\|_{L^2}^2 \, dt.$$

Then we have just to use the inequality $ab \leq \varepsilon a^2 + \frac{1}{4\varepsilon} b^2$ for all positive a, b, ε.

7. Set

$$F = \int_0^T \|Xv(t)\|_{L^2}^2 \, dt + \int_0^T \|Yv(t)\|_{L^2}^2 \, dt + \frac{k}{2} \int_0^T \|v(t)\|_{L^2}^2 \, dt.$$

We have shown in question 5 that,

$$F \leq C \left(\int_0^T \|\tilde{P}v(t)\|_{L^2}^2 \, dt + \int_0^T \|v(t)\|_{H^1} \|v(t)\|_{L^2} \, dt \right).$$

Using question 6b) we obtain,

$$F \leq C \int_0^T \|\tilde{P}v(t)\|_{L^2}^2 \, dt + C'\varepsilon \int_0^T \|Yv(t)\|_{L^2}^2 \, dt + C''(1 + kT) \int_0^T \|v(t)\|_{L^2}^2 \, dt.$$

Taking $\varepsilon > 0$ small enough so that $C'\varepsilon \leq \frac{1}{2}$, k large enough and T small enough so that $C''(1 + kT) \leq \frac{k}{4}$, we can absorb the two last terms in the right-hand side by F and we obtain,

$$\int_0^T \left(\|Xv(t)\|_{L^2}^2 + \|Yv(t)\|_{L^2}^2 + k\|v(t)\|_{L^2}^2 \right) dt \leq C_1 \int_0^T \|\tilde{P}v(t)\|_{L^2}^2 \, dt.$$

8. a) From question 6a) we have,

$$\|v(t)\|_{H^1}^2 \leq C \left(\|Yv(t)\|_{L^2}^2 + (1 + k^2 T^2) \|v(t)\|_{L^2}^2 \right).$$

Using question 7 we obtain, if k is large enough,

$$\int_0^T \|Yv(t)\|_{L^2}^2 \, dt + (1 + k^2 T^2) \int_0^T \|v(t)\|_{L^2}^2 \, dt \leq C(1 + kT^2) \int_0^T \|\tilde{P}v(t)\|_{L^2}^2 \, dt, \tag{9.28}$$

which proves the result.

b) We have $D_t v = Xv - Av$ so that this inequality follows from a) and from question 7, since A is continuous from H^1 to L^2.

c) We have $I = \int_0^T \|\tilde{P}v(t)\|_{L^2}^2 \, dt$ so that the first inequality to be proved follows from question 7. To prove the second one we notice that

$$e^{\frac{k}{2}(t-T)^2} \partial_t u(t) = \partial_t v(t) - k(t - T)v,$$

so that,

$$\int_0^T e^{k(t-T)^2} \|\partial_t u(t)\|_{L^2}^2 \, dt \leq C \left(\int_0^T \|\partial_t v(t)\|_{L^2}^2 \, dt + k^2 T^2 \int_0^T \|v(t)\|_{L^2}^2 \, dt \right).$$

We have just to use (9.28) and question 8b).

9. a) Since $B^2 = Id - \Delta_x$, we have $Q = D_t^2 + B^2 = -\partial_t^2 - \Delta_x + Id = -\Delta_{x,t} + Id$.

b) Set

$$J = \int_0^T e^{k(t-T)^2} \left(\|u(t)\|_{H^1(\mathbf{R}^d)}^2 + \|D_t u(t)\|_{L^2(\mathbf{R}^d)}^2 \right) dt.$$

It follows from questions 8b) and 9a) that,

$$J \leq C(1 + kT^2) \int_0^T e^{k(t-T)^2} \|(D_t + iB)u(t)\|_{L^2(\mathbf{R}^d)}^2 \, dt,$$

$$\int_0^T e^{k(t-T)^2} \|(D_t + iB)u(t)\|_{L^2(\mathbf{R}^d)}^2 \, dt \leq \frac{C}{k} \int_0^T e^{k(t-T)^2} \|Qu(t)\|_{L^2(\mathbf{R}^d)}^2 \, dt.$$

The inequality to be proved follows from the above two inequalities.

10. Since $c_j, d \in L^\infty((0, T) \times \mathbf{R}^d)$ we have, setting $x_0 = t$,

$$\int_0^T e^{k(t-T)^2} \|Lu(t)\|_{L^2(\mathbf{R}^d)}^2 \, dt \leq \int_0^T e^{k(t-T)^2} \|Qu(t)\|_{L^2(\mathbf{R}^d)}^2 \, dt$$

$$+ C \left(\sum_{j=0}^d \int_0^T e^{k(t-T)^2} \|\partial_{x_j} u(t)\|_{L^2(\mathbf{R}^d)}^2 \, dt + \int_0^T e^{k(t-T)^2} \|u(t)\|_{L^2(\mathbf{R}^d)}^2 \, dt \right).$$

We obtain the desired result by using the inequality in question 9b) and taking $\frac{1}{k} + T^2$ small enough.

Solution 28

1. By the recalled formula and the fact that $1 + |\xi| \leq Ca$, we can write, setting $\sigma = T - t$ and $I = |\partial_x^\alpha \partial_\xi^\beta e^{-\sigma a}|$,

$$I \leq \sum_{j=0}^{|\alpha|+|\beta|} C_{j\alpha\beta} \sigma^j (1 + |\xi|)^{j-|\beta|} e^{-\sigma a} \leq C_{\alpha\beta} \sum_{j=0}^{|\alpha|+|\beta|} [(a\sigma)^j e^{-\sigma a}](1 + |\xi|)^{-|\beta|},$$

$$\leq C_{\alpha\beta} \sum_{j=0}^{|\alpha|+|\beta|} [(a\sigma)^{j+m} e^{-\sigma a}] \frac{1}{(\sigma a)^m} (1 + |\xi|)^{-|\beta|} \leq \frac{C'_{\alpha\beta}}{\sigma^m} (1 + |\xi|)^{-m-|\beta|},$$

since for all $\rho \in (0, +\infty)$ and $t > 0$ we have $t^\rho e^{-t} \leq C_\rho$.

2. Taking $m = 1 - \varepsilon$ in the previous question we see that the symbol e_t belongs to $S^{-(1-\varepsilon)}$ and that all its semi-norms are bounded by $\frac{C_{\alpha\beta}}{(T-t)^{1-\varepsilon}}$. Therefore, the operator $\mathrm{Op}(e_t)$ is continuous from $L^2(\mathbf{R}^d)$ in $H^{1-\varepsilon}(\mathbf{R}^d)$ and we have,

$$\| \mathrm{Op}(e_t)v \|_{H^{1-\varepsilon}} \leq \frac{C'}{(T-t)^{1-\varepsilon}} \|v\|_{L^2}, \quad \forall v \in L^2(\mathbf{R}^d).$$

Then we have $\partial_t e_t = e_t a \in S^{-1-\varepsilon} \times S^1 \subset S^{-\varepsilon}$ and its semi-norms are bounded by $\frac{C_{\alpha\beta}}{(T-t)^{1-\varepsilon}}$. On the other hand, by the symbolic calculus, $\mathrm{Op}(e_t)\,\mathrm{Op}(a) \in \mathrm{Op}(S^{-(1-\varepsilon)} \times S^1) \subset \mathrm{Op}(S^{-\varepsilon})$ and its semi-norms are bounded by $\frac{C_{\alpha\beta}}{(T-t)^{1-\varepsilon}}$. Eventually, again by the symbolic calculus,

$$\mathrm{Op}(\partial_t e_t) - \mathrm{Op}(e_t)\,\mathrm{Op}(a) = \mathrm{Op}(e_t a) - \mathrm{Op}(e_t)\,\mathrm{Op}(a) \in S^{-(1-\varepsilon)},$$

and its semi-norms are bounded by $\frac{C_{\alpha\beta}}{(T-t)^{1-\varepsilon}}$. By the same argument as in the previous estimate we have,

$$\| \operatorname{Op}(\partial_t e_t) - \operatorname{Op}(e_t)\operatorname{Op}(a))v\|_{H^{1-\varepsilon}} \leq \frac{K}{(T-t)^{1-\varepsilon}}\|v\|_{L^2}.$$

3. Since $\partial_t u = -\operatorname{Op}(a)u + f$ one can write,

$$\partial_t \left(\operatorname{Op}(e_t)u(t)\right) = \operatorname{Op}(\partial_t e_t)u(t) + \operatorname{Op}(e_t)\partial_t u(t),$$
$$= (\operatorname{Op}(\partial_t e_t) - \operatorname{Op}(e_t)\operatorname{Op}(a))\,u(t) + \operatorname{Op}(e_t)f(t).$$

Since $u(0) = 0$ and $e_T = 1$ integrating this inequality between 0 and T we obtain,

$$u(T) = \int_0^T (\operatorname{Op}(\partial_t e_t) - \operatorname{Op}(e_t)\operatorname{Op}(a))\,u(t)\,dt + \int_0^T \operatorname{Op}(e_t)f(t)\,dt.$$

4. Using question 3 we can write,

$$\|u(T)\|_{H^{1-\varepsilon}} \leq C \int_0^T \frac{dt}{(T-t)^{1-\varepsilon}} \left(\sup_{t\in[0,T]} \|u(t)\|_{L^2} + \sup_{t\in[0,T]} \|f(t)\|_{L^2} \right),$$

which proves that $u(T) \in H^{1-\varepsilon}(\mathbf{R}^d)$.

Solution 29

Part 1
Integrating by parts, and using the fact that $\widehat{\varphi}$ has compact support in $\{\eta : |\eta| \leq 1\}$ we obtain,

$$y_j^N (R_\lambda\varphi)(y) = (2\pi)^{-d} \int y_j^N e^{iy\cdot\eta} r_\lambda(y,\eta)\widehat{\varphi}(\eta)\,d\eta,$$
$$= (2\pi)^{-d} \int (D_{\eta_j}^N e^{iy\cdot\eta}) r_\lambda(y,\eta)\widehat{\varphi}(\eta)\,d\eta,$$
$$= (2\pi)^{-d} \sum_{N_1+N_2=N} \binom{N}{N_1} \int e^{iy\cdot\eta} D_{\eta_j}^{N_1} r_\lambda(y,\eta) D_{\eta_j}^{N_2}\widehat{\varphi}(\eta)\,d\eta.$$

It follows from the estimates on r_λ that for all $N \in \mathbf{N}$, all $y \in \mathbf{R}^d$ and $\lambda \geq 1$ we have,

$$|y_j|^N |(R_\lambda \varphi)(y)| \leq C(\varphi)\lambda^{m-\frac{1}{2}}(1+|y|).$$

Since $(1 + \sum_{j=1}^d y_j^2)^k \leq C(1 + \sum_{j=1}^d y_j^{2k})$ we deduce from the above inequality that for all $k \in \mathbf{N}$ we have,

$$|(R_\lambda \varphi)(y)| \leq C'(\varphi)\lambda^{m-\frac{1}{2}} \frac{(1+|y|)}{(1+|y|^2)^k}.$$

If $2(2k-1) > d$, the right-hand side belongs to $L^2(\mathbf{R}^d)$.

Part 2
1. We have, setting $y = \lambda^{\frac{1}{2}}(x - x^0)$,

$$\widehat{u_\lambda}(\xi) = \lambda^{\frac{d}{4}} \int e^{-ix\cdot(\xi - \lambda\xi^0)} \varphi(\lambda^{\frac{1}{2}}(x - x^0))\, dx,$$

$$= \lambda^{-\frac{d}{4}} e^{-ix^0\cdot(\xi-\lambda\xi^0)} \int e^{-i\lambda^{-\frac{1}{2}}y\cdot(\xi-\lambda\xi^0)}\varphi(y)\, dy,$$

$$= \lambda^{-\frac{d}{4}} e^{-ix^0\cdot(\xi-\lambda\xi^0)} \widehat{\varphi}(\lambda^{-\frac{1}{2}}(\xi - \lambda\xi^0)).$$

2. Using question 1 and setting $\eta = \lambda^{-\frac{1}{2}}(\xi - \lambda\xi^0)$, so $d\xi = \lambda^{\frac{d}{2}}d\eta$, we obtain,

$$\|u_\lambda\|_{H^{m-\delta}}^2 = \lambda^{-\frac{d}{2}} \int (1+|\xi|^2)^{m-\delta}|\widehat{\varphi}(\lambda^{-\frac{1}{2}}(\xi - \lambda\xi^0))|^2\, d\xi,$$

$$= \int (1 + |\lambda\xi^0 + \lambda^{\frac{1}{2}}\eta|^2)^{m-\delta}|\widehat{\varphi}(\eta)|^2\, d\eta.$$

On the support of $\widehat{\varphi}$ we have $|\eta| \leq 1$ so that, since $|\xi^0| = 1$ we have,

$$|\lambda\xi^0 + \lambda^{\frac{1}{2}}\eta| \geq \lambda - \lambda^{\frac{1}{2}}|\eta| \geq \lambda - \lambda^{\frac{1}{2}} \geq \frac{1}{2}\lambda \quad \text{if } \lambda \geq 4.$$

It follows that,

$$\|u_\lambda\|_{H^{m-\delta}}^2 \geq c\lambda^{2(m-\delta)} \int |\widehat{\varphi}(\eta)|^2\, d\eta.$$

3. Using question 1 one can write,

$$Pu_\lambda(x) = (2\pi)^{-d}\lambda^{-\frac{d}{4}}\int e^{ix\cdot\xi}e^{-ix^0\cdot(\xi-\lambda\xi^0)}p(x,\xi)\widehat{\varphi}(\lambda^{-\frac{1}{2}}(\xi-\lambda\xi^0))\,d\xi,$$

$$= e^{i\lambda x^0\cdot\xi^0}\lambda^{-\frac{d}{4}}(2\pi)^{-d}\int e^{i(x-x^0)\cdot\xi}p(x,\xi)\widehat{\varphi}(\lambda^{-\frac{1}{2}}(\xi-\lambda\xi^0))\,d\xi.$$

Setting $\eta = \lambda^{-\frac{1}{2}}(\xi - \lambda\xi^0)$, so $d\xi = \lambda^{\frac{d}{2}}d\eta$ we obtain,

$$Pu_\lambda(x) = e^{i\lambda x\cdot\xi^0}\lambda^{\frac{d}{4}}(2\pi)^{-d}\int e^{i\lambda^{\frac{1}{2}}(x-x^0)\cdot\eta}p(x,\lambda\xi^0+\lambda^{\frac{1}{2}}\eta)\widehat{\varphi}(\eta)\,d\eta.$$

Set,

$$Q_\lambda(y, D_y)\varphi(y) = (2\pi)^{-d}\int e^{iy\cdot\eta}p(x^0+\lambda^{-\frac{1}{2}}y,\lambda\xi^0+\lambda^{\frac{1}{2}}\eta)\widehat{\varphi}(\eta)\,d\eta,$$

then,

$$e^{-i\lambda x\cdot\xi^0}Pu_\lambda(x) = \lambda^{\frac{n}{4}}(Q_\lambda(y, D_y)\varphi)(\lambda^{\frac{1}{2}}(x-x^0)).$$

4. Let $q_\lambda(y, \eta) = p(x^0+\lambda^{-\frac{1}{2}}y, \lambda\xi^0+\lambda^{\frac{1}{2}}\eta)$. Using the Taylor formula with integral reminder at the order 1 we can write,

$$q_\lambda(y, \eta) = p(x_0, \lambda\xi^0) + r_\lambda(y, \eta), \quad r_\lambda = r_\lambda^1 + r_\lambda^2,$$

$$r_\lambda^1(y, \eta) = \lambda^{-\frac{1}{2}}\int_0^1 (y\cdot\partial_x p)(X_t)\,dt, \quad X_t = (x_0 + t\lambda^{-\frac{1}{2}}y, \lambda\xi^0 + t\lambda^{\frac{1}{2}}\eta),$$

$$r_\lambda^2 = \lambda^{\frac{1}{2}}\int_0^1 (\eta\cdot\partial_\xi p)(X_t)\,dt.$$

Let us show that r_λ satisfies the estimates in Part 1. We have,

$$\partial_\eta^\alpha r_\lambda^1(y, \eta) = \lambda^{-\frac{1}{2}}\int_0^1 (t\lambda^{\frac{1}{2}})^{|\alpha|}y\cdot(\partial_\xi^\alpha\partial_x p)(X_t)\,dt.$$

Since $p \in S^m$ we obtain,

$$|\partial_\eta^\alpha r_\lambda^1(y, \eta)| \le \lambda^{-\frac{1}{2}}\lambda^{\frac{|\alpha|}{2}}|y|\int_0^1 (1+|\lambda\xi^0+t\lambda^{\frac{1}{2}}\eta|)^{m-|\alpha|}\,dt.$$

Now, since $|\xi^0| = 1$, $|\eta| \leq 1$ and $0 \leq t \leq 1$ we have,

$$1 + |\lambda\xi^0 + t\lambda^{\frac{1}{2}}\eta| \leq 1 + \lambda + \lambda^{\frac{1}{2}} \leq 3\lambda \quad \text{if } \lambda \geq 1,$$

$$1 + |\lambda\xi^0 + t\lambda^{\frac{1}{2}}\eta| \geq \lambda - t\lambda^{\frac{1}{2}}|\eta| \geq \lambda - \lambda^{\frac{1}{2}} \geq \frac{1}{2}\lambda, \quad \text{if } \lambda \geq 4. \tag{9.29}$$

It follows that,

$$|\partial_\eta^\alpha r_\lambda^1(y, \eta)| \leq C\lambda^{m-\frac{1}{2}}\lambda^{-\frac{|\alpha|}{2}}|y| \leq C\lambda^{m-\frac{1}{2}}|y|.$$

Now by the Leibniz formula we have,

$$\partial_\eta^\alpha r_\lambda^2(y, \eta) = \lambda^{\frac{1}{2}} \int_0^1 (t\lambda^{\frac{1}{2}})^{|\alpha|}\eta \cdot (\partial_\xi^\alpha \partial_\xi p)(X_t)\, dt$$

$$+ C\lambda^{\frac{1}{2}} \sum_{|\beta|\leq|\alpha|-1} \int_0^1 (t\lambda^{\frac{1}{2}})^{|\beta|}(\partial_\xi^\beta \partial_\xi p)(X_t)\, dt.$$

Thanks to the fact that $p \in S^m$ and to (9.29), the first term above can be bounded by $C\lambda^{\frac{1}{2}}\lambda^{\frac{|\alpha|}{2}}\lambda^{m-|\alpha|-1}$ or by $\lambda^{m-\frac{1}{2}}\lambda^{-\frac{|\alpha|}{2}}$ thus by $C\lambda^{m-\frac{1}{2}}$. The second term can be bounded, in a similar way, by $C \sum_{|\beta|\leq|\alpha|-1} \lambda^{\frac{1}{2}}\lambda^{\frac{|\beta|}{2}}\lambda^{m-|\beta|-1}$ or by $C'\lambda^{m-\frac{1}{2}}$. Gathering the estimates obtained for r_λ^1 and r_λ^2 we obtain eventually,

$$|\partial_\eta^\alpha r_\lambda(y, \eta)| \leq C\lambda^{m-\frac{1}{2}}(1 + |y|).$$

5. It follows from questions 3 and 4 that

$$e^{-i\lambda x \cdot \xi^0} P u_\lambda(x) = \lambda^{\frac{n}{4}} p(x^0, \lambda\xi^0)\varphi(\lambda^{\frac{1}{2}}(x - x^0)) + \lambda^{\frac{n}{4}} (R_\lambda(y, D_y)\varphi)(\lambda^{\frac{1}{2}}(x - x^0)). \tag{9.30}$$

Since $\|f(\lambda^{\frac{1}{2}}(x - x^0)\|_{L^2} = \lambda^{-\frac{d}{4}}\|f\|_{L^2}$ we deduce from (9.30) and from the result in Part 1 that,

$$\|Pu_\lambda\|_{L^2} \leq |p(x^0, \lambda\xi^0)| \|\varphi\|_{L^2} + C(\varphi)\lambda^{m-\frac{1}{2}}.$$

Using the fact that $\|\varphi\|_{L^2} = 1$, the estimate (6.7) and question 2 we can write,

$$c_0\lambda^{m-\delta} \leq C|p(x^0, \lambda\xi^0)| + C(\varphi)\lambda^{m-\frac{1}{2}}.$$

Since $\delta < \frac{1}{2}$ there exists $\lambda_0 > 0$ such that for $\lambda \geq \lambda_0$ we have $c_0\lambda^{m-\frac{1}{2}} - C(\varphi)\lambda^{m-\delta} \geq \frac{c_0}{2}\lambda^{m-\delta}$. It follows that,

$$C|p(x^0, \lambda\xi^0)| \geq \frac{c_0}{2}\lambda^{m-\delta}, \quad \forall \lambda \geq \lambda_0.$$

6. Since $p - p_m \in S^{m-1}$ and $\lambda \geq 1$ we have $|(p-p_m)(x^0, \lambda\xi^0)| \leq C(1+\lambda)^{m-1} \leq C_1\lambda^{m-1}$. It follows from the previous question and from the fact that $\delta < \frac{1}{2}$ that there exists $\lambda_3 > 0$ such that for $\lambda \geq \lambda_3$ we have,

$$|p_m(x^0, \lambda\xi^0)| \geq |p(x^0, \lambda\xi^0)| - |(p-p_m)(x^0, \lambda\xi^0)| \geq C_2\lambda^{m-\delta} - C_1\lambda^{m-1}$$
$$\geq \frac{1}{2}C_2\lambda^{m-\delta}.$$

7. Since p_m is homogeneous of degree m, taking $\lambda = \lambda_3$ we obtain from the previous question,

$$|p_m(x^0, \xi^0)| \geq \frac{1}{2}C_2\lambda^{-m}\lambda^{m-\delta} = \frac{1}{2}C_2\lambda^{-\delta}.$$

Let $x \in \mathbf{R}^d$ and $\xi \in \mathbf{R}^d, \xi \neq 0$. We can take in the above inequality $\xi^0 = \frac{\xi}{|\xi|} \in S^{d-1}$. Using again the homogeneity of p_m we obtain,

$$|p_m(x, \xi)| \geq \frac{1}{2}C_2\lambda^{-\delta}|\xi|^m.$$

Solution 30

Part 1
1. The functions a_j being real and C^1 with respect to all variables, the Cauchy–Lipschitz theorem ensures that this system has a unique solution in a neighborhood of $(t, y) = (0, 0)$. Moreover, the solution has a C^1 dependence with respect to the initial data.
2. a) We have

$$\frac{\partial x_1}{\partial s}(0,0) = 1, \quad \frac{\partial x_1}{\partial y_j}(0,0) = 0,$$

$$\frac{\partial x_j}{\partial s}(0,0) = \dot{x}_j(0,0) = a_j(0,0), \quad \frac{\partial x_j}{\partial y_k}(0,0) = \delta_{jk}, \quad j,k \geq 2.$$

It follows that $\det \frac{\partial\varphi(s,y)}{\partial(s,y)} = 1$.

b) This is a consequence of the inverse function theorem.

c) We have

$$\frac{\partial}{\partial s}(u(\varphi(s, y))) = \frac{\partial}{\partial s}(u(s, x_2(s, y), \ldots, x_d(s, y))),$$

$$= \frac{\partial u}{\partial x_1}(\varphi(s, y))\dot{x}_1(s, y) + \sum_{j=2}^{d} \frac{\partial u}{\partial x_j}(\varphi(s, y))\dot{x}_j(s, y),$$

$$= \frac{\partial u}{\partial x_1}(\varphi(s, y)) + \sum_{j=2}^{d} a_j(\varphi(s, y))\frac{\partial u}{\partial x_j}(\varphi(s, y)) = Xu(\varphi(s, y)).$$

3. Set $\widetilde{b}(s, y) = b(\varphi(s, y))$, $\widetilde{f}(s, y) = f(\varphi(s, y))$. Then the equation $Xu(t, x) + b(t, x)u(t, x) = f(t, x)$ is equivalent to,

$$\frac{\partial \widetilde{u}}{\partial s}(s, y) + \widetilde{b}(s, y)\widetilde{u}(s, y) = \widetilde{f}(s, y). \tag{9.31}$$

Now since $\varphi(0, y) = (0, y)$ we have $\widetilde{u}(0, y) = u(0, y) = u_0(y)$. The equation (9.31) can be solved easily. Indeed it is equivalent to,

$$\frac{d}{ds}\left(\exp\left(\int_0^s \widetilde{b}(\sigma, y)\,d\sigma\right)\widetilde{u}(s, y)\right) = \exp\left(\int_0^s \widetilde{b}(\sigma, y)\,d\sigma\right)\widetilde{f}(s, y),$$

which according to the initial data provides a unique solution given by,

$$\widetilde{u}(s, y) = \exp\left(-\int_0^s \widetilde{b}(\sigma, y)\,d\sigma\right)u_0(y) + \int_0^s \exp\left(\int_s^\sigma \widetilde{b}(z, y)\,dz\right)\widetilde{f}(\sigma, y)\,ds.$$

This solution is C^1 with respect to (s, y) in V so u is C^1 for $(t, x) \in U$.

Part 2

1. Setting $\widetilde{b}_j(x) = b_j(x + x^0)$ likewise for c, g, u, we are led to the case where $x^0 = 0$.

By hypothesis there exists j_0 such that $\widetilde{b}_{j_0}(0) \neq 0$. Then there exists a neighborhood $U_1 \subset U$ of 0 such that $b_{j_0}(x) \neq 0$ for all $x \in V_1$.

On the other hand, we can change the indices so that $j_0 = 1$ (setting $t = z_1 = x_{j_0}, z_k = x_k$ if $k \neq j_0$.). In the new coordinates $Y = \widetilde{b}_{j_0}(t, z)\frac{\partial}{\partial t} + \sum_{k=2}^{d} \widetilde{b}_k(t, z)\frac{\partial}{\partial z_k}$ so that setting $\overline{b}_j = \frac{\widetilde{b}_j}{\widetilde{b}_{j_0}}, \overline{c} = \frac{\widetilde{c}}{\widetilde{b}_{j_0}}, \overline{g} = \frac{\widetilde{g}}{\widetilde{b}_{j_0}}$ the initial problem can be written as,

$$\left(\frac{\partial}{\partial t} + \sum_{k=2}^{d} \overline{b}_j(t, z)\frac{\partial}{\partial z_k}\right)\overline{u} + \overline{c}\,\overline{u} = \overline{g}, \quad \overline{u}|_{t=0} = \widetilde{u}^0.$$

Then we apply the result in Part 1. It follows that this problem has a C^1 unique solution $\tilde{u}$ in a neighborhood W of 0.

Solution 31

Part 1

1. In $\mathbf{R}^d$, let $P = D_1^2 - \sum_{j=2}^d D_j^2$ be the wave operator. Then $p(\xi) = \xi_1^2 - \sum_{j=2}^n \xi_j^2$ and $\xi^0 = (\xi_1^0 = 1, \omega^0)$, where $\omega^0 \in \mathbf{S}^{d-2}$, satisfies $p(\xi^0) = 0$, $\frac{\partial p}{\partial \xi_1}(\xi^0) \neq 0$.

2. If P is hypoelliptic we have $E \subset C^\infty(V)$. The spaces E and $C^\infty(V)$ are Fréchet spaces. Let us consider the map $E \to C^\infty(V) : u \mapsto u$. It has a closed graph. Indeed, if a sequence (u_n, u_n) converges to (u, v) in $E \times C^\infty(V)$ then (u_n) converges to u in E so in $C^0(V)$ and (u_n) converges to u in $C^\infty(V)$ so in $C^0(V)$. It follows that $u = v$. It follows from the closed graph theorem that this map is continuous. Since the semi-norms of $C^\infty(V)$ are of the form $q_{F,M}(u) = \sum_{|\alpha| \leq M} \sup_F |D^\alpha u|$ where F is a compact subset of V and $M \in \mathbf{N}$, taking for F a compact containing the point x^0 there exists a compact K in V and $N_0 \in \mathbf{N}$ such that,

$$|\mathrm{grad} u(x^0)| \leq C \sum_{|\alpha| \leq 1} \sup_F |D^\alpha u| \leq C' \left(\sup_K |u| + \sum_{|\alpha| \leq N_0} \sup_K |D^\alpha Pu| \right)$$

for all u in E which contains $C^\infty(V)$.

3. We have $D_k(e^{i\lambda\varphi}w) = e^{i\lambda\varphi}(D_k w + i\lambda(D_k\varphi)w)$. On the other hand, we have,

$$D_j D_k(e^{i\lambda\varphi}w) = e^{i\lambda\varphi}\Big(D_j D_k w + i\lambda(D_j\varphi)D_k w + i\lambda(D_k\varphi)D_j w$$
$$+ i\lambda(D_j D_k\varphi)w - \lambda^2(D_k\varphi)(D_j\varphi)w \Big).$$

We deduce that,

$$e^{-i\lambda\varphi}P(e^{i\lambda\varphi}w) = \lambda^2 p(x, \mathrm{grad}\varphi(x))w$$
$$- 2i\lambda \sum_{j,k=1}^d a_{jk}(x)(\partial_j\varphi)\partial_k w + i\lambda((P-c)\varphi)w + Pw.$$

By $(\star)$ we have $p(x, \mathrm{grad}\varphi(x)) = 0$. Moreover, we have $\frac{\partial p}{\partial \xi_k} = 2\sum_{j=1}^{d} a_{jk}\xi_j$. It follows that,

$$e^{-i\lambda\varphi} P(e^{i\lambda\varphi} w) = -i\lambda \left(\sum_{k=1}^{d} \frac{\partial p}{\partial \xi_k} (x, \mathrm{grad}\varphi(x)) \partial_k w - ((P-c)\varphi)w \right) + Pw.$$

We have just to set $b = -(P-c)\varphi$.

4. a) We argue by induction on $|\alpha| = n$. The formula is true if $n = 0$. Assume it is true for n and let $|\beta| = n+1$. Then $\partial^\beta = \partial_\ell \partial^\alpha$ where $|\alpha| = n$. Then we have,

$$\partial^\beta (e^{i\lambda\varphi} f) = \partial_\ell \partial^\alpha (e^{i\lambda\varphi} f) = \left(\sum_{k=0}^{n} (\partial_\ell c_{k\alpha})\lambda^k + i\sum_{k=0}^{n} c_{k\alpha}(\partial_\ell \varphi)\lambda^{k+1} \right) e^{i\lambda\varphi},$$

$$= \sum_{j=0}^{n+1} c_{j\beta}\lambda^j e^{i\lambda\varphi}.$$

b) Since $\lambda \geq 1$ we have $\lambda^k \leq \lambda^{|\alpha|}$ if $k \leq |\alpha|$, so that,

$$\sup_K |\partial^\alpha (e^{i\lambda\varphi} f)| \leq \lambda^{|\alpha|} \sum_{k=0}^{|\alpha|} \sup_K |c_{k\alpha}|.$$

5. By (H), $(\star)$ and Problem 30, these problems have C^∞ solutions in a neighborhood $V_1 \subset V$ of x^0.

6. a) We have,

$$e^{-i\lambda\varphi} Pv = e^{-i\lambda\varphi} \sum_{j=0}^{N_0} \lambda^{-j} P(e^{i\lambda\varphi} w_j) = \sum_{j=0}^{N_0} \lambda^{-j}(-i\lambda L w_j + P w_j),$$

$$= -i\lambda L w_0 - i \sum_{j=1}^{N_0} \lambda^{-(j-1)} L w_j + \sum_{j=0}^{N_0} \lambda^{-j} P w_j,$$

$$= -i \sum_{j=0}^{N_0-1} \lambda^{-j} L w_{j+1} + \sum_{j=0}^{N_0-1} \lambda^{-j} P w_j + \lambda^{-N_0} P w_{N_0} = \lambda^{-N_0} P w_{N_0},$$

since $L w_0 = 0$ and $L w_{j+1} = -i P w_j$ for $0 \leq j \leq N_0 - 1$.

b) Since φ is real and $\lambda \geq 1$, we have $\sup_K |v| \leq \sum_{j=0}^{N_0} \sup_K |w_j| \leq C_K$ and we deduce from questions 4b) and 6a) that $\sup_K |\partial^\alpha Pv| \leq C\lambda^{|\alpha|-N_0}$.

c) Since by $(\star)$ we have $\operatorname{grad}\varphi(x_0) = \xi^0 \neq 0$ we have,

$$e^{-i\lambda\varphi(x_0)}\operatorname{grad} v(x_0) = \sum_{j=0}^{N_0}\lambda^{-j}\left(i\lambda w_j(x_0)\xi^0 + \operatorname{grad} w_j(x_0)\right),$$

$$= i\lambda w_0(x_0)\xi^0 + i\sum_{j=0}^{N_0-1}\lambda^{-j}w_{j+1}(x_0)\xi^0 + \sum_{j=0}^{N_0}\lambda^{-j}\operatorname{grad} w_j(x_0).$$

Since $\lambda \geq 1$ we have $|i\sum_{j=0}^{N_0-1}\lambda^{-j}w_{j+1}(x_0)\xi^0 + \sum_{j=0}^{N_0}\lambda^{-j}\operatorname{grad} w_j(x_0)| \leq C$. On the other hand, by construction we have $w_0(x_0) = 1$. It follows that,

$$|\operatorname{grad} v(x_0)| \geq \lambda|\xi^0| - C \geq \frac{1}{2}\lambda|\xi^0|,$$

if λ is large enough.

Then we use the inequality (6.9). By question 6b) the right-hand side is bounded while by question 6c) the left-hand side tends to $+\infty$ when λ goes to $+\infty$. Therefore, we have a contradiction. This implies that P cannot be hypoelliptic.

Part 2

1. The matrix $A(x_0)$ being real and symmetric it has d real eigenvalues $\lambda_1, \ldots, \lambda_d$. Moreover, there exists an orthogonal matrix O such that $OA(x_0)O = D$ where $D = \operatorname{diag}(\lambda_j)$. Therefore, setting $\eta = O\xi$, we have,

$$p(x_0, \xi) = \langle A(x_0)\xi, \xi\rangle = \langle DO\xi, O\xi\rangle = \langle D\eta, \eta\rangle = \sum_{j=1}^{d}\lambda_j\eta_j^2 := q(\eta).$$
$$(9.32)$$

If all the λ_j are equal to zero then $p(x_0, \xi) = 0$ for all $\xi \in \mathbf{R}^d$ which contradicts condition (6.10) in the statement. Now if all the eigenvalues are ≤ 0 (resp. ≥ 0) then $p(x_0, \xi) \leq 0$ (resp. $p(x_0, \xi) \geq 0$) for all $\xi \in \mathbf{R}^d$ by (9.32); this contradicts (6.10) in the statement. Thus there exists $\lambda_{j_0} > 0$ and $\lambda_{j_1} < 0$. We may assume that $j_0 < j_1$.

2. Let $\eta^0 = (\eta_j^0)$ where $\eta_{j_0}^0 = 1$, $\eta_{j_1}^0 = \sqrt{\frac{\lambda_{j_0}}{-\lambda_{j_1}}}$ and $\eta_j^0 = 0$ otherwise. Then,

$$q(\eta^0) = \lambda_{j_0}(\eta_{j_0}^0)^2 + \lambda_{j_1}(\eta_{j_1}^0)^2 = \lambda_{j_0} + \lambda_{j_1}\frac{\lambda_{j_0}}{-\lambda_{j_1}} = 0.$$

Now, $\frac{\partial q}{\partial \eta_{j_0}} = 2\lambda_{j_0}\eta_{j_0}^0 = 2\lambda_{j_0} \neq 0$.

3. Set $\xi^0 = O^{-1}\eta^0$. By (9.32) we have $p(x_0, \xi^0) = 0$. On the other hand, $d_\xi p(x_0, \xi^0) = d_\eta q(\eta^0) \circ O$. So $d_\xi p(x_0, \xi^0) \neq 0$.

4. We can state the following theorem.

Theorem *Let P be an operator of the form (6.8) in the statement, where the coefficients a_{jk} are real. If P is hypoelliptic in Ω then for any point $x^0 \in \Omega$ we have,*

$$\text{either} \quad \left(p(x_0, \xi) \geq 0, \ \forall \xi \in \mathbf{R}^d\right) \quad \text{or} \quad \left(p(x_0, \xi) \leq 0, \ \forall \xi \in \mathbf{R}^d\right).$$

Indeed, otherwise the condition (6.10) in the statement would be satisfied, which ensures by Part 2 that the condition (H) is satisfied, which by Part 1 implies that P is not hypoelliptic.

Solution 32

1. Let $x \in \mathbf{R}^d$ be such that $x_d \neq 0$. There exists a neighborhood V of this point in which u is equal to 0 or 1 so is C^∞ in V. It follows that for $\xi \in \mathbf{R}^d \setminus \{0\}$ we have $(x, \xi) \notin \mathrm{WF}(u)$.

2. a) If $\xi_0' \neq 0$ we have $\frac{|\xi_{0,d}|}{|\xi_0|} < 1$. Let $\varepsilon > 0$ be such that $\alpha := \frac{|\xi_{0,d}|}{|\xi_0|} + \varepsilon < 1$ and
$$\Gamma_{\xi_0} =: \left\{ \xi \in \mathbf{R}^d \setminus \{0\} : \left| \frac{\xi}{|\xi|} - \frac{\xi_0}{|\xi_0|} \right| < \varepsilon \right\}. \text{ If } \xi \in \Gamma_{\xi_0} \text{ we have } \left| \frac{\xi_d}{|\xi|} - \frac{\xi_{0,d}}{|\xi_0|} \right| < \varepsilon$$
so $|\xi_d - \xi_{0,d} \frac{|\xi|}{|\xi_0|}| < \varepsilon |\xi|$. Then,

$$|\xi_d| \leq \left| \xi_d - \xi_{0,d} \frac{|\xi|}{|\xi_0|} \right| + \left| \frac{|\xi_{0,d}| |\xi|}{|\xi_0|} \right| \leq \left(\varepsilon + \frac{|\xi_{0,d}|}{|\xi_0|} \right) |\xi| = \alpha |\xi|.$$

It follows that, $|\xi| \leq |\xi_d| + |\xi'| \leq \alpha |\xi| + |\xi'|$ so $(1 - \alpha)|\xi| \leq |\xi'|$. We take $c_0 = 1 - \alpha$.

b) We have, $|\xi'| \leq |\xi| \leq \frac{1}{c_0} |\xi'|$.

c) We have,

$$\widehat{\varphi u}(\xi) = \int e^{-ix \cdot \xi} \varphi(x) u(x) \, dx = \int_0^{+\infty} \int_{\mathbf{R}^{d-1}} e^{-ix \cdot \xi} \varphi(x) \, dx.$$

We use the fact that $(Id - \Delta_{x'})^N e^{-ix \cdot \xi} = (1 + |\xi'|^2)^N e^{-ix \cdot \xi}$ where $\Delta_{x'} = \sum_{j=1}^{d-1} \frac{\partial^2}{\partial x_j^2}$. Since $\varphi \in C_0^\infty(\mathbf{R}^d)$ we can, setting $P = Id - \Delta_{x'}$, integrate by parts and write,

$$(1 + |\xi'|^2)^N \widehat{\varphi u}(\xi) = \int_0^{+\infty} \int_{\mathbf{R}^{d-1}} (P^N e^{-ix \cdot \xi}) \varphi(x) \, dx,$$

$$= \int_0^{+\infty} \int_{\mathbf{R}^{d-1}} e^{-ix \cdot \xi} (P^N \varphi)(x) \, dx.$$

It follows that,

$$(1 + |\xi'|^2)^N |\widehat{\varphi u}(\xi)| \leq \int_0^{+\infty} \int_{\mathbb{R}^{d-1}} |(P^N \varphi)(x)| \, dx,$$

and by question 2b), $(1 + |\xi|^2)^N |\widehat{\varphi u}(\xi)| \leq C_N$ for $\xi \in \Gamma_{\xi_0}$, which shows that $(x_0, \xi_0) \notin \mathrm{WF}(u)$. Therefore, $\mathrm{WF}(u) \subset \{(x, \xi) : x_d = 0, \xi' = 0\}$.

3. Let V be a neighborhood of the point $(x_0', 0)$. Take $\varphi \in C_0^\infty(V), \varphi = 1$ near $(x_0', 0)$ with $\int \varphi(x', 0) \, dx = c \neq 0$; (we can take for instance $\varphi(x) = \varphi_1(x')\varphi_2(x_d)$). Set $I = i\, \xi_{0,d}\, \widehat{\varphi u}(0, \xi_{0,d})$. We can write,

$$I = \int_{\mathbb{R}^{d-1}} \int_0^{+\infty} -\partial_{x_d} \left(e^{-i x_d\, \xi_{0,d}} \right) \varphi(x', x_d) \, dx_d \, dx'$$

$$= \int_{\mathbb{R}^{d-1}} \varphi(x', 0) \, dx' + \int_{\mathbb{R}^{d-1}} \int_0^{+\infty} e^{-i x_d\, \xi_{0,d}} (\partial_{x_d}\varphi)(x', x_d) \, dx_d \, dx', = c + A.$$

Now we have,

$$i\, \xi_{0,d}\, A = \int_{\mathbb{R}^{d-1}} \int_0^{+\infty} -(\partial_{x_d} e^{-i x_d\, \xi_{0,d}})(\partial_{x_d}\varphi)(x', x_d) \, dx_d \, dx',$$

$$= \int_{\mathbb{R}^{d-1}} \int_0^{+\infty} e^{-i x_d\, \xi_{0,d}} (\partial_{x_d}^2 \varphi)(x', x_d) \, dx_d \, dx',$$

since $(\partial_{x_d}\varphi)(x', 0) = (\partial_{x_d}\varphi)(x', +\infty) = 0$. It follows that $|\xi_{0,d}\, A| \leq C$.
For $|\xi_{0,d}| \to +\infty$ we can write,

$$\widehat{\varphi u}(0, \xi_{0,d}) = \frac{c}{i \xi_{0,d}} + O\left(\frac{1}{|\xi_{0,d}|^2} \right).$$

We deduce that,

$$|\widehat{\varphi u}(0, \xi_{0,d})| \geq \frac{c}{2} \frac{1}{|\xi_{0,d}|},$$

for $|\xi_{0,d}|$ large enough. This shows that $\widehat{\varphi u}$ does not decay rapidly in any cone Γ_{ξ_0} and therefore that $(x_0', 0, 0, \xi_{0,d}) \in \mathrm{WF}(u)$.

Solution 33

1. a) The equality is obviously true for $k = 0$ with $P_0(\eta) = 1$. Assume it is true up to the order $k \geq 0$ then we have,

$$\left(\frac{d}{d\eta}\right)^{k+1}\left(\frac{1}{1+\eta^2}\right) = \frac{d}{d\eta}\left(\frac{P_k(\eta)}{(1+\eta^2)^{k+1}}\right) = \frac{P_k'(\eta)}{(1+\eta^2)^{k+1}} - \frac{2(k+1)\eta P_k(\eta)}{(1+\eta^2)^{k+2}},$$

$$= \frac{(1+\eta^2)P_k'(\eta) - 2(k+1)\eta P_k(\eta)}{(1+\eta^2)^{k+2}},$$

and $P_{k+1}(\eta) = (1+\eta^2)P_k'(\eta) - 2(k+1)\eta P_k(\eta)$ is a polynomial of degree $\leq k+1$.

   b) According to a) we can write,

$$\left(\frac{d}{d\eta}\right)^{k+1}\left(\frac{1}{1+\eta^2}\right) = \frac{1}{(1+\eta)^{k+2}}\frac{(1+\eta)^{k+2}P_k(\eta)}{(1+\eta^2)^{k+1}}.$$

The function $v_k(\eta) = \frac{(1+\eta)^{k+2}P_k(\eta)}{(1+\eta^2)^{k+1}}$ is continuous on $[0, +\infty)$ and tends to a finite limit when η goes to $+\infty$. It is therefore bounded by a constant C_k.

2. a) Set $f(\eta) = \frac{1}{1+\eta^2}$. The equality is true for $N = 1$ with $a_{0,N} = 0$.

   Assume it is true up to the order $N \geq 1$ and let us prove it at the order $N + 1$. We have $e^{ix\eta} = \frac{1}{ix}\frac{d}{d\eta}e^{ix\eta}$ so that integrating by parts we obtain,

$$\int_0^{+\infty} e^{ix\eta}f^{(N-1)}(\eta)\,d\eta = \frac{1}{ix}\left[e^{ix\eta}f^{(N-1)}(\eta)\right]_0^{+\infty} - \frac{1}{ix}\int_0^{+\infty}e^{ix\eta}f^{(N)}(\eta)\,d\eta,$$

$$= -\frac{f^{(N-1)}(0)}{ix} - \frac{1}{ix}\int_0^{+\infty}e^{ix\eta}f^{(N)}(\eta)\,d\eta,$$

since from question 1b) we have $\lim_{\eta\to+\infty}|f^{(N-1)}(\eta)| = 0$. Using the induction hypothesis at the order N we can write,

$$u(x) = \sum_{j=0}^{N-1}\frac{a_{j,N}}{x^j} + \frac{i^N f^{(N-1)}(0)}{x^N} + \frac{i^N}{x^N}\int_0^{+\infty}e^{ix\eta}f^{(N)}(\eta))\,d\eta,$$

which proves the desired inequality at the order $N + 1$.

   b) Let us show that u is C^∞ in $\mathbf{R} \setminus \{0\}$. Let $p \in \mathbf{N}$. Using question a) with $N = p + 1$ it is sufficient to prove that the function $x \to \int_0^{+\infty}e^{ix\eta}f^{(N-1)}(\eta)\,d\eta$ is C^p. But this results from a classical theorem of differentiability of the integrals. Indeed for $j \le p$ we have,

$$\left| \left(\frac{d}{dx} \right)^j \left(e^{ix\eta} f^{(N-1)}(\eta) \right) \right| = |\eta^j f^{(N-1)}(\eta)| \le \frac{C_N \eta^j}{(1+\eta)^{N+1}} \le \frac{C_N (1+\eta)^p}{(1+\eta)^{N+1}}$$

$$\le \frac{C_N}{(1+\eta)^{N+1-p}} \in L^1(\mathbf{R}),$$

since $N + 1 - p \ge 2$. Therefore, the integral is C^p for all $p \in \mathbf{N}$ so it is C^∞ on $\mathbf{R}$. It follows that u is C^∞ on $\mathbf{R} \setminus \{0\}$. We deduce that for all $x_0 \ne 0$ and all $\xi \in \mathbf{R} \setminus \{0\}$ we have, $(x_0, \xi) \notin WF(u)$ that is $WF(u) \subset \{(0, \xi), \xi \ne 0\}$.

3. a) We have $\widehat{\varphi u}(\xi) = \int_{\mathbf{R}} e^{-ix\xi} \varphi(x) \left(\int_0^{+\infty} \frac{e^{ix\eta}}{1+\eta^2} d\eta \right) dx$. The function $(x, \eta) \to \frac{e^{-ix(\xi-\eta)}}{1+\eta^2} \varphi(x)$ is in $L^1(\mathbf{R} \times (0, +\infty))$, so we can use the Fubini theorem and deduce that,

$$\widehat{\varphi u}(\xi) = \int_0^{+\infty} \left(\int_{\mathbf{R}} e^{-ix(\xi-\eta)} \varphi(x) \, dx \right) \frac{1}{1+\eta^2} = \int_0^{+\infty} \frac{\widehat{\varphi}(\xi-\eta)}{1+\eta^2} \, d\eta.$$

b) Since $\widehat{\varphi} \in S(\mathbf{R})$, for all $N \in \mathbf{N}$ the function $|\zeta|^N \widehat{\varphi}(\zeta)$ is bounded on $\mathbf{R}$ by a constant C_N. Then for $\xi < 0$ and $\eta \ge 0$ we have $|\xi| + \eta = -\xi + \eta = -(\xi - \eta) = |\xi - \eta|$. Therefore,

$$|\xi|^N |\widehat{\varphi}(\xi)| \le \int_0^{+\infty} |\xi|^N \frac{|\widehat{\varphi}(\xi-\eta)|}{1+\eta^2} \, d\eta \le \int_0^{+\infty} (|\xi| + \eta)^N \frac{|\widehat{\varphi}(\xi-\eta)|}{1+\eta^2} \, d\eta,$$

$$\le \int_0^{+\infty} |\xi - \eta|^N \frac{|\widehat{\varphi}(\xi-\eta)|}{1+\eta^2} \, d\eta \le C_N \int_0^{+\infty} \frac{1}{1+\eta^2} \, d\eta.$$

4. a) We have $\varphi(0) = \int \psi(y) \psi(-y) \, dy$. So $\varphi(0) = \int \psi(y)^2 \, dy = 1$ since ψ is even. Now, $\varphi'(0) = \int \psi(y) \psi'(-y) \, dy = -\int \psi(y) \psi'(y) \, dy$ since ψ' is odd, so $\varphi'(0) = -\frac{1}{2} \int \frac{d}{dy}[\psi(y)^2] \, dy = 0$. Now, since ψ is even $\widehat{\psi}$ is real and we have $\widehat{\varphi} = (\widehat{\psi})^2 \ge 0$. Eventually $\widehat{\varphi}(0) = (\widehat{\psi}(0))^2 = \left(\int \psi(y) \, dy \right)^2 \ne 0$ since ψ has positive values and L^2 norm equal to 1.

b) Since $\widehat{\varphi}$ is positive we have from question 3a), $\widehat{\varphi u}(\xi) \ge \int_A \widehat{\varphi}(\xi - \eta) \frac{1}{1+\eta^2} \, d\eta$. Now we write $\widehat{\varphi}(\zeta) = \widehat{\varphi}(0) + \zeta \int_0^1 (\widehat{\varphi})'(\lambda \zeta) \, d\lambda = \widehat{\varphi}(0) + \zeta \chi(\zeta)$ where $\sup_{\mathbf{R}} |\chi| \le C$. Then,

$$\widehat{\varphi u}(\xi) \ge \widehat{\varphi}(0) \int_A \frac{1}{1+\eta^2} \, d\eta + \int_A (\xi - \eta) \chi(\xi - \eta) \frac{1}{1+\eta^2} \, d\eta = (1) + (2).$$

We have $|(2)| \leq C\varepsilon \int_A \frac{1}{1+\eta^2} \, d\eta$ so,

$$\widehat{\varphi u}(\xi) \geq (\widehat{\varphi}(0) - C\varepsilon) \int_A \frac{1}{1+\eta^2} \, d\eta.$$

We fix $\varepsilon > 0$ such that $C\varepsilon = \frac{1}{2}\widehat{\varphi}(0)$. For $\xi > 0$ large, we have $A = \{\eta : \xi - \varepsilon \leq \eta \leq \xi + \varepsilon\}$. Then,

$$\int_{\xi-\varepsilon}^{\xi+\varepsilon} \frac{1}{1+\eta^2} \, d\eta \geq \frac{2\varepsilon}{1 + (\xi + \varepsilon)^2}.$$

Summing up we have,

$$\widehat{\varphi u}(\xi) \geq \frac{\varepsilon \, \widehat{\varphi}(0)}{1 + (\xi + \varepsilon)^2} \quad \text{so} \quad \xi^2 \widehat{\varphi u}(\xi) \geq \varepsilon \, \widehat{\varphi}(0) \frac{\xi^2}{1 + (\xi + \varepsilon)^2}.$$

This proves our claim since the right-hand side tends to $\varepsilon \widehat{\varphi}(0) > 0$ when $\xi \to +\infty$.

To prove that $\{(0, \xi) : \xi > 0\} \subset WF(u)$ we have to prove a lower bound, similar to that proved above, for all functions $\varphi_1 \in C_0^\infty$ equal to 1 near zero (not only for one function !).

c) Let φ_1 be given in the statement. We have $\theta(0) = \varphi(0) - \varphi_1(0) = 1 - 1 = 0$ and $\theta'(0) = \varphi'(0) - \varphi_1'(0) = 0 - 0 = 0$. Since $\theta \in C_0^\infty(\mathbf{R})$ the Taylor formula with integral reminder shows that $\theta(x) = x^2 \theta_1(x)$ and $\theta_1 \in C_0^\infty(\mathbf{R})$. Then $\widehat{\theta}(\xi) = -(\frac{d^2}{d\xi^2}\widehat{\theta_1})(\xi)$. It follows from question 3a) that,

$$\widehat{\theta u}(\xi) = \int_0^{+\infty} \theta(\xi - \eta) \frac{1}{1+\eta^2} \, d\eta = -\int_0^{+\infty} \left(\frac{d^2}{d\xi^2}\widehat{\theta_1}\right)(\xi - \eta) \frac{1}{1+\eta^2} \, d\eta,$$

$$= -\int_0^{+\infty} \frac{d^2}{d\eta^2}[\widehat{\theta_1}(\xi - \eta)] \frac{1}{1+\eta^2} \, d\eta.$$

Integrating two times by parts we obtain, since $(\widehat{\theta_1})' \in \mathcal{S}(\mathbf{R})$,

$$\widehat{\theta u}(\xi) = -(\widehat{\theta_1})'(\xi) - \int_0^{+\infty} \widehat{\theta_1}(\xi - \eta) \frac{d^2}{d\eta^2}\left(\frac{1}{1+\eta^2}\right) \, d\eta.$$

Then according to question 1b) with $k = 2$ we have,

$$|\widehat{\theta u}(\xi)| \leq |(\widehat{\theta_1})'(\xi)| + C_1 \int_0^{+\infty} |\widehat{\theta_1}(\xi - \eta)| \frac{1}{(1+\eta)^4} \, d\eta.$$

Let $\varepsilon > 0$ be small. Writing $|\xi|^{3-\varepsilon} \le C_2(|\xi - \eta|^{3-\varepsilon} + |\eta|^{3-\varepsilon})$ we get,

$$|\xi|^{3-\varepsilon}|\widehat{\theta u}(\xi)| \le |\xi|^{3-\varepsilon}|(\widehat{\theta_1})'(\xi)| + C_3 \int_0^{+\infty} |\xi - \eta|^{3-\varepsilon} \frac{|\widehat{\theta_1}(\xi - \eta)|}{(1+\eta)^4} \, d\eta$$

$$+ C_3 \int_0^{+\infty} |\widehat{\theta_1}(\xi - \eta)| \frac{|\eta|^{3-\varepsilon}}{(1+\eta)^4} \, d\eta,$$

$$|\xi|^{3-\varepsilon}|\widehat{\theta u}(\xi)| \le C_4 + C_5 \left(\int_0^{+\infty} \frac{1}{(1+\eta)^4} \, d\eta + \int_0^{+\infty} \frac{1}{(1+\eta)^{1+\varepsilon}} \, d\eta \right).$$

d) Then we have for $\xi \ge 1$, $\xi^2|\widehat{\theta u}(\xi)| \le \frac{C_6}{|\xi|^{1-\varepsilon}}$. We write,

$$\xi^2 \widehat{\varphi_1 u}(\xi) = \xi^2 \widehat{\varphi u}(\xi) - \xi^2 \widehat{\theta u}(\xi).$$

It follows from question 4b) that for $\xi \ge 1$,

$$\xi^2 \widehat{\varphi_1 u}(\xi) \ge C - \frac{C_6}{|\xi|^{1-\varepsilon}} \ge C_7 > 0,$$

if ξ is large enough. We deduce that $\{(0, \xi), \xi > 0\} \subset WF(u)$.

Solution 34

1. It follows from the hypotheses that $(x_1^2 + x_2^2)u \in C^\infty(\mathbf{R}^2)$, so $u \in C^\infty(\mathbf{R}^2 \setminus \{0\})$ which implies that $WF(u) \subset \{(x, \xi) \in \mathbf{R}^2 \times (\mathbf{R}^2 \setminus \{0\}) : x = 0\}$.
2. a) By hypothesis we have $x_j \varphi u \in C_0^\infty(\mathbf{R}^2)$, so $\widehat{x_j \varphi u} = i\frac{\partial}{\partial \xi_j}(\widehat{\varphi u}) \in \mathcal{S}(\mathbf{R}^2)$, $j = 1, 2$.
   b) Let $f(t) = |t\omega + (1-t)\omega_0|^2 = t^2 + (1-t)^2 + 2t(1-t)(\omega \cdot \omega_0)$. We have $\omega \cdot \omega_0 = \cos\theta$ where $\theta \in (0, \pi)$ is the angle between ω and ω_0. By hypothesis $0 \le \theta < \pi$.
      We have $f(t) = 2(1-\cos\theta)t^2 - 2(1-\cos\theta)t + 1$. If $\theta = 0$ we have $f(t) = 1$. If $0 < \theta < \pi$ we have $1 + \cos\theta > 0$ and $f'(t) = 2t(1 - \cos\theta)(2t - 1)$. We see then that f has a minimum at $t = \frac{1}{2}$ and that $f(\frac{1}{2}) = \frac{1}{2}(1 + \cos\theta)$. Therefore, $f(t) \ge \frac{1}{2}(1 + \cos\theta) = c^2 > 0$.
   c) Assume that the angle (ω, ω_0) is strictly smaller than π. There exists $\varepsilon > 0$ such that for all $\omega' \in \mathbf{S}^1$ with $|\omega - \omega'| < \varepsilon$ the angle (ω', ω_0) is strictly smaller than π. We write,

$$\widehat{\varphi u}(\lambda \omega') = \widehat{\varphi u}(\lambda \omega_0) + \sum_{j=1}^2 \lambda(\omega'_j - \omega_{0,j}) \int_0^1 \frac{\partial}{\partial \xi_j}(\widehat{\varphi u})(\lambda(t\omega' + (1-t)\omega_0)) \, dt.$$

$$(9.33)$$

By hypothesis we have $|\widehat{\varphi u}(\lambda \omega_0)| \leq C_N \lambda^{-N}$ for all $N \in \mathbf{N}$. Now, since from question a) we have $\frac{\partial}{\partial \xi_j}(\widehat{\varphi u}) \in S(\mathbf{R}^2)$, for all $N \in \mathbf{N}$ there exists $C_N > 0$ such that,

$$\left| \frac{\partial}{\partial \xi_j}(\widehat{\varphi u})(\lambda(t\omega' + (1-t)\omega_0)) \right| \leq C_N |\lambda(t\omega' + (1-t)\omega_0)|^{-N}.$$

It follows from question b) applied to ω' that,

$$\left| \frac{\partial}{\partial \xi_j}(\widehat{\varphi u})(\lambda(t\omega' + (1-t)\omega_0)) \right| \leq C_N c^{-N} \lambda^{-N}.$$

We deduce from formula (9.33) that for all $N \in \mathbf{N}$ there exists $D_N > 0$ such that,

$$|\widehat{\varphi u}(\lambda \omega')| \leq D_N \lambda^{-N}.$$

Therefore, $(0, \lambda\omega) \notin WF(u)$.

d) Since $\mathbf{S}^1$ can be covered by three sectors making an angle $< \pi$, if there exists ω_0 such that $(0, \lambda\omega_0) \notin WF(u)$, then for all $\omega \in \mathbf{S}^1$ we have, from question c), $(0, \lambda\omega) \notin WF(u)$. Therefore, $0 \notin \text{supp sing}(u)$ which means that u is C^∞ in a neighborhood of zero. Since from question 1 we know that u is C^∞ in $\mathbf{R}^2 \setminus \{0\}$ we have $u \in C^\infty(\mathbf{R}^2)$ which contradicts our hypothesis. Therefore, all points $(0, \lambda\omega)$ belong to $WF(u)$, i.e., $WF(u) = \{0, \xi), \xi \neq 0\}$.

A simple example of such a distribution is given by the Dirac mass at the origin $\delta_{(0,0)}$.

3. Assume that we show,

$$\left(\exists \xi^0 \neq 0 : (0, \xi^0) \notin WF(u) \right) \implies (\forall \xi \neq 0 : (0, \xi) \notin WF(u)). \qquad (9.34)$$

This would imply that $u \in C^\infty$ near zero, therefore $u \in C^\infty(\mathbf{R}^2)$, which is not true by hypothesis. So $(0, \xi^0) \in WF(u)$ for every $\xi^0 \neq 0$.

Let us prove (9.34). Let $(0, \xi)$ with $\xi \neq 0, \xi \neq \xi^0$. We may assume that $\xi_1 \neq \xi_1^0$. Set $\lambda = \frac{\xi_2 - \xi_2^0}{\xi_1 - \xi_1^0}$ and consider the operator $P = x_1 + \lambda x_2$. Since by hypothesis $Pu \in C^\infty(\mathbf{R}^2)$ the bicharacteristic issued from $(0, \xi^0)$ does not meet $WF(u)$. Now this bicharacteristic is given for all $t \in \mathbf{R}$ by $x(t) = 0, \xi_1(t) = -t + \xi_1^0, \xi_2(t) = -\lambda t + \xi_2^0$. Now if $t = \xi_1^0 - \xi_1$ we have $\xi_1(t) = \xi_1, \xi_2(t) = \xi_2$. Thus $(0, \xi) \notin WF(u)$.

Solution 35

1. Let us denote by $\Sigma = \{(x, y, \xi, \eta) \in \mathbf{R}^2 \times (\mathbf{R}^2 \setminus \{0\}) : p(x, y, \xi, \eta) = 0\} = \{(x, y, \xi, \eta) \in \mathbf{R}^2 \times (\mathbf{R}^2 \setminus \{0\}) : \xi^2 = x\eta^2\}$ the characteristic set of P. If $u \in \mathcal{D}'(\mathbf{R}^2)$ we know that,

$$\mathrm{WF}(u) \subset \mathrm{WF}(Pu) \cup \Sigma.$$

Since $Pu \in C^\infty(\mathbf{R}^2)$ we have $\mathrm{WF}(Pu) = \emptyset$. The points (x, y, ξ, η) such that $x < 0$ do not belong to Σ so neither to $\mathrm{WF}(u)$. This implies that u is C^∞ near these points, so $u \in C^\infty(H)$.

2. Denote by $(x(t), y(t), \xi(t), \eta(t))$ the bicharacteristic starting at $t = 0$ from a point $(x_0, y_0, \xi_0, \eta_0)$. Its equations are,

$$\dot{x}(t) = 2\xi(t), \quad \dot{y}(t) = -2x(t)\eta(t), \quad \dot{\xi}(t) = \eta(t)^2, \quad \dot{\eta}(t) = 0.$$

The last equation gives $\eta(t) = \eta_0$, the one before gives $\xi(t) = \eta_0^2 t + \xi_0$, the first one implies that $x(t) = \eta_0^2 t^2 + 2\xi_0 t + x_0$ eventually the second one gives $y(t) = -\frac{2}{3}\eta_0^3 t^3 - 2\eta_0\xi_0 t^2 - 2x_0\eta_0 t + y_0$.

3. We know from question 1 that u is C^∞ for $x < 0$. On the other hand, by hypothesis u is C^∞ in a neighborhood of $x = 0$. It is therefore sufficient to show that u is C^∞ for $x > 0$.

   By the propagation of singularities theorem it is sufficient to show that, starting from a point $(x_0, y_0, \xi_0, \eta_0)$ where $x_0 > 0$ and (ξ_0, η_0) is to be chosen, the *null* bicharacteristic starting at this point meets, at a time $t > 0$, the line $\{x = 0\}$ and therefore a neighborhood of this line (where u is C^∞). From question 2 we have $x(t) = \eta_0^2 t^2 + 2\xi_0 t + x_0$ where $\xi_0^2 = x_0\eta_0^2$. Notice that $\eta_0 \neq 0$ otherwise $\xi_0 = \eta_0 = 0$, which is excluded. The reduced discriminant of this second order polynomial in t is $\xi_0^2 - x_0\eta_0^2 = 0$ so that we have $x(t) = \eta_0^2(t - \frac{\xi_0}{\eta_0^2})^2$.

   Take $\xi_0 > 0$, then at the time $t = \frac{\xi_0}{\eta_0^2}$ we have $x(t) = 0$.

Solution 36

1. If K is a compact subset of $\mathbf{R}^d$ we have $\int_K |u_k(x)|^2\, dx = \int_K |b(x)|^2\, dx < +\infty$. Moreover, if $f \in L^2_c(\mathbf{R}^d) \subset L^1(\mathbf{R}^d)$ we have,

$$\int_{\mathbf{R}^d} e^{ikx\cdot\xi_0} f(x)\, dx = \widehat{f}(-k\xi_0),$$

and the right-hand side tends to zero, when $k \to +\infty$, since the Fourier transform sends $L^1(\mathbf{R}^d)$ to $C^0_{\to 0}(\mathbf{R}^d)$.

2. We have

$$\widehat{\chi u_k}(\xi) = \int_{\mathbf{R}^d} e^{-iy\xi} \chi(y) b(y) e^{iky\cdot\xi_0} \, dy = \widehat{\chi b}(\xi - k\xi_0).$$

Therefore, if we set $(1) = (2\pi)^d A(\chi u_k)(x)$ we obtain,

$$(1) = \int_{\mathbf{R}^d} e^{ix\xi} a(x, \xi) \widehat{\chi u_k}(\xi) \, d\xi = \int_{\mathbf{R}^d} e^{ix\xi} a(x, \xi) \widehat{\chi b}(\xi - k\xi_0) \, d\xi,$$

$$= e^{ikx\cdot\xi_0} \int_{\mathbf{R}^d} e^{ix\cdot\eta} \widehat{\chi b}(\eta) \, d\eta.$$

Since $\overline{\chi u_k(x)} = e^{-ikx\cdot\xi_0} \overline{\chi b(x)}$ it follows that

$$I_k = (2\pi)^{-d} \iint_{\mathbf{R}^d \times \mathbf{R}^d} e^{ix\cdot\eta} a(x, \eta + k\xi_0) \widehat{\chi b}(\eta) \chi(x) \overline{b(x)} \, d\eta \, dx.$$

3. a) For $|\eta| \geq \frac{1}{2}k$ we can write,

$$|f_k(x, \eta)| = (2\pi)^{-d} |\eta|^{-1} |a(x, \eta + k\xi_0)| (|\eta| |\widehat{\chi b}(\eta)| |\chi(x) b(x)|,$$

$$\leq \frac{C}{k} (|\eta| |\widehat{\chi b}(\eta)| |\chi(x) b(x)|.$$

Then the conclusion follows from the fact that the function

$$(|\eta| |\widehat{\chi b}(\eta)| |\chi(x) b(x)|$$

belongs to $L^1(\mathbf{R}^d \times \mathbf{R}^d)$.

b) If $|\eta| \leq \frac{1}{2}k$ we write, $\eta + k\xi_0 = \frac{k}{2}(2\xi_0 + \frac{2}{k}\eta)$. Now

$$|2\xi_0 + \frac{2}{k}\eta| \geq 2 - \frac{2}{k}|\eta| \geq 2 - 1 = 1.$$

Since $\lambda = \frac{k}{2} \geq 1$ and a is homogeneous of degree zero for $|\xi| \geq 1$ we have,

$$a(x, \eta + k\xi_0) = a\left(x, \frac{k}{2}(2\xi_0 + \frac{2}{k}\eta)\right) = a\left(x, 2\xi_0 + \frac{2}{k}\eta\right),$$

$$= a(x, 2\xi_0) + \frac{2}{k}\int_0^1 \eta \cdot \partial_\eta a\left(x, 2\xi_0 + \frac{2t}{k}\eta\right) dt$$

$$= a(x, \xi_0) + \frac{1}{k} c_k(x, \eta).$$

Since a is a symbol of order zero we have,

$$|c_k(x, \eta)| \le C_1 |\eta| \int_0^1 \left\langle 2\xi_0 + \frac{2t}{k}\eta \right\rangle^{-1} \le C_2 |\eta|.$$

c) It follows that,

$$\left| \iint_{|\eta| \le \frac{1}{2}k} e^{ix \cdot \eta} \frac{1}{k} c_k(c, \eta) \widehat{\chi b}(\eta) \chi(x) \overline{b(x)} \, d\eta \, dx \right|$$

$$\le \frac{C}{k} \iint_{\mathbf{R}^d \times \mathbf{R}^d} |\eta| |\widehat{\chi b}(\eta)| |\chi(x) b(x)| \, d\eta \, dx,$$

and the right-hand side tends to zero when k goes to $+\infty$.

d) Set $A_k = \iint_{|\eta| \ge \frac{1}{2}k} e^{ix \cdot \eta} a(x, \xi_0) \widehat{\chi b}(\eta) \chi(x) \overline{b(x)} \, d\eta \, dx$. We have,

$$|A_k| \le \iint_{|\eta| \ge \frac{1}{2}k} |\eta|^{-1} |a(x, \xi_0)| |\eta| |\widehat{\chi b}(\eta)| |\chi(x) b(x)| \, d\eta \, dx,$$

$$\le \frac{C}{k} \iint_{\mathbf{R}^d \times \mathbf{R}^d} |\eta| |\widehat{\chi b}(\eta)| |\chi(x) b(x)| \, d\eta \, dx \to 0.$$

It follows that,

$$(2\pi)^{-d} \iint_{|\eta| \le \frac{1}{2}k} e^{ix \cdot \eta} a(x, \xi_0) \widehat{\chi b}(\eta) \chi(x) \overline{b(x)} \, d\eta \, dx$$

$$= (2\pi)^{-d} \iint_{\mathbf{R}^d \times \mathbf{R}^d} e^{ix \cdot \eta} a(x, \xi_0) \widehat{\chi b}(\eta) \chi(x) \overline{b(x)} \, d\eta \, dx + o(1),$$

$$= \int_{\mathbf{R}^d} a(x, \xi_0) \chi(x) \overline{b(x)} \left((2\pi)^{-d} \int_{\mathbf{R}^d} e^{ix \cdot \eta} \widehat{\chi b}(\eta) \, d\eta \right) dx + o(1),$$

$$= \int_{\mathbf{R}^d} a(x, \xi_0) \chi(x)^2 |b(x)|^2 \, dx + o(1) = \int_{\mathbf{R}^d} a(x, \xi_0) |b(x)|^2 \, dx + o(1),$$

since χ is equal to one on the support in x of a.

e) It follows from the previous questions that,

$$\lim_{k \to +\infty} (A(\chi u_k), \chi u_k)_{L^2(\mathbf{R}^d)} = \int_{\mathbf{R}^d} a\left(x, \frac{\xi_0}{|\xi_0|} \right) |b(x)|^2 \, dx,$$

since $a(x, \xi_0) = a\left(x, \frac{\xi_0}{|\xi_0|} \right)$.

4. a) The estimates of (1) and (2) follow from the Cauchy–Schwarz inequality and
the fact that $A \in \text{Op } S^0$ thus is continuous on L^2. Now since $a \in S^0$ it is
bounded so we have,

$$|(4)| \leq C \int_{\mathbf{R}^d} (|b_n(x)| + |\widetilde{b}(x)|)(|b_n(x)) - \widetilde{b}(x)|) \, dx.$$

Using the Cauchy–Schwarz inequality we obtain,

$$|(4)| \leq C(\|b_n\|_{L^2} + \|\widetilde{b}\|_{L^2})\|b_n - \widetilde{b}\|_{L^2}.$$

b) Let $\varepsilon > 0$. We fix n so large that $|(1)| + |(2)| + |(4)| \leq \frac{\varepsilon}{2}$. Now n being fixed
we can use the previous questions with $b = b_n \in C_0^\infty(\mathbf{R}^d)$. It follows that,
$\lim_{k \to +\infty}(3) = 0$ so there exists k_0 such that for $k \geq k_0$ we have $|(4)| \leq \frac{\varepsilon}{2}$.
Therefore, for $k \geq k_0$ we have $|I_k| \leq \varepsilon$. To conclude we have just to notice that
since χ_1 is equal to one on the support in x of a we have $|\widetilde{b}(x)|^2 a \left(x, \frac{\xi_0}{|\xi_0|}\right) =$
$|b(x)|^2 a \left(x, \frac{\xi_0}{|\xi_0|}\right)$.

Solution 37

Part 1
1. We have $F(x, p) = p_{11}p_{22} - p_{12}^2 - K(x)$. Since,

$$p_{11}^0 = \partial_{x_1}^2 \widetilde{u}(0, 0) = g''(0), \ p_{22}^0 = \partial_{x_2}^2 \widetilde{u}(0, 0) = A, \ p_{12}^0 = \partial_1 \partial_{x_2} \widetilde{u}(0, 0) = h'(0),$$

the condition $F(0, p^0) = 0$ is equivalent to $Ag''(0) - h'(0)^2 - K(0, 0) = 0$.
Therefore, it is sufficient to take $A = \frac{h'(0)^2 + K(0,0)}{g''(0)}$. On the other hand, since
$\varphi(x) = x_2$ we have $\partial_{x_1}\varphi(0, 0) = 0, \partial_{x_2}\varphi(0, 0) = 1$. Therefore,

$$\sum_{j,k=1}^{2} \frac{\partial F}{\partial p_{jk}}(0, p^0)(\partial_{x_j}\varphi(0, 0)\partial_{x_k}\varphi(0, 0)) = \frac{\partial F}{\partial p_{22}}(0, p^0) = p_{11}^0 = g''(0) \neq 0.$$

2. We can therefore apply the Cauchy–Kovalevski theorem. Notice that F is a
polynomial in p. Then for every K, every g, h real analytic near the origin with
g satisfying the condition $g''(0) \neq 0$ the problem (8.1) has a unique real analytic
solution u near the origin.

Part 2

1. Recall that $\widetilde{u}(x) = g(x') + h(x')x_d + \frac{1}{2}Ax_d^2$ where, $g(x') = a_0 + \sum_{j=1}^{d-1} a_j x_j + \sum_{j=1}^{d-1} \lambda_j x_j^2 + O(|x'|^3)$ and $p^0 = \left(\widetilde{u}(0), (\partial_{x_j}\widetilde{u}(0))_{j=1,\dots,d}, (\partial_{x_j}\partial_{x_k}\widetilde{u}(0))_{j,k=1,\dots,d}\right)$. Then,

if $|\alpha| = 0$, $p_0^0 = g(0)$; if $|\alpha| = 1$, $p_j^0 = \partial_{x_j}g(0)$, $j = 1\dots, d-1$, $p_d^0 = h(0)$,

if $|\alpha| = 2$, $p_{jk}^0 = 0$, $1 \le j \ne k \le d-1$, $p_{jj}^0 = 2\lambda_j$, $1 \le j \le d-1$,

$p_{jd}^0 = \partial_{x_j}h(0)$, $1 \le j \le d-1$, $p_{dd}^0 = A$.

We have $F(0, p^0) = \det(p_{jk}^0) - K(0)$. Developing the determinant with respect to the last line we get, $\det(p_{jk}^0) = A \prod_{j=1}^{d-1}(2\lambda_j) + F(\lambda_j, \partial_{x_j}h(0))_{1\le j \le d-1}$. Since by condition (8.3) we have $\lambda_j \ne 0$, $j = 1, \dots, d-1$ one can find A such that $F(0, p^0) = 0$. On the other hand, since $\varphi(x) = x_d$ we have,

$$\sum_{|\alpha|=2} \frac{\partial F}{\partial p_\alpha}(0, p^0)(d\varphi(0))^\alpha = \frac{\partial F}{\partial p_{dd}}(0, p^0) = \det((p_{jk}^0)_{1\le j,k\le d-1}),$$

$$= \prod_{j=1}^{d-1}(2\lambda_j) \ne 0.$$

Thus the conditions in the Cauchy–Kovalevski theorem are satisfied. Then we have the same conclusion as in Part 1.

2. a) The Hessian matrix of g at zero is a real and symmetric matrix. It is therefore diagonalizable in an orthonormal basis, that means that we can find $M' \in O(d-1)$ such that ${}^tM'\text{Hess}(g(0))M' = \text{diag}(\lambda_j)$. In particular, since $|\det M'| = 1$ it follows from (8.3) that $\det(\text{Hess}(g(0))) = \prod_{j=1}^{d-1}\lambda_j \ne 0$.

   b) This follows from a simple computation since ${}^tM'M'$ is the $(d-1) \times (d-1)$ identity matrix. Moreover, since $m_{jd} = m_{dj} = 0$ for $1 \le j \le d-1$ and $m_{dd} = 1$ we have, $M(y', y_d) = (M'y', y_d)$.

   c) Using the Taylor formula we can write in a neighborhood of the origin,

$$v(y + hy_0) = v(y) + h\nabla_y v(y) \cdot y_0 + \frac{h^2}{2}\langle \text{Hess}(v)(y)y_0, y_0 \rangle + O(|h|^3),$$

$$u(My + hMy_0) = u(My) + h\nabla_x u(My_0) \cdot (My_0)$$
$$+ \frac{h^2}{2}\langle \text{Hess}(u)(My)My_0, My_0 \rangle + O(|h|^3).$$

We have by definition $v(y + hy_0) = u(My + hMy_0)$, $v(y) = u(My)$. Then dividing by h and taking $h = 0$ we get $\nabla_y v(y) \cdot y_0 = \nabla_x u(My_0) \cdot (My_0)$.

By the same way we obtain $\langle \text{Hess}(v)(y)y_0, y_0 \rangle = \langle \text{Hess}(u)(My)My_0, My_0 \rangle$ which can be written as,

$$\langle \text{Hess}(v)(y)y_0, y_0 \rangle = \langle {}^t M \text{Hess}(u)(My)My_0, y_0 \rangle .$$

This equality for any $y_0 \in \mathbf{R}^d$ implies that $\text{Hess}(v)(y) = {}^t M \text{Hess}(u)(My)M$.

d) This follows from the fact that $|\det M| = 1$.

e) Applying the Taylor formula to g we can write,

$$\widetilde{g}(y') = g(0) + \nabla_x g(0) \cdot (M'y') + \frac{1}{2} \langle \text{Hess}(g)(0)M'y', M'y' \rangle + O(|y'|^3),$$

$$= g(0) + \nabla_x g(0) \cdot (M'y') + \frac{1}{2} \langle {}^t M' (\text{Hess}(g))(0)M'y', y' \rangle + O(|y'|^3),$$

$$= g(0) + \nabla_x g(0) \cdot (M'y') + \frac{1}{2} \sum_{j=1}^{d-1} \lambda_j y_j^2 + O(|y'|^3),$$

by question 2a).

f) Summing up we have proved that the function $v(y) = u(My) = u(M'y', y_d)$ satisfies,

$$\det(\text{Hess}(v)(y)) = K(My), \quad v|_{y_d=0} = \widetilde{g}(y'), \quad \frac{\partial v}{\partial y_d}|_{y_d=0} = h(M'y'),$$

$$(9.35)$$

with,

$$\widetilde{g}(y') = g(0) + \nabla_x g(0) \cdot (M'y') + \frac{1}{2} \sum_{j=1}^{d-1} \lambda_j y_j^2 + O(|y'|^3).$$

Therefore, we can apply the result obtained in Case 1 and deduce that the problem (9.35) has a unique real analytic solution in a neighborhood of the origin. It follows that $u(x) = v(M^{-1}x)$ is the unique real analytic solution of problem (8.2).

Solution 38

1. a) We have $|X_1 - X_2| \le |X_1 - X_0| + |X_0 - X_2| \le r$. Let $Y \in B_n(X_2, r)$ then $|Y - X_1| \le |Y - X_2| + |X_2 - X_1| \le 2r$. Now by the mean value formula we

have, since $u \geq 0$,

$$u(X_2) = \frac{1}{|B_n(X_2,r)|} \int_{B_n(X_2,r)} u(Y)\,dY \leq \frac{1}{|B_n(X_2,r)|} \int_{B_n(X_1,2r)} u(Y)\,dY,$$

$$\leq \frac{|B_n(X_1,2r)|}{|B_n(X_2,r)|} \frac{1}{|B_n(X_1,2r)|} \int_{B_n(X_1,2r)} u(Y)\,dY = \frac{|B_n(X_1,2r)|}{|B_n(X_2,r)|} u(X_1).$$

Now $|B_n(X_2,r)| = c_n r^n$ and $|B_n(X_1,2r)| = c_n(2r)^n$. It follows that $u(X_2) \leq 2^n u(X_1)$.

b) Now since u is continuous on the closed ball $B_n(X_0, \frac{r}{2})$ we have $\inf_{B_n(X_0,\frac{r}{2})} u = u(X_1)$ and $\sup_{B_n(X_0,\frac{r}{2})} u = u(X_2)$ for some points $X_1, X_2 \in B_n(X_0, \frac{r}{2})$. Applying the above inequality with these two points we obtain,

$$\sup_{B_n(X_0,\frac{r}{2})} u \leq 2^n \inf_{B_n(X_0,\frac{r}{2})} u.$$

2. a) We have $\partial_t^2 \psi = e^{t\sqrt{\lambda}} \lambda \varphi = -e^{t\sqrt{\lambda}} \Delta_x \varphi = -\Delta_x \psi$ so $\Delta_{t,x} \psi = 0$.
Let $X_0 = (\frac{3r}{2}, x_0)$. We have $X = (t,x) \in B_{d+1}(X_0, \frac{r}{2})$ if and only if

$$\max\left(|t - \frac{3r}{2}|, |x - x_0|\right) \leq \frac{r}{2} \Leftrightarrow |t - \frac{3r}{2}| \leq \frac{r}{2}, |x - x_0| \leq \frac{r}{2},$$

which is equivalent to,

$$X \in (r, 2r) \times B_d(x_0, \frac{r}{2}).$$

b) According to the previous question we have,

$$\sup_{B_{d+1}(X_0,\frac{r}{2})} \psi = e^{2r\sqrt{\lambda}} \sup_{B_d(x_0,\frac{r}{2})} \varphi \geq e^{2r\sqrt{\lambda}} \inf_{B_d(x_0,\frac{r}{2})} \varphi = e^{r\sqrt{\lambda}}\left(e^{r\sqrt{\lambda}} \inf_{B_d(x_0,\frac{r}{2})} \varphi\right),$$

$$= e^{r\sqrt{\lambda}} \inf_{B_{d+1}(X_0,\frac{r}{2})} \psi.$$

c) Applying the previous question and question 1 in $\Omega = B_{d+1}(X_0, 4r)$ to the positive harmonic function ψ we can write,

$$e^{r\sqrt{\lambda}} \inf_{B_{d+1}(X_0,\frac{r}{2})} \psi \leq \sup_{B_{d+1}(X_0,\frac{r}{2})} \psi \leq M \inf_{B_{d+1}(X_0,\frac{r}{2})} \psi.$$

It follows that $e^{r\sqrt{\lambda}} \leq M$ which implies that $r \leq \frac{C}{\sqrt{\lambda}}$ where C depends only on the dimension d.

3. By the conclusion of question 2c) there exists $C > 0$ (depending only on d) such that,

$$r > \frac{C}{\sqrt{\lambda}} \implies B_d(x_0, 4r) \cap Z(\varphi) \neq \emptyset. \tag{9.36}$$

We proceed by contradiction. If the claim (8.6) is not true we have,

$$\forall M > 0, \exists \lambda > 0, \exists x \in \omega : d(x, Z(\varphi)) > \frac{M}{\sqrt{\lambda}}. \tag{9.37}$$

Take $M = 5(C + 1)$ and the corresponding λ, x. Take $r = \frac{C+1}{\sqrt{\lambda}} > \frac{C}{\sqrt{\lambda}}$. By (9.36) we have $B_d(x, 4r) \cap Z(\varphi) \neq \emptyset$. But if $z \in Z(\varphi)$ one can write,

$$|z - x| \geq d(x, Z(\varphi)) > \frac{5(C + 1)}{\sqrt{\lambda}} = 5r$$

which gives a contradiction.

Solution 39

Part 1
1. Recall that in the polar coordinates $(r, \omega) \in (0, +\infty) \times \mathbf{S}^2$, the Laplacian can be written as $\Delta = \frac{\partial^2}{\partial r^2} + \frac{2}{r}\frac{\partial}{\partial r} + \frac{1}{r^2}\Delta_\omega$, where Δ_ω is an operator on $\mathbf{S}^2$ (the Laplace–Beltrami operator). The function u_k is C^∞ on $\overline{B}$ and depends only on r. It follows that,

$$\Delta u_k = \left(\frac{\partial^2}{\partial r^2} + \frac{2}{r}\frac{\partial}{\partial r}\right)\frac{\sin(k\pi r)}{r}.$$

An elementary computation shows that $-\Delta u_k = (k\pi)^2 u_k$. On the other hand, $u_k|_{r=1} = \sin(k\pi) = 0$. Thus u_k is an eigenfunction corresponding to the eigenvalue $(k\pi)^2$.

2. Using the polar coordinates we get,

$$\|u_k\|_{L^2(B)}^2 = \int_{\mathbf{S}^2}\int_0^1 \sin^2(k\pi r)\,dr\,d\omega = \frac{|\mathbf{S}^2|}{2}\int_0^1 (1 - \cos(2k\pi r))\,dr = \frac{|\mathbf{S}^2|}{2}.$$

Now since $|\frac{\sin x}{x}| \leq 1$ we have $|u_k(x)| \leq k\pi$. On the other hand, $u_k(0) = k\pi$. Therefore, $\|u_k\|_{L^\infty(B)} = k\pi$.

3. Let $e_k = \frac{u_k}{\|u_k\|_{L^2(B)}}$. Then e_k is also an eigenfunction such that $\|e_k\|_{L^2(B)} = 1$ and
$\|e_k\|_{L^\infty(B)} = \sqrt{\frac{2}{|S^2|}}k\pi = Ck = Ck^{\frac{3-1}{2}}$. Therefore, the estimate (7.7) cannot be improved in that case.

Part 2
1. We have $\int_\Omega a(y)\,dy = \sum_{i=1}^m \int_\Omega |v_i(y)|^2\,dy = m$ since $\|v_i\|_{L^2(\Omega)} = 1$. On the other hand, since the function a is continuous on $\overline{\Omega}$ there exists $y_0 \in \overline{\Omega}$ such that $\sup_{\overline{\Omega}} a = a(y_0)$. Therefore, by the above equality we get $m \le |\Omega|\,a(y_0)$.
2. Since (v_i) is an orthonormal basis of V we have $\|u_{y_0}\|_{L^2(\Omega)}^2 = \sum_{i=1}^m |v_i(y_0)|^2 = a(y_0)$. On the other hand, we have $\|u_{y_0}\|_{L^\infty(\Omega)} \ge |u_{y_0}(y_0)| = a(y_0)$.
3. Let $U = \frac{u_{y_0}}{\|u_{y_0}\|_{L^2(\Omega)}}$. Then $U \in V$ and $\|U\|_{L^2(\Omega)} = 1$. By the previous question we have,

$$\|U\|_{L^\infty(\Omega)} = \frac{\|u_{y_0}\|_{L^\infty(\Omega)}}{\|u_{y_0}\|_{L^2(\Omega)}} \ge \frac{a(y_0)}{\sqrt{a(y_0)}} = \sqrt{a(y_0)}.$$

It follows from question 1 that,

$$\sup_{\substack{u \in V \\ \|u\|_{L^2(\Omega)}=1}} \|u\|_{L^\infty(\Omega)} \ge \|U\|_{L^\infty(\Omega)} \ge |\Omega|^{-\frac{1}{2}} m^{\frac{1}{2}}.$$

4. The estimate (7.7) shows that for $u \in V$ we have $\|u\|_{L^\infty(\Omega)} \le C\lambda^{\frac{d-1}{4}}$. It follows then from question 3 that,

$$|\Omega|^{-\frac{1}{2}} m^{\frac{1}{2}} \le C\lambda^{\frac{d-1}{4}},$$

which proves (8.7).

Solution 40

1. If $f(x) = (x_1 + ix_2)^j = z^j$ we have

$$\Delta_{\mathbf{R}^{d+1}} f = \left(\frac{\partial^2}{\partial x_1^2} + \frac{\partial^2}{\partial x_2^2}\right) f = \frac{1}{4}\frac{\partial}{\partial z}\frac{\partial}{\partial \overline{z}} z^j = 0.$$

On the other hand, $f(r\omega) = r^j(\omega_1 + i\omega_2)^j$ so that, using formula (8.8) in the statement, we obtain for $r > 0$,

$$
\begin{aligned}
0 = \Delta_{\mathbf{R}^{d+1}} f \\
= r^{j-2} \left[j(j-1)e_j + d\, je_j + \Delta_\omega e_j \right] \\
= r^{j-2} \left[\Delta_\omega e_j + j(j+d-1)e_j \right],
\end{aligned}
$$

which proves our result.

2. We have $|e_j(\omega)|^2 = (\omega_1^2 + \omega_2^2)^j = (1 - |\omega'|^2)^j$, so that, using the hint given in the statement, we can write, when $d \geq 3$,

$$
\begin{aligned}
\|e_j\|_{L^2(\mathbf{S}^d)}^2 &= \int_{\{\omega' \in \mathbf{R}^{d-1}: |\omega'| \leq 1\}} \int_0^{2\pi} (1 - |\omega'|^2)^j (1 - |\omega'|^2)^{\frac{1}{2}}\, d\alpha\, d\omega', \\
&= 2\pi \int_0^1 \int_{\mathbf{S}^{d-2}} (1 - \rho^2)^{j+\frac{1}{2}} \rho^{d-2}\, d\rho\, d\Theta, \\
&= 2\pi |\mathbf{S}|^{d-2} \int_0^{\sqrt{j}} \left(1 - \frac{t^2}{j}\right)^{j+\frac{1}{2}} \left(\frac{t}{\sqrt{j}}\right)^{d-2} \frac{dt}{\sqrt{j}}.
\end{aligned}
$$

Since by the Lebesgue dominated convergence theorem we have,

$$
\lim_{j \to +\infty} \int_0^{\sqrt{j}} \left(1 - \frac{t^2}{j}\right)^{j+\frac{1}{2}} t^{d-2}\, dt = \int_0^{+\infty} e^{-t^2} t^{d-2}\, dt := c
$$

we obtain,

$$
\lim_{j \to +\infty} j^{\frac{d-1}{2}} \|e_j\|_{L^2(\mathbf{S}^d)}^2 = 2\pi |\mathbf{S}^{d-2}| c := c_0.
$$

If $d = 2$ we have $\omega' \in (-1, 1)$ so that,

$$
\|e_j\|_{L^2(\mathbf{S}^2)}^2 = \int_{-1}^1 \int_0^{2\pi} (1 - |\omega'|^2)^j (1 - |\omega'|^2)^{\frac{1}{2}}\, d\alpha\, d\omega'.
$$

Then we set $\omega' = \frac{t}{\sqrt{j}}$ and we conclude as above.

3. We have $\|e_j\|_{L^\infty(\mathbf{S}^d)} = \sup_{\omega \in \mathbf{S}^d} (1 - |\omega'|^2)^j = 1$ so that,

$$
\|u_j\|_{L^\infty(\mathbf{S}^d)} = \frac{1}{\|e_j\|_{L^2(\mathbf{S}^d)}} \sim c_1 j^{\frac{d-1}{4}}, \quad j \to +\infty.
$$

Solution 41

Part 1

1. Recall that we have, $V' = H^{-1} + \{f \in S' : \langle x \rangle^{-1} f \in L^2\}$. Since $u \in H^1$ we have $-\Delta u \in H^{-1}$; moreover $\langle x \rangle^{-1} (\langle x \rangle^2 u) = \langle x \rangle u \in L^2 - $. Therefore, $-\Delta u + \langle x \rangle^2 u \in V'$. The operator $-\Delta + \langle x \rangle^2$ is bijective by the existence and uniqueness result proved in Problem 1 question 3. It is continuous since,

$$\|(-\Delta + \langle x \rangle^2)u\|_{V'} \leq \|\Delta u\|_{H^{-1}} + \| \langle x \rangle^{-1} \langle x \rangle^2 u\|_{L^2},$$

$$\leq \|u\|_{H^1} + \| \langle x \rangle u\|_{L^2} \leq \|u\|_V.$$

The inverse operator is therefore continuous by the Banach theorem. Summing up, $P = -\Delta + \langle x \rangle^2$ is an isomorphism from V to V'.

2. a) If $f \in L^2$ we deduce from (2.1) that,

$$(\nabla_x u, \nabla_x w)_{L^2} + (\langle x \rangle u, \langle x \rangle w)_{L^2} = \left(\overline{f}, w\right)_{L^2}, \quad \forall w \in V.$$

Taking $w = D_h v$ where $v \in V$ we obtain,

$$(\nabla_x u, \nabla_x D_h v)_{L^2} + \left(\langle x \rangle^2 u, D_h v\right)_{L^2} = \left(\overline{f}, D_h v\right)_{L^2}, \quad \forall v \in V. \qquad (9.38)$$

Now we have,

$$(\nabla_x u, \nabla_x D_h v)_{L^2} = (\nabla_x D_{-h} u, \nabla_x v)_{L^2(\mathbf{R}^d)}, \quad \left(\overline{f}, D_h v\right)_{L^2} = \left(D_{-h}\overline{f}, v\right)_{L^2},$$

and,

$$\left(\langle x \rangle^2 u, D_h v\right)_{L^2} = \left(\langle y \rangle^2 D_{-h} u, v\right)_{L^2}$$

$$+ \frac{1}{h} \left([\langle y + he_k \rangle^2 - \langle y \rangle^2]u(y + he_k), v\right)_{L^2}.$$

We have, $\frac{1}{h} \left(\langle y + he_k \rangle^2 - \langle y \rangle^2\right) = 2y_k + h$. It follows from (9.38) that,

$$(\nabla_x D_{-h} u, \nabla_x v)_{L^2} + \left(\langle y \rangle^2 D_{-h} u, v\right)_{L^2} = \left(D_{-h}\overline{f}, v\right)_{L^2}$$

$$- \int_{\mathbf{R}^d} (2y_k + h)u(y + he_k)\overline{v(y)}\, dy, \quad \forall v \in V. \qquad (9.39)$$

b) We can take in (9.39) $v = D_{-h}u \in V$. Using the Cauchy–Schwarz inequality, setting $x = y + he_k$ in the integral and using the fact that $|2x_k - h| \leq C \langle x \rangle$, we obtain,

$$\|\nabla_x D_{-h}u\|_{L^2}^2 + \| \langle x \rangle D_{-h}u\|_{L^2}^2 \leq |(D_{-h}\overline{f}, D_{-h}u)_{L^2}| + C\| \langle x \rangle u\|_{L^2}\|D_{-h}u\|_{L^2}.$$

Now using Problem 1 question 3c) we have,

$$\| \langle x \rangle u\|_{L^2}\|D_{-h}u\|_{L^2(\mathbf{R}^d)} \leq \| \langle x \rangle u\|_{L^2}\|\nabla_x u\|_{L^2} \leq \|u\|_V^2,$$

$$|(D_{-h}\overline{f}, D_{-h}u)_{L^2}| \leq \|D_{-h}f\|_{H^{-1}}\|D_{-h}u\|_{H^1} \leq C\|f\|_{L^2}\|D_{-h}u\|_{H^1},$$

$$\leq \varepsilon\|D_{-h}u\|_{H^1}^2 + \frac{C'}{4\varepsilon}\|f\|_{L^2}^2.$$

Taking $\varepsilon = \frac{1}{2}$ we obtain,

$$\|\nabla_x D_{-h}u\|_{L^2}^2 + \| \langle x \rangle D_{-h}u\|_{L^2}^2 \leq C\|\nabla_x D_{-h}u\|_{L^2}^2 + \| \langle x \rangle D_{-h}u\|_{L^2}^2.$$

c) Let (h_n) be a sequence in $\mathbf{R}$ which converges to zero. From the above inequality we see that the sequence $(\nabla_x D_{-h_n}u)$ is uniformly bounded in L^2. Therefore, it has a subsequence which converges weakly in L^2, thus in $\mathcal{D}'$, to a $v \in L^2$. On the other hand, since $\nabla_x u \in L^2$ Problem 1 question 3b) shows that $(D_{-h_n}\nabla_x u)$ converges to $\partial_{x_k}\nabla_x u$ in H^{-1}, thus in $\mathcal{D}'$. Therefore, $\partial_{x_k}\nabla_x u = v \in L^2$. It follows that $\Delta u \in L^2$ and the equation shows that $\langle x \rangle^2 u \in L^2$.

d) The operator P is clearly continuous from D to L^2. It is bijective; indeed if $f \in L^2 \subset V'$ there exists a unique $u \in V$ such that $-\Delta u + \langle x \rangle^2 u = f$. Using (8.10) we see that $u \in D$. Its inverse is continuous by the Banach isomorphism theorem since D and L^2 are Banach spaces.

3. If $f \in L^2$ is such that $P^{-1}f = 0 \in D$ then $f = 0$ by the previous question. Moreover, we have $P^{-1}(L^2) = D$.

   Now, since a composition of continuous map and compact map is compact it follows from Problem 1 question 2 that P^{-1} is compact from L^2 to itself.

   On the other hand, by question 2c) P is an isomorphism from D to L^2. Therefore, if $f \in L^2$ then $P^{-1}f = u \in D$ and this is equivalent to $(-\Delta + \langle x \rangle^2)u = f$. Therefore, we have,

$$\left(P^{-1}f, f\right)_{L^2(\mathbf{R}^d)} = \left(u, (-\Delta + \langle x \rangle^2)u\right)_{L^2}.$$

Now for $u \in H^2$ we have $(u, -\Delta u)_{L^2} = \|\nabla_x u\|_{L^2}^2$. Therefore,

$$\left(P^{-1}f, f\right)_{L^2} = \|\nabla_x u\|_{L^2}^2 + \| \langle x \rangle u\|_{L^2}^2 \geq 0.$$

This means that P^{-1} is a positive operator.

By the same argument we see that for $f, g \in L^2$ setting $P^{-1}g = v$ we have,

$$\left(P^{-1}f, g\right)_{L^2} = \left(u, (-\Delta + \langle x \rangle^2)v\right)_{L^2} = (\nabla_x u, \nabla_x v)_{L^2} + (\langle x \rangle u, \langle x \rangle v)_{L^2},$$

$$\left(f, P^{-1}g\right)_{L^2} = \left((-\Delta + \langle x \rangle^2)u, v\right)_{L^2} = (\nabla_x u, \nabla_x v)_{L^2} + (\langle x \rangle u, \langle x \rangle v)_{L^2}.$$

Therefore, we have $\left(P^{-1}f, g\right)_{L^2} = \left(f, P^{-1}g\right)_{L^2}$, which proves that P^{-1} is self-adjoint.

4. This is obvious.

5. a) We have $u \in V$ and $-\Delta u + \langle x \rangle^2 u = (\lambda + 1)u$. Our claim is true when $k = 2$ since $u \in D$. Now assume the claim true at the order $k \geq 2$ and let $\alpha \in \mathbf{N}^d, |\alpha| = k - 1$. Then $\partial^\alpha u \in V$. Indeed, by the induction, we have $\partial^\alpha u \in H^1$ and $\langle x \rangle \partial^\alpha u = \partial^\alpha (\langle x \rangle u) - \sum_{\beta \neq 0} \binom{\alpha}{\beta} \partial^\beta \langle x \rangle \partial^{\alpha - \beta} u \in L^2$ since $\langle x \rangle u \in H^{k-1}$, $\partial^\beta \langle x \rangle$ is bounded and $|\alpha - \beta| \leq k - 2$. Moreover, differentiating the equation we obtain,

$$(-\Delta + \langle x \rangle^2)\partial^\alpha u = (\lambda + 1)\partial^\alpha u - \sum_{\beta \neq 0} \binom{\alpha}{\beta} \partial^\beta \langle x \rangle^2 \, \partial^{\alpha - \beta} u =: g.$$

By the induction we have $(\lambda + 1)\partial^\alpha u \in L^2$, then for $|\beta| = 1$, $\langle x \rangle \partial^{\alpha - \beta} u \in L^2$, for $|\beta| = 2$, $\partial^\beta \langle x \rangle^2$ is bounded, $\partial^{\alpha - \beta} u \in L^2$ and for $|\beta| \geq 3$, $\partial^\beta \langle x \rangle^2 = 0$, so $g \in L^2$. Therefore, we can apply (8.10) to $\partial^\alpha u$ and deduce that $\partial^\alpha u \in H^2$ and $\langle x \rangle \nabla_x \partial^\alpha u \in L^2$. Therefore, $u \in H^{k+1}$ and $\langle x \rangle u \in H^k$. Eventually we deduce from the equation that $\langle x \rangle^2 u = \Delta u + (\lambda + 1)u \in H^{k-1}$.

b) We have,

$$\left(-\Delta u + |x|^2 u, u\right)_{L^2} = \lambda \|u\|_{L^2}^2.$$

Now $u \in \cap_{k \in \mathbf{N}} H^k \subset C_{\to 0}^\infty$. Therefore, we can integrate by parts in the above formula and deduce that,

$$\|\nabla_x u\|_{L^2}^2 + \||x|u\|_{L^2}^2 = \lambda \|u\|_{L^2}^2.$$

c) Integrating by parts we can write,

$$\int_{\mathbf{R}^d} \partial_{x_j} u(x) x_j \overline{u(x)} \, dx = -\int_{\mathbf{R}^d} |u(x)|^2 dx - \int_{\mathbf{R}^d} x_j u(x) \overline{\partial_{x_j} u(x)} \, dx.$$

Therefore, by the Cauchy–Schwarz inequality,

$$\|u\|_{L^2}^2 = -2\mathrm{Re} \int_{\mathbf{R}^d} \partial_{x_j} u(x) x_j \overline{u(x)} \, dx \leq 2\|\partial_{x_j} u\|_{L^2} \|x_j u\|_{L^2}.$$

Summing in j from 1 to d and using the Cauchy–Schwarz inequality we obtain,

$$d\|u\|_{L^2}^2 \leq 2\left(\sum_{j=1}^{d}\|\partial_{x_j}u\|_{L^2}^2\right)^{\frac{1}{2}}\left(\sum_{j=1}^{d}\|x_ju\|_{L^2}^2\right)^{\frac{1}{2}} = 2\|\nabla_x u\|_{L^2}\||x|u\|_{L^2},$$

$$\leq \|\nabla_x u\|_{L^2}^2 + \||x|u\|_{L^2}^2.$$

d) Follows immediately from questions b) and c) since $\|u\|_{L^2} \neq 0$.

6. a) If $j, k \in \mathbf{N}$ we have,

$$(e_j, e_k)_V = (\nabla_x e_j, \nabla_x e_k)_{L^2} + (\langle x\rangle^2 e_j, e_k)_{L^2} = ((-\Delta + \langle x\rangle^2)e_j, e_k)_{L^2},$$

$$= (\lambda_j + 1)(e_j, e_k)_{L^2} = 0.$$

Therefore, $(e_j, e_k)_V = 0$ if $j \neq k$ and $\|e_j\|_V^2 = (\lambda_j + 1)$.

b) It follows that $(u, u)_V = \sum_{j,k} u_j \overline{u_k}(e_j, e_k)_V = \sum_j (\lambda_j + 1)|u_j|^2$.

c) We have,

$$u \in D \Leftrightarrow u \in V \text{ and } (-\Delta + \langle x\rangle^2)u \in L^2 \Leftrightarrow u = \sum_n u_n e_n \text{ and } \sum_n \lambda_n^2|u_n|^2 < +\infty.$$

7. a) Since $\lambda \neq \lambda_n$ for all $n \in \mathbf{N}$ the operator $-\Delta + |x|^2 - \lambda$ is injective. To see that it is surjective let $f = \sum_{n \in \mathbf{N}} f_n e_n \in L^2$. Set $u = \sum_n \frac{f_n}{\lambda_n - \lambda} e_n$. Then by question 6c) $u \in D$ and $(-\Delta + |x|^2 - \lambda)u = f$.

b) Thus $(-\Delta + |x|^2 - \lambda) : D \to L^2$ is not invertible if and only if $\lambda = \lambda_n$ for some $n \in \mathbf{N}$.

Part 2

Case 1. $d = 1$

1. The first identity is obvious. Let us prove the second by induction on n. If $n = 1$ we have,

$$(L^-L^+ - L^+L^-)u = (\frac{d}{dx} + x)(-u' + xu) - (-\frac{d}{dx} + x)(u' + xu) =$$

$$= (-u'' + u + xu' - xu' + x^2u) - (-u'' - u - xu' + xu' + x^2u) = 2u.$$

Assume the identity true at the order n. Then using the induction we can write,

$$L^-(L^+)^{n+1} = L^-(L^+)^n L^+ = (L^+)^n L^- L^+ + 2n(L^+)^n.$$

Using the case $n = 1$ above we deduce that,

$$L^-(L^+)^{n+1} = (L^+)^{n+1}L^- + 2(L^+)^n + 2n(L^+)^n = (L^+)^{n+1}L^- + 2(n+1)(L^+)^n.$$

2. a) This follows from the fact that $L^-\phi_0 = 0$ and question 1. This shows that $\lambda = 1$ is an eigenvalue of $\mathcal{H}$.

   b) We argue by induction on n. The claim is obviously true for $n = 0$. Assume it is true for n and let us prove it for $n + 1$. We have

   $$\phi_{n+1}(x) = (L^+\phi_n)(x) = (-\frac{d}{dx} + x)\left(P_n(x)\exp\left(-\frac{x^2}{2}\right)\right),$$

   $$= (-P_n'(x) + 2x\,P_n(x))\exp\left(-\frac{x^2}{2}\right),$$

   and $-P_n' + 2x\,P_n$ is a polynomial of order $n + 1$ whose coefficient of x^{n+1} is equal to 2^n.

   c) We have, $\mathcal{H}\phi_n = (L^+L^- + 1)\phi_n = L^+L^-(L^+)^n\phi_0 + \phi_n$. We deduce from question 1 that,

   $$\mathcal{H}\phi_n = (L^+)^{n+1}L^-\phi_0 + 2n(L^+)^n\phi_0 + \phi_n = (2n + 1)\phi_n,$$

   since $L^-\phi_0 = 0$.

   d) Notice that for $\varphi, \psi \in \mathcal{S}(\mathbf{R})$ we have $(\varphi, L^+\psi)_{L^2} = (L^-\varphi, \psi)_{L^2}$. It follows that for any $m \geq 0$ we have $(\varphi, (L^+)^m\psi)_{L^2} = ((L^-)^m\varphi, \psi)_{L^2}$. Therefore, we have for $n, m \in \mathbf{N}$ with $n < m$,

   $$(\phi_n, \phi_m)_{L^2} = ((L^-)^m(L^+)^n\phi_0, \phi_0)_{L^2}.$$

   We shall prove, by induction on n that for every $n \geq 1$ and every $m < n$ we have $I_{m,n} := ((L^-)^m(L^+)^n\phi_0, \phi_0)_{L^2} = 0$. If $n = 1, m = 0$, since $L^-\phi_0 = 0$ we have,

   $$I_{0,1} = (L^+\phi_0, \phi_0)_{L^2} = (\phi_0, L^-\phi_0)_{L^2} = 0.$$

   Assume that our claim is true at the order $n \geq 1$ for all $m < n$. Let $m < n+1$ and consider $I_{m,n+1}$. If $m = 0$ we have,

   $$I_{0,n+1} = ((L^+)^{n+1}\phi_0, \phi_0)_{L^2} = (\phi_0, (L^-)^{n+1}\phi_0)_{L^2} = 0.$$

If $m \geq 1$ we write using question 1,

$$I_{m,n+1} = \left((L^-)^{m-1} L^- (L^+)^{n+1} \phi_0, \phi_0 \right)_{L^2},$$

$$= \left((L^-)^{m-1} (L^+)^{n+1} L^- \phi_0, \phi_0 \right)_{L^2} + (n+1) \left((L^-)^{m-1} (L^+)^n \phi_0, \phi_0 \right)_{L^2} = 0,$$

using $L^- \phi_0 = 0$ and the induction since $m - 1 < n$.

e) Recall that $\phi_j(x) = P_j(x) e^{-\frac{x^2}{2}}$ where the polynomials P_j are of exact order j. Then $(P_0, \ldots, P_n)$ form a basis of the space of polynomials of order $\leq n$ for every $n \in \mathbf{N}$ which implies that $\int x^j e^{-\frac{x^2}{2}} f(x) \, dx = 0$ for every $j \in \mathbf{N}$.

For $\xi \in \mathbf{R}$ the function $x \mapsto e^{-ix\xi} e^{-\frac{x^2}{2}} f(x)$ belongs to $L^1(\mathbf{R})$ since $f \in L^2(\mathbf{R})$ and $e^{-\frac{x^2}{2}} \in L^2(\mathbf{R})$. Then we have,

$$F(\xi) = \int e^{-ix\xi} e^{-\frac{x^2}{2}} f(x) \, dx = \int_{-\infty}^{+\infty} \sum_{j=0}^{\infty} \frac{(-ix\xi)^j}{j!} e^{-\frac{x^2}{2}} f(x) \, dx.$$

Since,

$$\int \sum_{j=0}^{+\infty} \frac{(|x\xi|)^j}{j!} e^{-\frac{x^2}{2}} |f(x)| \, dx = \int e^{|x\xi|} e^{-\frac{x^2}{2}} |f(x)| \, dx < +\infty,$$

we can apply the Fubini theorem and write,

$$F(\xi) = \sum_{j=0}^{+\infty} \frac{(-i\xi)^j}{j!} \int x^j e^{-\frac{x^2}{2}} f(x) \, dx = 0, \quad \forall \xi \in \mathbf{R}.$$

This shows that the Fourier transform of $e^{-\frac{x^2}{2}} f(x)$ vanishes identically. This implies that $e^{-\frac{x^2}{2}} f(x) \equiv 0$ thus $f \equiv 0$.

3. a) This follows immediately from the fact that (e_n) is an orthonormal basis of L^2.

b) This follows from question 3 of Part 2 where we have proved that $(\mathcal{H}u, \varphi)_{L^2} = (u, \mathcal{H}\varphi)_{L^2}$ for $u, \varphi \in D$.

c) Taking $\varphi = e_k$ in question b) we obtain, since $\mathcal{H}e_k = (2k+1)e_k$, $(u, (2k+1-\lambda)e_k)_{L^2} = 0 = (2k+1-\lambda)(u, e_k)_{L^2}$. If $2k+1-\lambda \neq 0$ for all $k \in \mathbf{N}$ then $(u, e_k)_{L^2} = 0$ for all $k \in \mathbf{N}$ which implies that $u = 0$ which contradicts our assumption $u \neq 0$. Therefore, there exists $k \in \mathbf{N}$ such that $\lambda = 2k + 1$.

4. Let $k \in \mathbf{N}$ be fixed. For any $n \neq k$ we have

$$(\mathcal{H}u, e_n)_{L^2} = (2k+1)(u, e_n)_{L^2},$$

and

$$(\mathcal{H}u, e_n)_{L^2} = (u, \mathcal{H}e_n)_{L^2} = (2n+1)\,(u, e_n)_{L^2}.$$

Therefore, $2(k-n)\,(u, e_n)_{L^2} = 0$. This implies that $(u, e_n)_{L^2} = 0$. Now, (e_n) being an orthonormal basis of L^2 we have

$$u = \sum_{n \in \mathbf{N}} (u, e_n)_{L^2}\, e_n = (u, e_k)_{L^2}\, e_k.$$

This shows that u is proportional to e_k.

Case 2. $d \geq 2$ The proof is straightforward using the path given by questions 1 to 4 above.

Solution 42

Part 0
1. The first claim follows from the Hardy inequality (4.9). For the second we use the Cauchy Schwarz and the Hardy inequalities. We obtain,

$$\int \frac{|\psi(x)|^2}{|x|}\,dx \leq \left(\int |\psi(x)|^2\,dx\right)^{\frac{1}{2}} \left(\int \frac{|\psi(x)|^2}{|x|^2}\,dx\right)^{\frac{1}{2}} \leq C\|\psi\|_{L^2}\|\nabla_x\psi\|_{L^2}.$$

Part 1
1. Using the Hint and setting $\varepsilon x = y$, $\varepsilon > 0$ in the integral we obtain,

$$F(\psi) = \varepsilon\left(\varepsilon\int |\nabla_x\chi(y)|^2\,dy - \int \frac{|\chi(y)|^2}{|y|}\,dy\right) < 0$$

if $\varepsilon > 0$ is sufficiently small and $\chi \not\equiv 0$. Thus the infimum E is strictly negative.
2. a) This follows from the definition of an infimum. Indeed in the case where $E > -\infty$ for any $n \geq 1$ $E + \frac{1}{n}$ is no more an infimum that is there exists $\psi_n \in H^1$ with $\|\psi_n\|_{L^2} = 1$ such that $E \leq F(\psi_n) \leq E + \frac{1}{n}$. If $E = -\infty$ then for any $n > 0$ there exists ψ_n as above such that $F(\psi_n) \leq -n$. In all cases $F(\psi_n) \to E$.
   b) We have $\|\psi_n\|_{L^2} = 1$. On the other hand, since $F(\psi_n) < 0$, for n large enough, using question 1 we obtain,

$$\|\nabla_x\psi_n\|_{L^2}^2 \leq \left\|\frac{\psi_n}{|x|^{\frac{1}{2}}}\right\|^2 \leq C\|\psi_n\|_{L^2}\|\nabla_x\psi_n\|_{L^2} = C\|\nabla_x\psi_n\|_{L^2}.$$

Therefore, $\|\nabla_x\psi_n\|_{L^2} \leq C$.

c) Since $\|\psi_n\|_{L^2} = 1$ there exists a subsequence $(\psi_{\sigma_1(n)})$ which converges
weakly in L^2 to $\tilde{\psi}$. Thus $(\nabla_x \psi_{\sigma_1(n)})$ converges to $\nabla_x \tilde{\psi}$ in $\mathcal{D}'$. Since
$\|\nabla_x \psi_{\sigma_1(n)}\|_{L^2}$ is uniformly bounded (by C) there exists a subsequence $(\psi_{\sigma(n)})$
such that $(\nabla_x \psi_{\sigma(n)})$ converges weakly in L^2 (thus in $\mathcal{D}'$) to v. Then $v = \nabla_x \tilde{\psi}$.

Now for $\varphi \in L^2$ we have, $|(\psi_{\sigma(n)}, \varphi)_{L^2}| \leq \|\varphi\|_{L^2}$. The left-hand side con-
verges to $|(\tilde{\psi}, \varphi)_{L^2}|$ so we obtain, $\|\tilde{\psi}\|_{L^2} = \sup\{|(\tilde{\psi}, \varphi)_{L^2}| : \|\varphi\|_{L^2} = 1\}$
≤ 1. The same argument can be applied to $\nabla_x \psi_{\sigma(n)}$.

d) Set $\theta_n = \psi_{\sigma(n)} - \tilde{\psi}$. Let $\varepsilon > 0$. Fix $R > 0$ such that $\frac{2}{R}(1 + \|\tilde{\psi}\|_{L^2}^2) \leq \frac{\varepsilon}{2}$.
Then we have,

$$
I_n =: \int_{\{|x| > R\}} \frac{|\theta_n(x)|^2}{|x|} \, dx \leq \frac{2}{R}(\|\psi_{\sigma(n)}\|_{L^2}^2 + \|\tilde{\psi}\|_{L^2}^2) \leq \frac{\varepsilon}{2}.
$$

Now by the Hardy inequality and the fact that (ψ_n) is bounded in H^1 by K
we have,

$$
J_n =: \int_{\{|x| \leq R\}} \frac{|\theta_n(x)|^2}{|x|} \, dx \leq \|\theta_n\|_{L^2(|x| \leq R)} \left(\int \frac{|\theta_n(x)|^2}{|x|^2} \, dx \right)^{\frac{1}{2}},
$$

$$
\leq C\|\theta_n\|_{L^2(|x| \leq R)} \|\nabla_x \theta_n\|_{L^2} \leq C\|\theta_n\|_{L^2(|x| \leq R)}(K + \|\tilde{\psi}\|_{H^1}).
$$

Let $\chi \in C_0^\infty$ be such that $\chi(x) = 1$ if $|x| \leq 1$, $\chi(x) = 0$ if $|x| \geq 2$ and
$0 \leq \chi \leq 1$. Then,

$$
\|\theta_n\|_{L^2(|x| \leq R)} \leq \left\| \chi(\frac{x}{R})\theta_n \right\|_{L^2}.
$$

Now by question c) we know that (θ_n) converges weakly to 0 in H^1 which
implies that $(\chi(\frac{x}{R})\theta_n)$ converges as well weakly to 0 in H^1. Since the
map $u \mapsto u$ from $\{u \in H^1 : \text{supp } u \subset \{|x| \leq 2R\}\}$ to L^2 is compact we
deduce that $(\chi(\frac{x}{R})\theta_n)$ converges strongly (that is for the norm) to zero in L^2.
Therefore, there exists $N \in \mathbf{N}$ such that for $n \geq N$, $J_n \leq C\|\theta_n\|_{L^2(|x| \leq R)}(K + \|\tilde{\psi}\|_{H^1}) \leq \frac{\varepsilon}{2}$. Therefore, $I_n + J_n \leq \varepsilon$ for $n \geq N$.

3. a) Using questions c) and d) we can write,

$$
F(\tilde{\psi}) = \|\nabla_x \tilde{\psi}\|_{L^2}^2 - \int \frac{|\tilde{\psi}(x)|^2}{|x|} \, dx
$$

$$
\leq \liminf_{n \to +\infty} \|\nabla_x \psi_{\sigma(n)}\|_{L^2}^2 - \lim_{n \to +\infty} \int \frac{|\psi_{\sigma(n)}|^2}{|x|} \, dx,
$$

$$
\leq \liminf_{n \to +\infty} \left(\|\nabla_x \psi_{\sigma(n)}\|_{L^2}^2 - \int \frac{|\psi_{\sigma(n)}|^2}{|x|} \, dx \right) = \liminf_{n \to +\infty} F(\psi_{\sigma(n)}) = E.
$$

b) Since $\theta = \dfrac{\tilde{\psi}}{\|\tilde{\psi}\|_{L^2}}$ belongs to H^1 and has an L^2 norm equal to one we have
$F(\theta) \geq E$.

c) It follows from questions a) and b) that $E(1 - \|\tilde{\psi}\|_{L^2}^2) \geq 0$. Since $E < 0$ we deduce that $\|\tilde{\psi}\|_{L^2}^2 \geq 1$. Using question 2c) we deduce that $\|\tilde{\psi}\|_{L^2}^2 = 1$ and from questions 3a) and b) that $F(\tilde{\psi}) = E$.

4. a) When $n \to +\infty$ we have $\|\nabla_x \varphi_n - \nabla_x \psi\|_{L^2} \to 0$ and by question 1 $\left\| \dfrac{\varphi_n - \psi}{|x|} \right\|_{L^2} \to 0$ so $F(\varphi_n) \to F(\psi)$.

b) Since $H^2 \subset H^1$ we have $E_1 \leq E_2$.

Now let $\psi \in H^1$ with $\|\psi\|_{L^2} = 1$ and $(\varphi_n) \subset H^2$ such that $\varphi_n \to \psi$ in H^1. Let $\varepsilon \in (0,1)$. There exists $n \in \mathbf{N}$ such that $\|\varphi_n\|_{L^2} \geq \|\psi\|_{L^2} - \varepsilon = 1 - \varepsilon > 0$ and by question a) $F(\varphi_n) \leq F(\psi) + \varepsilon$. Then we can write,

$$E_2 \leq F\left(\frac{\varphi_n}{\|\varphi_n\|_{L^2}} \right) \leq \frac{1}{(1-\varepsilon)^2}(F(\psi) + \varepsilon).$$

Taking that infimum on all $\psi \in H^1$ with $\|\psi\|_{L^2} = 1$ we obtain,

$$E_2 \leq \frac{1}{(1-\varepsilon)^2}(E_1 + \varepsilon).$$

Since this holds for all $\varepsilon > 0$ we obtain $E_2 \leq E_1$.

Part 2

1. a) Set $f = \tilde{\psi} \pm \varepsilon\varphi$ and $G(f) = F(f) - E\|f\|_{L^2}^2 \geq 0$. Since $G(\tilde{\psi}) = 0$ we obtain,

$$\pm 2\varepsilon \mathrm{Re} \left(\int \nabla_x \tilde{\psi}(x) \cdot \overline{\nabla_x \varphi(x)}\, dx - \int \frac{\tilde{\psi}(x)\overline{\varphi(x)}}{|x|}\, dx \right.$$
$$\left. - E \int \tilde{\psi}(x)\overline{\varphi(x)}\, dx \right) \geq -\varepsilon^2 G(\varphi).$$

Dividing both members by $2\varepsilon > 0$ and letting ε go to zero we obtain,

$$\mathrm{Re} \left(\int \nabla_x \tilde{\psi}(x) \cdot \overline{\nabla_x \varphi(x)}\, dx - \int \frac{\tilde{\psi}(x)\overline{\varphi(x)}}{|x|}\, dx - E \int \tilde{\psi}(x)\overline{\varphi(x)}\, dx \right) = 0,$$

which implies that,

$$\mathrm{Re} \left\langle -\Delta\tilde{\psi} - \frac{\tilde{\psi}}{|x|} - E\tilde{\psi}, \varphi \right\rangle = 0,$$

in the sense of distributions.

b) Changing φ to $i\varphi$ we see that the imaginary part of the quantity above in the brackets is equal to zero. This proves our claim.

c) Since $\tilde{\psi} \in H^1$, question 1 Part 1 shows that $\frac{\tilde{\psi}}{|x|} \in L^2$. It follows from the equation that $\Delta\tilde{\psi} \in L^2$ which implies that $u \in H^2$. Thus $\tilde{\psi} \in D$.

2. If $u \neq 0$ there exists $x_0 \in \mathbf{R}^3$ such that $u(x_0) \neq 0$. Since $H^2(\mathbf{R}^3) \subset C^0$, u is continuous, so there exists a neighborhood of x_0 in which u does not vanish. Therefore, $\|u\|_{L^2} \neq 0$. On the other hand, if $u \in D$ we have,

$$E\|u\|_{L^2}^2 = \left(-\Delta u - \frac{u}{|x|}, u\right)_{L^2} = \|\nabla_x u\|_{L^2}^2 - \left\|\frac{u}{|x|^{\frac{1}{2}}}\right\|_{L^2}^2 = F(u). \qquad (9.40)$$

Therefore, $\frac{u}{\|u\|_{L^2}}$ is a minimizer.

Part 3

1. Obviously $|u| \in L^2$. Now we have $\varepsilon \leq \sqrt{u^2 + \varepsilon^2} \leq |u| + \varepsilon$. Therefore, $0 \leq u_\varepsilon \leq |u|$ and for almost all $x \in \mathbf{R}^3$ $\lim_{\varepsilon \to 0} u_\varepsilon(x) = |u(x)|$. It follows from the Lebesgue dominated convergence that

$$\lim_{\varepsilon \to 0} \int u_\varepsilon(x)\varphi(x)\,dx = \int |u(x)|\varphi(x)\,dx,$$

for all $\varphi \in C_0^\infty$. This implies that $u_\varepsilon \to |u|$ in $\mathcal{D}'$. It follows that $\partial_j u_\varepsilon \to \partial_j |u|$ in $\mathcal{D}'$, $1 \leq j \leq 3$. Now $\partial_j u_\varepsilon = \frac{u\partial_j u}{\sqrt{u^2+\varepsilon^2}}$. It follows that $\partial_j u_\varepsilon(x)$ converges to zero if $u(x) = 0$ and to $\frac{u(x)}{|u(x)|}\partial_j u(x)$ if $u(x) \neq 0$. Moreover, for $\varphi \in C_0^\infty$ we have $|\partial_j u_\varepsilon(x)\varphi(x)| \leq |\partial_j u(x)\varphi(x)| \in L^1$. It follows from the Lebesgue dominated convergence that

$$\partial_j u_\varepsilon \to 1_{\{u(x)\neq 0\}}\frac{u(x)}{|u(x)|}\partial_j u \quad \text{in} \quad \mathcal{D}'.$$

Therefore, $\partial_j |u| = 1_{\{u(x)\neq 0\}}\frac{u(x)}{|u(x)|}\partial_j u \in L^2$.

2. a) It follows from question 1 that $|\tilde{\psi}| \in H^1$ and $\|\nabla_x|\tilde{\psi}|\| \leq |\nabla_x\tilde{\psi}|$. This shows that $F(|\tilde{\psi}|) \leq F(\tilde{\psi})$. Since $F(\tilde{\psi})$ is the infimum we must have the equality. It follows from question 1a) in Part 2 that we have the claimed equality.

b) This follows from question 1c) in Part 2.

c) Let $G(x) = \frac{e^{-\omega|x|}}{4\pi|x|}$. Then $G \in L^1 \cap L^2$ and $(-\Delta + \omega^2)G = \delta_0$ (the Dirac distribution at zero). It follows that,

$$|\tilde{\psi}(x)| = (\delta_0 \star |\tilde{\psi}|)(x) = ((-\Delta + \omega^2)G \star |\tilde{\psi}|)(x) = (G \star (-\Delta + \omega^2)|\tilde{\psi}|)(x),$$

$$= \left(G \star \frac{|\tilde{\psi}|}{|x|}\right)(x) = \int \frac{e^{-\omega|x-y|}}{4\pi|x-y|}\frac{|\tilde{\psi}(y)|}{|y|}\,dy.$$

d) If there is a point x_0 such that $|\widetilde{\psi}(x_0)| = 0$, question c) shows that $\widetilde{\psi}(y) = 0$ almost everywhere, which is impossible since $\|\widetilde{\psi}\|_{L^2} = 1$. Therefore (8.13) is proved.

3. (8.14) is true for $k = 0$ since $|\widetilde{\psi}| \in H^2 \subset L^2 \cap L^\infty$. Assume it is true up to the order $k \geq 0$ then since $|x|^{k+1} \leq C_k(|x - y|^{k+1} + |y|^{k+1})$ we have,

$$|x|^{k+1}|\widetilde{\psi}| \leq C \left(\int |x - y|^k e^{-\omega|x-y|} \frac{|\widetilde{\psi}(y)|}{|y|} \, dy + \int \frac{e^{-\omega|x-y|}}{|x - y|} |y|^k |\widetilde{\psi}(y)| \, dy \right).$$

Now on one hand,

$$|x|^k e^{-\omega|x|} \in L^1 \cap L^2, \qquad \frac{|\widetilde{\psi}(y)|}{|y|} \in L^2,$$

and on the other hand, by the induction,

$$\frac{e^{-\omega|x|}}{|x|} \in L^1 \cap L^2, \qquad |y|^k |\widetilde{\psi}| \in L^2.$$

By the Young inequality we have $L^1 \star L^2 \subset L^2$ and $L^2 \star L^2 \subset L^\infty$. Therefore, $|x|^{k+1}|\widetilde{\psi}| \in L^2 \cap L^\infty$.

4. a) Let $u \in \text{Ker}(\mathcal{H} - E)$. If $u \neq 0$ it follows from question 2. Part 2 that $\frac{u}{\|u\|_{L^2}}$ is a minimizer and from question 2d) Part 3 that $|u(x)| > 0$ for all $x \in \mathbf{R}^3$.

b) We have obviously $u \in \text{Ker}(\mathcal{H} - E)$ and $u(x_0) = 0$. It follows from question a) above that $u \equiv 0$. Therefore, $u_1 = \lambda u_2, \lambda \in \mathbf{C}$ which shows that $\dim \text{Ker}(\mathcal{H} - E) = 1$.

5. Since $|Ax| = |x|$ we have,

$$-\Delta\widetilde{\psi}_A(x) - \frac{\widetilde{\psi}_A(x)}{|x|} - E\widetilde{\psi}_A(x) = (-\Delta\widetilde{\psi})(Ax) - \frac{\widetilde{\psi}(Ax)}{|Ax|} - E\widetilde{\psi}(Ax)$$

$$= ((\mathcal{H} - E)\widetilde{\psi})(Ax) = 0.$$

Moreover, $\|\widetilde{\psi}_A\|_{L^2} = \|\widetilde{\psi}\|_{L^2} = 1$. Therefore, by question 2 Part 2, $\widetilde{\psi}_A$ is a minimizer. It follows from the previous question that there exists $c \neq 0$ such that $\widetilde{\psi}_A = c\,\widetilde{\psi}$. Since the L^2 norms are equal to one and $\widetilde{\psi} > 0$ we have $c = 1$. Thus $\widetilde{\psi}$ is radial.

Part 4

1. This follows from (9.40) in question 2 Part 2.
2. a) Since $\theta \in H^2 \subset C^0$ the function u is continuous on $[0, +\infty)$. Now since $\theta(x) = u(r)$ is radial we have in dimension $d = 3$, $\Delta\theta = u'' + \frac{2}{r}u'$.

   b) By hypothesis $r^k u \in L^\infty((0, +\infty))$ for every $k \in \mathbf{N}$. This implies the same properties for v and the fact that $v \in L^1((0, +\infty))$.

c) This follows from the fact that

$$v'' = ru'' + 2u' = r\left(u'' + \frac{2}{r}u'\right) = r\left(-\frac{1}{r}u - \lambda u\right) = -\frac{v}{r} - \lambda v.$$

3. a) By the equation satisfied by v we see that $T = 0$ on $(0, +\infty)$ and, by definition of w, on $(-\infty, 0)$. Therefore, we have supp $T \subset \{0\}$ and the result follows.

  b) We have,

$$\langle T, \varphi_\varepsilon \rangle = \left(w, -(r\varphi_\varepsilon)'' - \varphi_\varepsilon - \lambda r\varphi_\varepsilon\right),$$

$$= -\int_0^{+\infty} v(r)(r\varphi_\varepsilon(r))'' \, dr - \int_0^{+\infty} v(r)\left[\varphi\left(\frac{r}{\varepsilon}\right) + \lambda r\varphi\left(\frac{r}{\varepsilon}\right)\right] dr$$

$$= I + J.$$

Setting $r = \varepsilon s$ in the integral we see that $|J| \leq \varepsilon \|v\|_{L^\infty} \int_0^{+\infty}(1 + \varepsilon|\lambda|s)|\varphi(s)| \, ds$. Therefore, $J \to 0$ when $\varepsilon \to 0$. On the other hand, setting $r = \varepsilon s$ in the integral we have,

$$I = -\int_0^{+\infty} v(r)\left[\frac{r}{\varepsilon^2}\varphi''\left(\frac{r}{\varepsilon}\right) + \frac{2}{\varepsilon}\varphi'\left(\frac{r}{\varepsilon}\right)\right] dr$$

$$= -\int_0^{+\infty} v(\varepsilon s)\left(s\varphi''(s) + 2\varphi'(s)\right) ds.$$

Since u is continuous on $[0, +\infty)$ and $v = ru$ we have $v(0) = 0$. So using the dominated convergence theorem we see that $\lim_{\varepsilon \to 0} I = 0$.

  c) By question a) we have, $\langle T, \varphi_\varepsilon \rangle = \sum_{k=0}^{N}(-1)^k c_k \varepsilon^{-k}\varphi^{(k)}(0)$ and by question b) $\langle T, \varphi_\varepsilon \rangle \to 0$ when $\varepsilon \to 0$. This implies $T = 0$. Indeed we show that $c_N = c_{N-1} = \cdots c_0 = 0$ as follows. We have $\varepsilon^N \langle T, \varphi_\varepsilon \rangle \to 0$ thus $c_N = 0$. Then $\varepsilon^{N-1}\langle T, \varphi_\varepsilon \rangle \to 0$ which implies $c_{N-1} = 0$ and we continue up to c_0.

4. a) The fact that f is C^∞ on $\mathbf{R}$ follows from the fact that $r^k v \in L^1(\mathbf{R})$ for every $k \in \mathbf{N}$ and the Lebesgue theorem on differentiability of integrals. Now for Im $z < 0$ and $r \in (0, +\infty)$ we have $|e^{izr}v(r)| = e^{-\text{Im } zr}|v(r)| \leq |v(r)| \in L^1$. Therefore, we can set $f(z) = \int_0^{+\infty} e^{izr}v(r) \, dr$. Again by the Lebesgue theorem we see that f is holomorphic in the set Im $z < 0$.

  b) Recall that $-rw'' - w - \lambda rw = 0$ in $S'(\mathbf{R})$ and that $\hat{w} = f$. Taking the Fourier transform of both members and using the fact that

$$\mathcal{F}(rw'') = \frac{1}{i}\frac{d}{d\xi}(\xi^2 f), \mathcal{F}(\lambda rw) = -\frac{\lambda}{i}\frac{d}{d\xi}f,$$

we deduce that $(\xi^2 + \mu^2)f' + (2\xi + i)f = 0$ in $S'(\mathbf{R})$.

5. a) Since G is holomorphic in Im $z < 0$ and vanishes when Im $z = 0$ the function $H(x, y) = G(x + iy)$ is a solution of the problem,

$$\bar{\partial} H = \frac{1}{2}\left(\frac{\partial}{\partial x} + i\frac{\partial}{\partial y}\right) H = 0 \text{ for } y < 0, \quad G|_{y=0} = 0.$$

The surface $S = \{(x, y) \in \mathbf{R}^2 : y = 0\}$ is non-characteristic for the operator $\bar{\partial}$ which has constant, thus analytic, coefficients. By the Holmgren theorem every point $x_0 \in S$ has a neighborhood in which $H = 0$. It follows that G vanishes in a small ball contained in the set Im $z < 0$. Since G is holomorphic there, the principle of isolated zeros implies that $G = 0$.

Alternative proof: it follows from Morera's theorem that if u is a holomorphic function in Im $z > 0$ and Im $z < 0$ and is continuous in $\mathbf{C}$ then u is holomorphic in $\mathbf{C}$. Here consider the function defined by $u(z) = G(z)$ if Im $z \leq 0$ and $u(z) = G(-z)$ for Im $z > 0$. Then u satisfies all the conditions above (since $G = 0$ for Im $z = 0$) therefore u is holomorphic in $\mathbf{C}$. Since $u = 0$ for Im $z = 0$ the principle of isolated zeros proves that $u = 0$.

b) The function $G(z) = (z^2 + \mu^2)f'(z) + (2z + i)f(z)$ is holomorphic in Im $z < 0$. By question 4 G vanishes when Im $z = 0$. It follows from the previous question that $G = 0$ in Im $z < 0$.

6. a) By the general theory (see the appendix) the integral $\frac{1}{2i\pi}\int_B \frac{f'(z)}{f(z)} dz$ is equal to $n - p$ where n is the number of zeroes of f and p the number of its poles. Since f is holomorphic in Im $z < 0$ it has no pole. Therefore, the integral is equal to $n \in \mathbf{N}$.

b) By the question 5b) we have $\frac{f'(z)}{f(z)} = -\frac{2z+i}{z^2+\mu^2}$. Inside B it has a simple pole at $z = -i\mu$. The residue at this pole is equal to $\lim_{z\to-i\mu}(z + i\mu)\frac{f'(z)}{f(z)} = -\lim_{z\to-i\mu}\frac{2z+i}{z-i\mu} = \frac{1}{2\mu} - 1$. Therefore, by the residue theorem we have $\frac{1}{2i\pi}\int_B \frac{f'(z)}{f(z)} dz = \frac{1}{2\mu} - 1$. It follows from question a) that $\frac{1}{2\mu} - 1 = n$ so $\mu = \frac{1}{2(n+1)}$.

c) It follows from question b) that for all $\theta \in D$, $\theta > 0$ satisfying $(\mathcal{H} - \lambda)\theta = 0$ we must have $\lambda = -\frac{1}{4(n+1)^2}$. The minimum value of λ is then reached for $n = 0$. This minimum energy is precisely E. Therefore, we must have $E = -\frac{1}{4}$.

d) It follows that $\frac{f'(z)}{f(z)} = -\frac{2z+i}{z^2+\frac{1}{4}} = -2(z - \frac{i}{2})^{-1}$. This implies that $f(z) = C(z - \frac{i}{2})^{-2}$, where $C \in \mathbf{C}$.

7. a) We have,

$$\int_{\mathbf{R}} e^{-ir\xi}\varphi(r)\, dr = \int_0^{+\infty} e^{-r(\frac{1}{2}+i\xi)}\, dr = \frac{1}{\frac{1}{2} + i\xi}.$$

It follows that $\mathcal{F}(r\varphi)(\xi) = -\frac{1}{i}(\frac{d}{d\xi}\widehat{\varphi})(\xi) = (\xi - \frac{i}{2})^{-2}$.

b) The previous question shows that $r\varphi(r) = \mathcal{F}^{-1}\left((\xi - \frac{1}{2})^{-2}\right)$. This shows
that $w(r) = Cr\varphi(r) = C1_{r>0}re^{-\frac{r}{2}}$. This implies that $v(r) = Cre^{-\frac{r}{2}}$ so
$u(r) = Ce^{-\frac{r}{2}}$. Since the set of minimizers is a vector space of one dimension
we conclude that it is generated by the function $\tilde{\psi}(x) = Ce^{-\frac{1}{2}|x|}$ with $C = \frac{1}{\sqrt{8\pi}}$ to have $\|\tilde{\psi}\|_{L^2} = 1$.

Solution 43

Part 1

1. We have $u \in C^2([-h, 0], H^s(\mathbf{R}))$ for all $s \in \mathbf{R}$, $u|_{y=0} = 0$, $(\partial_y u)|_{y=-h} = 0$. An
integration by parts shows that,

$$0 = -\left(\Delta_{xy} u, u\right)_{L^2(\Omega)} = \iint_\Omega \left(|\partial_x u(x, y)|^2 + |\partial_y u(x, y)|^2\right) dx\, dy.$$

Therefore, u is constant and since $u|_{y=0} = 0$ we have $u \equiv 0$.

2. a) Applying a Fourier transformation in x we see that the equation is equivalent
   to $(\partial_y^2 - \xi^2)\hat{u}(\xi, y) = 0$ for all $\xi \in \mathbf{R}$. So $\hat{u}(\xi, y) = A(\xi)e^{y|\xi|} + B(\xi)e^{-y|\xi|}$.
   The boundary conditions show that,

$$A(\xi) + B(\xi) = \hat{\psi}(\xi), \quad |\xi|A(\xi)e^{-h|\xi|} - |\xi|B(\xi)e^{h|\xi|} = 0, \quad \forall \xi \in \mathbf{R},$$

so,

$$A(\xi) = \frac{e^{h|\xi|}}{e^{h|\xi|} + e^{-h|\xi|}}\hat{\psi}(\xi), \quad B(\xi) = \frac{e^{-h|\xi|}}{e^{h|\xi|} + e^{-h|\xi|}}\hat{\psi}(\xi), \quad \forall \xi \in \mathbf{R}.$$

   It follows that, $\hat{u}(\xi, y) = \frac{\text{ch}((y+h)|\xi|)}{\text{ch}(h|\xi|)}\hat{\psi}(\xi)$.

   b) We deduce from the above formula that $(\partial_y \hat{u})(\xi, 0) = |\xi|\frac{\text{sh}(h|\xi|)}{\text{ch}(h|\xi|)}\hat{\psi}(\xi) = |\xi|\text{th}(h|\xi|)\hat{\psi}(\xi)$.

   c) The symbol $a(\xi) = |\xi|\text{th}(h|\xi|)$ is C^∞ in $\mathbf{R} \setminus \{0\}$ and also near zero since
   $\text{th}(h|\xi|) = \sum_{k=0}^{+\infty} c_k(h|\xi|)^{2k+1}$. On the other hand, this symbol is even. It is
   therefore sufficient to estimate it for $\xi \geq 0$. The symbol ξ belongs to S^1 and
   $\text{th}(h|\xi|)$ is bounded (by 1) and has all its derivatives bounded, so it belongs to
   S^0. Therefore, $a \in S^1$.

   c) Since $|\text{th}\, x| \leq 1$ we have,

$$\|a(D_x)\psi\|^2_{H^{s-\frac{1}{2}}(\mathbf{R})} = \int \langle\xi\rangle^{2s-1} |\xi|^2(\text{th}\,(h|\xi|))^2|\hat{\psi}(\xi)|^2\, d\xi \leq \|\psi\|^2_{H^{s+\frac{1}{2}}(\mathbf{R})}.$$

Part 2

1. For $u \in C^{\infty,0}(\Omega)$ and $(x, y) \in \Omega)$ we can write $u(x, y) = -\int_y^{\eta(x)} \partial_y u(x, t) \, dt$ so that,

$$|u(x, y)|^2 \leq |\eta(x) + h| \int_{-h}^{\eta(x)} |\partial_y u(x, t)|^2 \, dt \leq \frac{3h}{2} \int_{-h}^{\eta(x)} |\partial_y u(x, t)|^2 \, dt.$$

Then,

$$\int_{\mathbf{R}} \int_{-h}^{\eta(x)} |u(x, y)|^2 \, dy \, dx \leq \left(\frac{3h}{2}\right)^2 \int_{\mathbf{R}} \int_{-h}^{\eta(x)} |\partial_y u(x, y)|^2 \, dy \, dx.$$

Since $C^{\infty,0}(\Omega)$ is dense in $\mathcal{H}^{1,0}(\Omega)$ for the $H^1(\Omega)$ norm this inequality extends to $H^1(\Omega)$.

2. a) It is obvious, by construction, that $\widetilde{\psi}|_{z=0} = \psi$, $\widetilde{\psi}|_{z=-h} = 0$. Now for fixed z, by the Parseval equality and the fact that $e^{z|\xi|} \leq 1$ for $z \leq 0$ we can write,

$$\|\widetilde{\psi}(\cdot, z)\|_{L^2(\mathbf{R})} = (2\pi)^{-\frac{1}{2}} |\chi(z)| \|e^{z|\xi|} \widehat{\psi}\|_{L^2(\mathbf{R})} \leq (2\pi)^{-\frac{1}{2}} |\chi(z)| \|\widehat{\psi}\|_{L^2(\mathbf{R})},$$

$$\leq |\chi(z)| \|\psi\|_{L^2(\mathbf{R})}.$$

Integrating in z we obtain, $\|\widetilde{\psi}\|_{L^2(\widetilde{\Omega})} \leq C\|\psi\|_{L^2(\mathbf{R})}$. Now, $\partial_x \widetilde{\psi} = \chi(z)e^{z|D_x|}\partial_x \psi$ and $\partial_z \widetilde{\psi} = \chi'(z)e^{z|D_x|}\psi + \chi(z)e^{z|D_x|}|D_x|\psi$. Therefore, we have,

$$\|\partial_{x,z}\widetilde{\psi}\|_{L^2(\widetilde{\Omega})}^2 \leq C \int_{\mathbf{R}} |\xi|^2 |\widehat{\psi}(\xi)|^2 \left(\int_{-h}^0 e^{2z|\xi|} \, dz\right) d\xi + C\|\psi\|_{L^2(\mathbf{R})}^2.$$

Since $\int_{-h}^0 2|\xi| e^{2z|\xi|} \, dz = \int_{-h}^0 \frac{d}{dz}\left(e^{2z|\xi|}\right) dz \leq 1$ we obtain,

$$\|\partial_{x,z}\widetilde{\psi}\|_{L^2(\widetilde{\Omega})}^2 \leq C' \int_{\mathbf{R}} |\xi| |\widehat{\psi}(\xi)|^2 \, d\xi + C\|\psi\|_{L^2(\mathbf{R})}^2 \leq C''\|\psi\|_{H^{\frac{1}{2}}(\mathbf{R})}^2.$$

   b) Set $\psi(x, y) = \widetilde{\psi}(x, y - \eta(x))$. We have $\psi|_{y=\eta(x)} = \psi$. On the other hand, if $y \leq -\frac{5h}{6}$ we have $y - \eta(x) \leq y + \frac{h}{2} \leq -\frac{h}{3}$, so $\psi(x, y) = 0$. Setting $y - \eta(x) = z$ in the integral in y we see that,

$$\|\psi\|_{L^2(\Omega)} + \|\partial_y \psi\|_{L^2(\Omega)} \leq C(\|\widetilde{\psi}\|_{L^2(\widetilde{\Omega})} + \|\partial_z \widetilde{\psi}\|_{L^2(\widetilde{\Omega})}).$$

Now, $(\partial_x \psi)(x, y) = (\partial_x \widetilde{\psi})(x, y - \eta(x)) - \partial_x \eta(x)(\partial_z \widetilde{\psi})(x, y - \eta(x))$. It follows that,

$$\|\partial_x \psi\|_{L^2(\Omega)} \leq C\|\partial_x \widetilde{\psi}\|_{L^2(\widetilde{\Omega})} + \|\partial_x \eta\|_{L^\infty(\mathbf{R})} \|\partial_z \widetilde{\psi}\|_{L^2(\widetilde{\Omega})}.$$

The desired estimate follows from question a).

3. We have,

$$a(u, v) = \iint_\Omega \left(|\partial_x u(x, y)|^2 + |\partial_y u(x, y)|^2 \right) dx\, dy.$$

The coercivity on $\mathcal{H}^{1,0}(\Omega)$ follows then from question 1.

4. a) By the Lax–Milgram lemma it is sufficient to prove that $\Delta_{xy}\underline{\psi}$ defines a continuous linear form on $\mathcal{H}^{1,0}(\Omega)$. For $\varphi \in C^{\infty,0}(\Omega)$ set,

$$\left\langle \Delta_{xy}\underline{\psi}, \varphi \right\rangle := \iint_\Omega \Delta_{xy}\underline{\psi}(x, y)\overline{\varphi(x, y)}\, dx\, dy.$$

Since $\varphi|_{y=\eta(x)} = 0$ and $\underline{\psi} \equiv 0$ for $y \leq -\frac{5h}{6}$, an integration by parts shows that,

$$\left\langle \Delta_{xy}\underline{\psi}, \varphi \right\rangle = -\iint_\Omega \nabla_{xy}\underline{\psi}(x, y) \cdot \overline{\nabla_{xy}\varphi(x, y)}\, dx\, dy,$$

so that,

$$\left| \left\langle \Delta_{xy}\underline{\psi}, \varphi \right\rangle \right| \leq \|\underline{\psi}\|_{H^1(\Omega)}\|\varphi\|_{H^1(\Omega)} \leq C(1 + \|\eta\|_{W^{1,\infty}(\mathbf{R})})\|\psi\|_{H^{\frac{1}{2}}(\mathbf{R})}\|\varphi\|_{H^1(\Omega)}.$$

This shows that $\Delta_{xy}\underline{\psi}$ defines a continuous linear form on $C^{\infty,0}(\Omega)$ endowed with the $H^1(\Omega)$ norm. By density it can be extended to a continuous linear form on $\mathcal{H}^{1,0}(\Omega)$. The Lax–Milgram lemma shows that there exists a unique $u \in \mathcal{H}^{1,0}(\Omega)$ such that,

$$a(u, v) = \left\langle \Delta_{xy}\underline{\psi}, v \right\rangle, \quad \forall v \in \mathcal{H}^{1,0}(\Omega).$$

If $\varphi \in C_0^\infty(\Omega)$ we have $a(u, \varphi) = \left\langle -\Delta_{xy}v, \varphi \right\rangle$.

b) This follows immediately from a).

5. We have $\rho(x, 0) = \eta(x)$, $\rho(x, -h) = -h$ and $\partial_z\rho(x, z) = \frac{1}{h}\eta(x) + 1 \geq 1 - \frac{1}{h}\|\eta\|_{L^\infty(\mathbf{R})} \geq \frac{1}{2}$. For fixed x, the map $z \mapsto \rho(x, z)$ from $(-h, 0)$ to $(-h, \eta(x))$ being strictly increasing, is bijective. Therefore, the map, $(x, z) \mapsto (x, \rho(x, z))$ is bijective from $\widetilde{\Omega}$ to Ω and its Jacobian, which is equal to $\partial_z\rho$, is different from zero. Therefore, it is a C^∞- diffeomorphism.

6. This follows immediately from the chain rule.

7. We have,

$$\partial_z U = \partial_z\Lambda_1\widetilde{\phi} - \partial_x\partial_z\rho \cdot \Lambda_2\widetilde{\phi} - \partial_x\rho \cdot \partial_z\Lambda_2\widetilde{\phi},$$

$$= (\partial_z\rho)\Lambda_1^2\widetilde{\phi} - \partial_x\partial_z\rho \cdot \Lambda_2\widetilde{\phi} - (\partial_z\rho)(\partial_x - \Lambda_2)\Lambda_2\widetilde{\phi},$$

$$= (\partial_z\rho)(\Lambda_1^2 + \Lambda_2^2)\widetilde{\phi} - \partial_x((\partial_z\rho)\Lambda_2\widetilde{\phi}) = -\partial_x((\partial_z\rho)\Lambda_2\widetilde{\phi}).$$

8. a) Recall that $\widetilde{\Omega} = \{(x, z) : x \in \mathbf{R}, -h < z < 0\}$. We have,

$$\|U\|_{L^2(\widetilde{\Omega})} \leq \|\Lambda_1\widetilde{\phi}\|_{L^2(\widetilde{\Omega})} + \|\partial_x\rho\|_{L^\infty(\widetilde{\Omega})}\|\Lambda_2\widetilde{\phi}\|_{L^2(\widetilde{\Omega})},$$

$$\leq \|\Lambda_1\widetilde{\phi}\|_{L^2(\widetilde{\Omega})} + C\|\partial_x\eta\|_{L^\infty(\mathbf{R})}\|\Lambda_2\widetilde{\phi}\|_{L^2(\widetilde{\Omega})}.$$

It follows from question 4b) that,

$$\|U\|_{L^2(\widetilde{\Omega})} \leq C(1 + \|\eta\|_{W^{1,\infty}(\mathbf{R})})\|\psi\|_{H^{\frac{1}{2}}}.$$

Now, from question 7 we have,

$$\|\partial_z U\|_{L^2((-h,0),H^{-1}(\mathbf{R}))} \leq \|\partial_z\rho\|_{L^\infty(\mathbf{R})}\|\Lambda_2\widetilde{\phi}\|_{L^2(\widetilde{\Omega})} \leq C(1 + \|\eta\|_{W^{1,\infty}(\mathbf{R})})\|\psi\|_{H^{\frac{1}{2}}}.$$

b) This follows immediately from Problem 5.

9. By definition we have the equality,

$$(\Lambda_1 - (\partial_x\rho)\Lambda_2)\widetilde{\phi}(x, z) = (\partial_x - (\partial_x\rho)\partial_y)\phi)(x, \rho(x, z)),$$

so that, since $\partial_x\rho|_{z=0} = \partial_x\eta$, we have,

$$U|_{z=0} = (\partial_x - (\partial_x\eta)\partial_y)\phi|_{y=\eta(x)}.$$

The surface Σ is given by the equation $y = \eta(x)$. Its unit normal is then the vector $n = (1 + (\partial_x\eta(x))^2)^{-\frac{1}{2}}(-\partial_x\eta(x), 1)$. The normal derivative is then,

$$\frac{\partial}{\partial n} = \frac{1}{\sqrt{1 + (\partial_x\eta(x))^2}}\left(\partial_y - \partial_x\eta(x)\partial_x\right).$$

Therefore, we have,

$$U|_{z=0} = \sqrt{1 + (\partial_x\eta(x))^2}\left(\frac{\partial\phi}{\partial n}\right)|_\Sigma.$$

Solution 44

Part 1

1. By the Sobolev embedding we have $H^1(\mathbf{R}^3) \subset L^p(\mathbf{R}^3)$ for $2 \leq p \leq 6$. Therefore, $u^3 \in L^2(\mathbf{R}^3)$, which implies that $(I - \Delta)u \in L^2(\mathbf{R}^3)$ so $u \in H^2(\mathbf{R}^3)$ using the Fourier transform.

2. The function $f(u) = u^3$ is a C^∞ function such that $f(0) = 0$. Since $k \geq 2 > \frac{3}{2}$ we can use (3.3) to infer that $u^3 \in H^k(\mathbf{R}^3)$. Then $(I - \Delta)u = u^3 \in H^k(\mathbf{R}^3)$

which implies, using the Fourier transform, that $u \in H^{k+2}(\mathbf{R}^3)$. Therefore, by induction, $u \in H^m(\mathbf{R}^3)$ for every $m \in \mathbf{N}$.

Part 2

1. Since E is radial so is $\widehat{E}$. Moreover, since $E \in L^1(\mathbf{R}^3)$, $\widehat{E}$ is a continuous function. Therefore, $\widehat{E}(A\xi) = \widehat{E}(\xi)$ for every $\xi \in \mathbf{R}^3$ and every orthogonal matrix A. Taking A such that $A\xi = (|\xi|, 0, 0)$ we can write,

$$\widehat{E}(\xi) = \int_{\mathbf{R}^3} e^{-i|\xi| x_1} \frac{e^{-|x|}}{4\pi |x|} dx.$$

To compute the integral we use the polar coordinates, $x_1 = r \cos \theta_1, x_2 = r \sin \theta_1 \cos \theta_2, x_3 = r \sin \theta_1 \sin \theta_2, r > 0, \theta_1 \in (0, \pi), \theta_2 \in (0, 2\pi)$. Then $dx = r^2 dr \, \sin \theta_1 d\theta_1 \, d\theta_2$. It follows that,

$$\widehat{E}(\xi) = \frac{1}{2} \int_0^{+\infty} \left(\int_0^\pi e^{-i|\xi| r \cos \theta_1} \sin \theta_1 d\theta_1 \right) e^{-r} r \, dr.$$

Setting $s = \cos \theta_1$ in the integral in θ_1 we obtain,

$$\widehat{E}(\xi) = \frac{1}{2} \int_0^{+\infty} \left(\int_{-1}^{+1} e^{-irs|\xi|} ds \right) e^{-r} r \, dr$$

$$= \frac{1}{2i|\xi|} \int_0^{+\infty} \left(e^{-r(1-i|\xi|)} - e^{-r(1+i|\xi|)} \right) dr = \frac{1}{1 + |\xi|^2}.$$

2. We deduce from the previous question that $(1 + |\xi|^2)\widehat{E}(\xi) = \widehat{(I - \Delta)E} = 1 = \widehat{\delta_0}$. By inverse Fourier transform we have $(I - \Delta)E = \delta_0$.

Part 3

1. From question 2 in the previous part one can write,

$$u = u \star \delta_0 = u \star (I - \Delta)E = (I - \Delta)u \star E = u^3 \star E.$$

Now since $E \in L^1(\mathbf{R}^3)$ and $u^3 \in L^\infty(\mathbf{R}^3)$ we deduce that,

$$u(x) = \int_{\mathbf{R}^3} E(x - y)u(y)^3 \, dy.$$

2. This follows from the fact that $u \in H^2(\mathbf{R}^3) \subset C^0_{\to 0}(\mathbf{R}^3)$ since $2 > \frac{3}{2}$, which means that u tends to zero when $|x| \to +\infty$.

3. (i) If $x \in \omega_n$ we have $|x| \geq 2^n$, so if $|y| \leq 2^{n-1}$ we have $|y| \leq \frac{1}{2}|x|$; this implies that $|x - y| \geq |x| - |y| \geq \frac{1}{2}|x| \geq 2^{n-1}$. Moreover, $u^3 \in L^1(\mathbf{R}^3)$. Therefore,

$$I_1 \leq \frac{1}{4\pi 2^{n-1}} e^{-2^{n-1}} \int_{\mathbf{R}^3} |u(y)|^3 \, dy \leq C(u) e^{-2^{n-1}}.$$

(ii) If $|x| \leq 2^{n+1}$ and $|y| \geq 2^{n+2}$ we have, $|x| \leq \frac{1}{2}|y|$ so $|x-y| \geq |y| - |x| \geq \frac{1}{2}|y| \geq 2^{n+1}$. Therefore, as in (i) we have,

$$I_2 \leq C(u) e^{-2^{n+1}}.$$

(iii) We have,

$$I_3 \leq \int_{\mathbf{R}^3} \frac{e^{-|x-y|}}{4\pi|x-y|} \sup_{y \in \omega_{n-1}} |u(y)|^3 \, dy \leq \frac{1}{4\pi} A_{n-1}^3 \int_{\mathbf{R}^3} \frac{e^{-|x|}}{|x|} \, dx \leq C A_{n-1}^3,$$

and the same proof holds for I_4 and I_5.

Since by question 1 $u(x) = \sum_{j=1}^5 I_j$ we obtain the estimate (8.20) in the statement.

4. Let $M \in \mathbf{N}$. By the previous question we have,

$$\sum_{n=N}^{N+M} A_n \leq C \left(\sum_{n=N}^{N+M} e^{-2^{n-1}} + A_{N-1}^3 + \sum_{n=N}^{N+M-1} A_n^3 + \sum_{n=N}^{N+M} A_n^3 \right.$$

$$\left. + \sum_{n=N+1}^{N+M} A_n^3 + A_{N+M+1}^3 \right) \leq C \left(\sum_{n=N}^{N+M} e^{-2^{n-1}} + A_{N-1}^3 + A_{N+M+1}^3 + 3 \sum_{n=N}^{N+M} A_n^3 \right).$$

Now we notice that, since $N - n - 1 \leq -1$,

$$e^{-2^{n-1}} = e^{-2^{N-2}} e^{-2^{n-1}(1-2^{N-n-1})} \leq e^{-2^{N-2}} e^{-2^{n-2}}.$$

It follows that,

$$\sum_{n=N}^{+\infty} e^{-2^{n-1}} \leq e^{-2^{N-2}} \sum_{n=N}^{+\infty} e^{-2^{n-2}} \leq C e^{-2^{N-2}}.$$

Now by question 2 we have $\lim_{n \to +\infty} A_n = 0$. Therefore, for any $\varepsilon > 0$ such that $3C\varepsilon \leq \frac{1}{2}$ one can find N_0 so large that for $n \geq N \geq N_0$ we have $A_n^2 \leq \varepsilon$. It follows that

$$3C \sum_{n=N}^{N+M} A_n^3 \leq 3C\varepsilon \sum_{n=N}^{N+M} A_n \leq \frac{1}{2} \sum_{n=N}^{N+M} A_n.$$

Summing up we obtain,

$$\frac{1}{2} \sum_{n=N}^{N+M} A_n \leq C \left(e^{-2^{N-2}} + A_{N-1}^3 + A_{N+M+1}^3 \right).$$

Now letting M go to $+\infty$ and using the fact that $\lim_{n\to+\infty} A_n = 0$, we obtain,

$$\sum_{n=N}^{+\infty} A_n \le C' \left(e^{-2^{N-2}} + A_{N-1}^3 \right).$$

5. a) In particular we have $A_N \le C' \left(e^{-2^{N-2}} + A_{N-1}^3 \right)$; so taking $N \ge N_1$ in order to have $A_N \le 1$ we obtain,

$$B_{N+1} \le (1+C') \left(e^{-2^{N-1}} + A_N^2 \right) \le (1+C') \left(e^{-2^{N-2}} + A_N \right)^2 = (1+C')B_N^2.$$

This implies, by induction on $N \ge N_1$ that,

$$B_N \le (1 + C')^{2^{N-N_1+1}-1}(B_{N_1})^{2^{N-N_1}} = \frac{1}{1+C'} \left((1+C')^2 B_{N_1} \right)^{2^{N-N_1}}.$$

b) Taking moreover N_1 so large that, $(1 + C')^2 B_{N_1} < 1$ we deduce from the previous question that,

$$B_N \le \frac{1}{1+C'} e^{2^{N-N_1} \ln((1+C')^2 B_{N_1})} = K e^{-\delta 2^N} \quad \text{with} \quad \delta > 0.$$

c) Since $A_N \le B_N$ we have $A_N \le K e^{-\delta 2^N}$ for $N \ge N_1$. Therefore, for $|x| \ge 2^{N_1}$ we have $x \in \omega_N$ for some $N \ge N_1$ which means that $2^N \le |x| \le 2^{N+1}$ so,

$$|u(x)| \le A_N \le K e^{-\delta 2^N} \le K e^{-\frac{1}{2}\delta|x|}.$$

Solution 45

Part 1
1. Since the Fourier transform (with respect to the variable x only) of $u_\ell(t, \cdot)$ is compactly supported, it is clear that this function is smooth.

Denote by $(\cdot, \cdot)_{H^s}$ the scalar product in $H^s(\mathbf{R}^d)$. By multiplying the equation $\partial_t u - \Delta u = 0$ by u_ℓ we find that,

$$(\partial_t u(t, \cdot), u_\ell(t, \cdot))_{H^s} + (-\Delta u(t, \cdot), u_\ell(t, \cdot))_{H^s} = 0.$$

Observe that P_ℓ is a projector, that is $P_\ell = P_\ell^2$. Consequently $u_\ell = P_\ell u_\ell$ and it follows from the Plancherel theorem that,

$$(\partial_t u(t, \cdot), u_\ell(t, \cdot))_{H^s} = (\partial_t u(t, \cdot), P_\ell u_\ell(t, \cdot))_{H^s} = (\partial_t P_\ell u(t, \cdot), u_\ell(t, \cdot))_{H^s}$$
$$= (\partial_t u_\ell(t, \cdot), u_\ell(t, \cdot))_{H^s}.$$

Since u_ℓ is C^1 in time with values in $H^s(\mathbf{R}^d)$, we can write,

$$(\partial_t u_\ell(t, \cdot), u_\ell(t, \cdot))_{H^s} = \frac{1}{2}\frac{d}{dt}\|u_\ell(t, \cdot)\|_{H^s}^2.$$

Similar arguments imply that,

$$(-\Delta u(t, \cdot), u_\ell(t, \cdot))_{H^s} = (\nabla u_\ell(t, \cdot), \nabla u_\ell(t, \cdot))_{H^s}.$$

This gives the wanted identity,

$$\frac{1}{2}\frac{d}{dt}\|u_\ell(t, \cdot)\|_{H^s}^2 + \|\nabla u_\ell(t, \cdot)\|_{H^s}^2 = 0.$$

2. By integrating in time the above identity we find that (u_ℓ) is bounded in $L^\infty((0, T); H^s(\mathbf{R}^d))$ and (∇u_ℓ) is bounded in $L^2((0, T); H^s(\mathbf{R}^d))$. This implies that (u_ℓ) is bounded in $L^2((0, T); H^{s+1}(\mathbf{R}^d))$ which in turn implies that u belongs to $L^2((0, T); H^{s+1}(\mathbf{R}^d))$.

Part 2

1. This is a consequence of the uniqueness result for solution of the heat equation in the space of tempered distribution.
2. The Bernstein inequality implies that there is $C > 0$ such that, for any function f,

$$\|\Delta_j f\|_{L^\infty} \le C 2^{j\frac{d}{2}} \|\Delta_j f\|_{L^2}.$$

So, it follows from the triangle inequality that,

$$\|u(t, \cdot)\|_{L^\infty} \le C \sum_{j \ge -1} 2^{j\frac{d}{2}} \|\Delta_j u(t, \cdot)\|_{L^2}.$$

Now recall that,

$$\hat{u}(t, \xi) = e^{-t|\xi|^2} \hat{g}(\xi),$$

and observe that $|\xi|$ is bounded from below by $c2^j$ on the support of the Fourier transform of $\Delta_j g$. Therefore, the Plancherel identity implies that,

$$\|\Delta_j u(t, \cdot)\|_{L^2} \leq e^{-c^2 t 2^{2j}} \|\Delta_j g\|_{L^2}.$$

By combining the previous estimates, we conclude that,

$$\|u(t, \cdot)\|_{L^\infty} \leq C \sum_{j \geq -1} 2^{j\frac{d}{2}} e^{-c^2 t 2^{2j}} \|\Delta_j g\|_{L^2},$$

which is the wanted estimate (up to replacing c by c^2).

3. a) The Cauchy–Schwarz inequality implies that,

$$|h_n|^2 \leq \left(\sum_{k \in \mathbf{Z}} |f_{n-k}|\right)\left(\sum_{k \in \mathbf{Z}} |f_{n-k}|\, |g_k|^2\right) \leq \|f\|_{\ell^1(\mathbf{Z})} \sum_{k \in \mathbf{Z}} |f_{n-k}|\, |g_k|^2.$$

So, summing on $n \in \mathbf{Z}$ and using the Fubini theorem for positive series,

$$\|h\|^2_{\ell^2(\mathbf{Z})} \leq \|f\|^2_{\ell^1(\mathbf{Z})} \|g\|^2_{\ell^2(\mathbf{Z})},$$

which implies the desired result.

b) Given $j \geq -1$ set,

$$A_j = \sum_{k=-1}^{+\infty} \frac{a_k}{2^{|k-j|}}.$$

The Cauchy–Schwarz inequality implies that,

$$\sum_{j=-1}^{\infty} \sum_{k=-1}^{+\infty} \frac{a_j a_k}{2^{|k-j|}} \leq \left(\sum_{j=-1}^{\infty} a_j^2\right)^{\frac{1}{2}} \left(\sum_{j=-1}^{\infty} A_j^2\right)^{\frac{1}{2}}.$$

It remains only to use the previous question which gives,

$$\sum_{j=-1}^{\infty} A_j^2 \leq C_0 \sum_{k=-1}^{\infty} a_k^2 \quad \text{with} \quad C_0 = \sum_{n \in \mathbf{Z}} 2^{-|n|} < +\infty.$$

c) We start from,

$$\|u(t, \cdot)\|_{L^\infty} \leq C \sum_{j=-1}^{\infty} 2^{j\frac{d}{2}} e^{-ct 2^{2j}} \|\Delta_j g\|_{L^2},$$

and take the square of both sides. We then integrate in time, using

$$\int_0^T e^{-ct2^{2j}} e^{-ct2^{2k}}\, dt \le \frac{1}{c(2^{2j}+2^{2k})},$$

to deduce that,

$$\|u\|_{L^2((0,T);L^\infty)}^2 \le \sum_{j=-1}^\infty \sum_{k=-1}^\infty \frac{1}{c(2^{2j}+2^{2k})} 2^{j\frac{d}{2}}\|\Delta_j g\|_{L^2} 2^{k\frac{d}{2}}\|\Delta_k g\|_{L^2}.$$

Set,

$$a_j = 2^{j(\frac{d}{2}-1)}\|\Delta_j g\|_{L^2}.$$

Then,

$$\|u\|_{L^2((0,T);L^\infty)}^2 \le \sum_{j=-1}^\infty \sum_{k=-1}^\infty \frac{2^{j+k}}{c(2^{2j}+2^{2k})} a_j a_k.$$

Write,

$$\frac{2^{j+k}}{c(2^{2j}+2^{2k})} = \frac{2^{j+k}}{c2^{2j}(1+2^{2(k-j)})} = \frac{2^{k-j}}{c(1+2^{2(k-j)})}.$$

Now by considering separately the case $k \ge j$ and $j \ge k$ we verify that,

$$\frac{2^{k-j}}{c(1+2^{2(k-j)})} \le \frac{1}{c2^{|k-j|}},$$

We thus have proved that,

$$\|u\|_{L^2((0,T);L^\infty)}^2 \le C \sum_j \sum_k \frac{a_j a_k}{c2^{|k-j|}}.$$

The result (8.22) then follows from the previous question.

Solution 46

1. We use the formula $\Delta = \operatorname{div} \nabla$ and elementary computations to write,

$$(\partial_t - \Delta)\operatorname{Log} h = \frac{1}{h}\partial_t h - \operatorname{div}\left(\frac{1}{h}\nabla h\right) = \frac{1}{h}\underbrace{(\partial_t h - \Delta h)}_{=0} + \frac{|\nabla h|^2}{h^2} = \frac{|\nabla h|^2}{h^2}.$$

Then,

$$(\partial_t - \Delta)(h\mathrm{Log}\,h) = h(\partial_t - \Delta)\mathrm{Log}\,h + \underbrace{(\partial_t h - \Delta h)}_{=0}\mathrm{Log}\,h - 2\nabla h \cdot \nabla\mathrm{Log}\,h$$

$$= h\frac{|\nabla h|^2}{h^2} - 2\frac{|\nabla h|^2}{h} = -\frac{|\nabla h|^2}{h}.$$

To obtain the third identity, we start again with the Leibniz rule,

$$(\partial_t - \Delta)\frac{|\nabla h|^2}{h} = \frac{1}{h}(\partial_t - \Delta)|\nabla h|^2 + |\nabla h|^2(\partial_t - \Delta)\frac{1}{h} - 2\nabla\left(|\nabla h|^2\right)\cdot\nabla h.$$

Since $(\partial_t - \Delta)\nabla h = 0$ we get,

$$(\partial_t - \Delta)|\nabla h|^2 = \left(\partial_t - \sum_j \partial_j^2\right)\sum_i(\partial_i h)^2$$

$$= 2\sum_i \partial_i h\underbrace{\left(\partial_t - \sum_j \partial_j^2\right)\partial_i h}_{=0} - 2\sum_{i,j}(\partial_i\partial_j h)^2 = -2\left|\nabla^2 h\right|^2.$$

On the other hand, since,

$$\Delta\frac{1}{h} = -\,\mathrm{div}\left(\frac{1}{h^2}\nabla h\right) = 2\frac{1}{h^3}|\nabla h|^2 - \frac{\Delta h}{h^2},$$

we have,

$$|\nabla h|^2(\partial_t - \Delta)\frac{1}{h} = -\frac{|\nabla h|^2}{h^2}\underbrace{(\partial_t h - \Delta h)}_{=0} + 2\frac{|\nabla h|^4}{h^3} = 2\frac{|\nabla h|^4}{h^3}.$$

We conclude the proof by verifying that,

$$-\frac{2}{h}\left|\nabla^2 h\right|^2 + 2\frac{|\nabla h|^4}{h^3} - 2\nabla\left(|\nabla h|^2\right)\cdot\nabla h = -\frac{2}{h}\left|\nabla^2 h - \frac{\nabla h\otimes\nabla h}{h}\right|^2.$$

2. Notice that if U is 2π-periodic with respect to x_j for $1 \le j \le d$ we have,

$$\int_{[0,2\pi]^d}\partial_j^2 U\,dx = 0,$$

therefore, $\int_{[0,2\pi]^d}\Delta U\,dx = 0$.

By integrating the identity,

$$(\partial_t - \Delta)(h \operatorname{Log} h) = -\frac{|\nabla h|^2}{h},$$

we obtain that

$$\frac{d}{dt} H = \int_{[0,2\pi]^d} \partial_t (h \operatorname{Log} h) \, dx = \int_{[0,2\pi]^d} (\partial_t - \Delta)(h \operatorname{Log} h) \, dx,$$

$$= -\int_{[0,2\pi]^d} \frac{|\nabla h|^2}{h} \, dx \le 0.$$

This is the first decay result. We also infer from the previous computation that,

$$\frac{d^2}{dt^2} H = -\frac{d}{dt} \int_{[0,2\pi]^d} \frac{|\nabla h|^2}{h} \, dx.$$

Since,

$$(\partial_t - \Delta)\frac{|\nabla h|^2}{h} \le 0 \quad \text{and} \quad \int_{[0,2\pi]^d} \Delta\left(\frac{|\nabla h|^2}{h}\right) dx = 0,$$

we immediately obtain the second inequality.

3. a) Notice that,

$$\frac{d}{dt} \int_{[0,2\pi]^d} h \, dx = \int_{[0,2\pi]^d} \partial_t h \, dx = \int_{[0,2\pi]^d} \Delta h \, dx = 0.$$

Since $\int_{[0,2\pi]^d} h(0, x) \, dx = 1$ we have $\int_{[0,2\pi]^d} h(t, x) \, dx = 1$ for all $t \ge 0$.

b) By integrating in time the equality,

$$(\partial_t - \Delta)\frac{|\nabla h|^2}{h} = -2h \left| \frac{\nabla^2 h}{h} - \frac{\nabla h \otimes \nabla h}{h^2} \right|^2$$

we immediately obtain that,

$$\frac{d}{dt} F + 2J = 0. \tag{9.41}$$

c) Since,

$$\int_{[0,2\pi]^d} \partial_{ii} h \, dx = 0, \qquad \int_{[0,2\pi]^d} h \, dx = 1,$$

by expanding $A(\lambda)$ as a polynomial in λ, we obtain,

$$A(\lambda) = J - 2\lambda F + \lambda^2 d.$$

The choice $\lambda = \frac{1}{d} F$ then yields the desired result. Indeed,

$$0 \le A(\lambda) = J - 2\frac{F^2}{d} + d\frac{F^2}{d^2} = J - \frac{F^2}{d},$$

so,

$$J \ge \frac{1}{d} F^2. \tag{9.42}$$

d) We have,

$$\frac{d}{dt} N = -\frac{2}{d} \exp\left(-\frac{2}{d}H\right) \frac{d}{dt} H,$$

so,

$$\frac{d^2}{dt^2} N = \frac{2}{d}\left(\frac{2}{d}\left(\frac{d}{dt}H\right)^2 - \frac{2}{d}\frac{d^2}{dt^2}H\right) \exp\left(-\frac{2}{d}H\right).$$

Now (9.41) and (9.42) imply that, $\frac{d^2}{dt^2} N(t) \le 0$.
e) The function $1/u$ satisfies,

$$\partial_t \left(\frac{1}{u}\right) = -\frac{\partial_t u}{u^2} \ge K.$$

By integrating in time this inequality we get,

$$\frac{1}{u(t)} - \frac{1}{u(0)} \ge Kt.$$

This immediately implies the desired result,

$$u(t) \le \frac{u(0)}{1 + Ku(0)t}.$$

f) By combining the results of the previous questions, we conclude that,

$$u(t) = \int_{[0,2\pi]^d} \frac{|\nabla h(t,x)|^2}{h(t,x)} \, dx,$$

satisfies the differential inequality $\partial_t u + K u^2 \leq 0$ with $K = \frac{2}{d}$, which gives $u(t) \leq \frac{du(0)}{d+2u(0)t}$, which implies the desired result.

Solution 47

1. Firstly integrating by parts we obtain,

$$\int_{[0,2\pi]^d} (\partial_t h)\kappa \, dx = \int_{[0,2\pi]^d} \nabla_x (\partial_t h) \cdot \frac{\nabla_x h}{\sqrt{1+|\nabla h|^2}} dx,$$

$$= \frac{d}{dt} \int_{[0,2\pi]^d} \sqrt{1+|\nabla h|^2} \, dx.$$

On the other hand, directly from the equation $\partial_t h + \sqrt{1+|\nabla h|^2}\kappa = 0$ one has,

$$\int_{[0,2\pi]^d} (\partial_t h)\kappa \, dx = -\int_{[0,2\pi]^d} \sqrt{1+|\nabla h|^2}\kappa^2 \, dx.$$

By combining these two results we conclude that,

$$\frac{d}{dt} \int_{[0,2\pi]^d} \sqrt{1+|\nabla h|^2} \, dx = -\int_{[0,2\pi]^d} \sqrt{1+|\nabla h|^2}\kappa^2 \, dx \leq 0.$$

2. If $n = 1$ then,

$$\kappa = -\partial_x \left(\frac{h_x}{(1+h_x^2)^{1/2}} \right) = -\frac{h_{xx}}{(1+h_x^2)^{1/2}} + \frac{h_x^2 h_{xx}}{(1+h_x^2)^{3/2}} = -\frac{h_{xx}}{(1+h_x^2)^{3/2}}.$$

So

$$\partial_t h + \sqrt{1+h_x^2}\kappa = \partial_t h - \frac{h_{xx}}{1+h_x^2}. \tag{9.43}$$

By integrating by parts, we have,

$$\frac{d}{dt} \int_0^{2\pi} h_x^2 \, dx = -2 \int_0^{2\pi} (\partial_t h) h_{xx} \, dx.$$

So, directly from (9.43), we conclude that

$$\frac{d}{dt} \int_0^{2\pi} h_x^2 \, dx = -2 \int_0^{2\pi} \frac{h_{xx}^2}{1+h_x^2} \, dx \leq 0.$$

3. a) This follows from the fact that,

$$\partial_t h + \sqrt{1 + h_x^2} \kappa = \partial_t h - \partial_x(\arctan(h_x)) = 0,$$

b) Multiplying the previous equation by h and integrating by parts we obtain,

$$\frac{1}{2} \frac{d}{dt} \int_0^{2\pi} h^2 \, dx = - \int_0^{2\pi} h_x \arctan(h_x) \, dx \le 0,$$

since $u \arctan u \ge 0$ for all $u \in \mathbf{R}$.

c) In view of the previous question, it is sufficient to prove that,

$$\frac{d}{dt} \int_0^{2\pi} h_x \arctan h_x \, dx \le 0.$$

To do so observe that,

$$\partial_t(h_x \arctan h_x) = \partial_t h_x \left(\arctan h_x + \frac{h_x}{1 + h_x^2} \right).$$

On the other hand,

$$\partial_t h_x = \partial_x \left(\frac{h_{xx}}{1 + h_x^2} \right).$$

Therefore, integrating by parts, we conclude that,

$$\frac{d}{dt} \int_0^{2\pi} h_x \arctan h_x \, dx = - \int_0^{2\pi} \frac{h_{xx}}{1 + h_x^2} \partial_x \left(\arctan h_x + \frac{h_x}{1 + h_x^2} \right) dx.$$

Now we use the chain rule to compute the derivative of the terms inside the brackets. We get,

$$\frac{d}{dt} \int_0^{2\pi} h_x \arctan h_x \, dx = -2 \int_0^{2\pi} \frac{h_{xx}}{1 + h_x^2} \cdot \frac{h_{xx}}{(1 + h_x^2)^2} \, dx \le 0,$$

which is the desired result.

d) Using question a) and differentiating both members with respect to t we can write,

$$\partial_t \dot{h} = \partial_x \left(\frac{\partial_x \dot{h}}{1 + (\partial_x h)^2} \right).$$

e) Multiplying this equation by $\dot{h}$ and integrating by parts in x we get,

$$\frac{d}{dt} \int_0^{2\pi} (\partial_t h)^2 \, dx \le 0.$$

To conclude, observe that, by using the equation for h, this is equivalent to

$$\frac{d}{dt} \int_0^{2\pi} (1 + h_x^2) \kappa^2 \, dx \le 0.$$

Solution 48

Part 1

1. This is proved by integrating by parts twice (using the periodicity assumption). Namely write,

$$\int_{[0,2\pi]^d} (\Delta \theta)^2 \, dx = \int_{[0,2\pi]^d} \Big(\sum_{1 \le j \le d} \partial_j^2 \theta \Big)^2 \, dx = \int_{[0,2\pi]^d} \sum_{1 \le j,k \le d} (\partial_j^2 \theta)(\partial_k^2 \theta) \, dx$$

$$= - \int_{[0,2\pi]^d} \sum_{1 \le j,k \le d} (\partial_j \theta)(\partial_j \partial_k^2 \theta) \, dx$$

$$= \int_{[0,2\pi]^d} \sum_{1 \le j,k \le d} (\partial_j \partial_k \theta)(\partial_j \partial_k^2 \theta) \, dx = \int_{[0,2\pi]^d} \left| \nabla^2 \theta \right|^2 \, dx,$$

which is the desired equality.

2. By an immediate computation one has,

$$I = 16 \int_{[0,2\pi]^d} \left| \nabla \theta^{1/2} \right|^4 \, dx$$

$$= \int_{[0,2\pi]^d} \theta^{-2} |\nabla \theta|^4 \, dx = - \int_{[0,2\pi]^d} \left(\nabla \theta^{-1} \cdot \nabla \theta \right) |\nabla \theta|^2 \, dx.$$

3. By integrating by parts one can rewrite I under the form,

$$I = \int_{[0,2\pi]^d} \theta^{-1} \Delta \theta |\nabla \theta|^2 \, dx + 2 \int_{[0,2\pi]^d} \theta^{-1} [(\nabla \theta \cdot \nabla) \nabla \theta] \cdot \nabla \theta \, dx. \qquad (9.44)$$

4. We begin by writing,

$$|(\nabla \theta \cdot \nabla) \nabla \theta|^2 = \sum_{1 \le j \le d} \Big(\sum_{1 \le k \le d} \partial_k \theta \partial_k \partial_j \theta \Big)^2,$$

and then use the Cauchy–Schwarz inequality in $\mathbf{R}^d$ for any fixed j,

$$\left(\sum_{1 \leq k \leq d} \partial_k \theta \partial_k \partial_j \theta \right)^2 \leq \left(\sum_{1 \leq k \leq d} (\partial_k \theta)^2 \right) \left(\sum_{1 \leq k \leq d} (\partial_k \partial_j \theta)^2 \right).$$

It follows that $|(\nabla \theta \cdot \nabla)\nabla \theta|^2 \leq |\nabla \theta|^2 \, |\nabla^2 \theta|^2$, which implies that,

$$|(\nabla \theta \cdot \nabla)\nabla \theta| \leq |\nabla \theta| \left| \nabla^2 \theta \right|.$$

Therefore, using the Cauchy–Schwarz inequality in $\mathbf{R}^d$ again,

$$\left| \int_{[0,2\pi]^d} \theta^{-1} [(\nabla \theta \cdot \nabla)\nabla \theta] \cdot \nabla \theta \, dx \right| \leq \int_{[0,2\pi]^d} \frac{|\nabla \theta|^2}{\theta} \left| \nabla^2 \theta \right| dx.$$

Since,

$$I = \int_{[0,2\pi]^d} \frac{|\nabla \theta|^4}{\theta^2} \, dx,$$

by using the Cauchy–Schwarz inequality in $L^2([0, 2\pi]^d)$ we get,

$$\left| \int_{[0,2\pi]^d} \theta^{-1} [(\nabla \theta \cdot \nabla)\nabla \theta] \cdot \nabla \theta \, dx \right| \leq I^{\frac{1}{2}} \left(\int_{[0,2\pi]^d} \left| \nabla^2 \theta \right|^2 dx \right)^{\frac{1}{2}}.$$

Remembering that,

$$\int_{[0,2\pi]^d} (\Delta \theta)^2 \, dx = \int_{[0,2\pi]^d} \left| \nabla^2 \theta \right|^2 dx,$$

this gives,

$$\left| \int_{[0,2\pi]^d} \theta^{-1} [(\nabla \theta \cdot \nabla)\nabla \theta] \cdot \nabla \theta \, dx \right| \leq I^{\frac{1}{2}} \left(\int_{[0,2\pi]^d} (\Delta \theta)^2 \, dx \right)^{\frac{1}{2}}. \tag{9.45}$$

Similarly,

$$\int_{[0,2\pi]^d} \theta^{-1} \Delta \theta |\nabla \theta|^2 \, dx \leq I^{\frac{1}{2}} \left(\int_{[0,2\pi]^d} (\Delta \theta)^2 \, dx \right)^{\frac{1}{2}}. \tag{9.46}$$

By combining (9.44), (9.45), and (9.46) we deduce that,

$$I \leq 3\,I^{1/2}\left(\int_{[0,2\pi]^d}(\Delta\theta)^2\,dx\right)^{1/2}.$$

Thus we get,

$$I \leq 9\int_{[0,2\pi]^d}(\Delta\theta)^2\,dx,$$

which immediately implies the desired inequality.

Part 2

1. Multiplying the equation by $(m+1)h^m$ and integrating by parts, we have

$$\frac{d}{dt}\int_{[0,2\pi]^d}h^{m+1}\,dx + m(m+1)\int_{[0,2\pi]^d}h^m|\nabla h|^2\,dx = 0. \tag{9.47}$$

Recalling that $h > 0$ by assumption, this implies that, for any $m \geq 0$,

$$\frac{d}{dt}\int_{[0,2\pi]^d}h^{m+1}\,dx \leq 0.$$

2. We multiply the Boussinesq equation by $\partial_t(h^2)$ and write the result under the form,

$$2h(\partial_t h)^2 - \frac{1}{2}\partial_t(h^2)\,\mathrm{div}(\nabla h^2).$$

Then, we integrate by parts in x we obtain,

$$\frac{1}{2}\int_{[0,2\pi]^d}\partial_t\nabla(h^2)\cdot\nabla(h^2)\,dx + \int_{[0,2\pi]^d}2h(\partial_t h)^2\,dx = 0.$$

Since $h > 0$ by assumption this immediately implies that,

$$\frac{1}{4}\frac{d}{dt}\int_{[0,2\pi]^d}\left|\nabla(h^2)\right|^2\,dx \leq 0,$$

which gives the desired inequality.

3. a) Starting from,

$$\partial_t(h^m\,|\nabla h|^2) = (\partial_t h^m)\,|\nabla h|^2 + 2h^m\nabla h\cdot\nabla\partial_t h,$$

and then using the equation,

$$\partial_t h^m - mh^{m-1}\mathrm{div}(h\nabla h) = 0,$$

we deduce that,

$$\frac{d}{dt}\int_{[0,2\pi]^d} h^m \, |\nabla h|^2 \, dx = \int_{[0,2\pi]^d} mh^{m-1}\mathrm{div}(h\nabla h)|\nabla h|^2 \, dx$$

$$+ \int_{[0,2\pi]^d} 2h^m \nabla h \cdot \nabla \mathrm{div}(h\nabla h) \, dx.$$

(9.48)

b) Directly from the Leibniz rule we have,

$$h^{m-1}\mathrm{div}(h\nabla h) \, |\nabla h|^2 = h^{m-1}|\nabla h|^4 + h^m (\Delta h) |\nabla h|^2,$$

and (after some elementary computations),

$$\mathrm{div}(h^m \nabla h)\,\mathrm{div}(h\nabla h) = \left(\mathrm{div}(h^{(m+1)/2}\nabla h)\right)^2 - \frac{(m-1)^2}{4}h^{m-1}|\nabla h|^4.$$

So, integrating by parts in the second term of the right-hand side of (9.48) we get,

$$\frac{d}{dt}\int_{[0,2\pi]^d} h^m \, |\nabla h|^2 \, dx = (m + \frac{1}{2}(m-1)^2) \int_{[0,2\pi]^d} h^{m-1}|\nabla h|^4 \, dx$$

$$+ m \int_{[0,2\pi]^d} h^m \Delta h|\nabla h|^2 \, dx - 2 \int_{[0,2\pi]^d} (\mathrm{div}(h^{(m+1)/2}\nabla h))^2 \, dx,$$

(9.49)

which is the desired identity.

c) Integrating by parts we have,

$$\int_{[0,2\pi]^d} h^m \, |\nabla h|^2 \, \Delta h \, dx = \int_{[0,2\pi]^d} \nabla h \cdot \left(h^m (\Delta h)\nabla h\right) \, dx,$$

$$= - \int_{[0,2\pi]^d} h \cdot \mathrm{div} \left(h^m (\Delta h)\nabla h\right) \, dx,$$

$$= - \int_{[0,2\pi]^d} mh^m (\Delta h) \, |\nabla h|^2 \, dx$$

$$- \int_{[0,2\pi]^d} h^{m+1}\nabla h \cdot \nabla \Delta h \, dx$$

$$- \int_{[0,2\pi]^d} h^{m+1}(\Delta h)^2 \, dx.$$

Then, we integrate by parts again to compute the contribution of $h^{m+1}\nabla h \cdot \nabla \Delta h$. We can write,

$$\int_{[0,2\pi]^d} h^{m+1}\nabla h \cdot \nabla \Delta h \, dx = -\int_{[0,2\pi]^d} \operatorname{div}\left(h^{m+1}\nabla h\right)\Delta h \, dx.$$

Now, we exploit an identity similar to the one used above:

$$\operatorname{div}\left(h^{m+1}\nabla h\right)\Delta h = \left(\operatorname{div}\left(h^{(m+1)/2}\nabla h\right)\right)^2 - \frac{(m+1)^2}{4}h^{m-1}|\nabla h|^4,$$

which can be verified by computing both sides. This yields,

$$\int_{[0,2\pi]^d} h^m |\nabla h|^2 \Delta h \, dx = -\int_{[0,2\pi]^d} mh^m(\Delta h)|\nabla h|^2 \, dx$$
$$+ \int_{[0,2\pi]^d} \left(\operatorname{div}\left(h^{(m+1)/2}\nabla h\right)\right)^2 dx$$
$$- \int_{[0,2\pi]^d} \frac{(m+1)^2}{4}h^{m-1}|\nabla h|^4 \, dx$$
$$- \int_{[0,2\pi]^d} h^{m+1}(\Delta h)^2 \, dx.$$

Simplifying we deduce that,

$$\int_{[0,2\pi]^d} h^m |\nabla h|^2 \Delta h \, dx = -\frac{1}{m+1}\int_{[0,2\pi]^d} h^{m+1}(\Delta h)^2 \, dx$$
$$+ \frac{1}{m+1}\int_{[0,2\pi]^d} \left(\operatorname{div}\left(h^{(m+1)/2}\nabla h\right)\right)^2 dx$$
$$- \frac{m+1}{4}\int_{[0,2\pi]^d} h^{m-1}|\nabla h|^4 \, dx.$$
$$(9.50)$$

d) By reporting the previous identity (9.50) in (9.49) we obtain,

$$\frac{d}{dt}\int_{[0,2\pi]^d} h^m |\nabla h|^2 \, dx$$
$$= \left(\frac{m^2+1}{2} - \frac{m(m+1)}{4}\right)\int_{[0,2\pi]^d} h^{m-1}|\nabla h|^4 \, dx$$

$$-\frac{m}{m+1}\int_{[0,2\pi]^d}h^{m+1}(\Delta h)^2\,dx$$

$$-\left(2-\frac{m}{m+1}\right)\int_{[0,2\pi]^d}(\mathrm{div}(h^{(m+1)/2}\nabla h))^2\,dx,$$

which is the desired result.

e) We will use the Sobolev inequality proved in the second part with,

$$\theta = h^{(m+3)/2}.$$

Then,

$$\begin{cases}\Delta\theta = \dfrac{m+3}{2}\,\mathrm{div}\left(h^{(m+1)/2}\nabla h\right),\\[2mm]\nabla\theta^{\frac12} = \nabla h^{(m+3)/4} = \dfrac{m+3}{4}h^{(m-1)/4}\nabla h.\end{cases}$$

Consequently, the inequality,

$$\int_{[0,2\pi]^d}|\nabla\theta^{1/2}|^4\,dx \le \frac{9}{16}\int_{[0,2\pi]^d}(\Delta\theta)^2\,dx,$$

reads,

$$\left(\frac{m+3}{4}\right)^4\int_{[0,2\pi]^d}h^{m-1}|\nabla h|^4\,dx \le \frac{9}{16}\int_{[0,2\pi]^d}\left(\frac{m+3}{2}\right)^2$$

$$\left(\mathrm{div}\left(h^{(m+1)/2}\nabla h\right)\right)^2\,dx.$$

This immediately implies the result.

f) This follows directly from questions 3d) and 3e).

g) This is an elementary computation.

4. The identity (9.47) applied with $m = 1$ shows that,

$$\frac{d}{dt}\int_{[0,2\pi]^d}h^2\,dx = -\int_{[0,2\pi]^d}h|\nabla h|^2\,dx \le 0.$$

To obtain the second statement, notice that, since $1 \le (1+\sqrt{7})/2$, the result of the previous question yields that,

$$\frac{d^2}{dt^2}\int_{[0,2\pi]^d}h^2\,dx = -\frac{d}{dt}\int_{[0,2\pi]^d}h|\nabla h|^2\,dx \ge 0.$$

This concludes the proof.

5. We now study the Boltzmann's entropy,

$$H(t) = \int_{[0,2\pi]^d} h(t,x) \mathrm{Log}\,(h(t,x))\,dx.$$

Directly from the above formula and from the equation we can write,

$$\frac{d}{dt}H = \int_{[0,2\pi]^d} (\partial_t h + (\mathrm{Log}\,h)\partial_t h)\,dx$$

$$= \int_{[0,2\pi]^d} (\mathrm{div}(h\nabla h) + (\mathrm{Log}\,h)\,\mathrm{div}(h\nabla h))\,dx.$$

Notice that $\int_{[0,2\pi]^d} \mathrm{div}(h\nabla h)\,dx = 0$; indeed $\int_{[0,2\pi]} \partial_i f(x)\,dx_i = 0$ for any C^1 function $f\colon \mathbf{R}^d \to \mathbf{R}$ which is 2π-periodic with respect to x_i. We compute the contribution of the second term by integrating by parts. This gives,

$$\frac{d}{dt}H = -\int_{[0,2\pi]^d} |\nabla h|^2\,dx.$$

This proves that $\frac{d}{dt}H \leq 0$. Furthermore, one can apply the result of the previous question 3 with $m = 0$ to obtain,

$$\frac{d^2}{dt^2}H = -\frac{d}{dt}\int_{[0,2\pi]^d} |\nabla h|^2\,dx \geq 0.$$

This concludes the proof.

Solution 49

1. Using the Cauchy–Schwarz inequality and the fact that $\mu(\mathbf{S}^2) = 4\pi$, we can write,

$$|u_1(t,x)|^2 \leq \frac{t^2}{(4\pi)^2}\int_{\mathbf{S}^2} |g(x-t\omega)|^2\,d\omega,$$

so that,

$$\int_0^{+\infty} |u_1(\cdot,x)|^2\,dt \leq \frac{1}{(4\pi)^2}\int_0^{+\infty}\int_{\mathbf{S}^2} |g(x-t\omega)|^2 t^2\,d\omega\,dt.$$

Setting $y = t\omega$ we have $dy = t^2 \, d\omega \, dt$ and we obtain,

$$\int_0^{+\infty} |u_1(\cdot, x)|^2 \, dt \le \frac{1}{(4\pi)^2} \int_{\mathbb{R}^3} |g(x - y)|^2 \, dy = \frac{1}{(4\pi)^2} \|g\|_{L^2(\mathbb{R}^3)}^2.$$

The estimate of u_3 is similar using the fact that $|\nabla_x u \cdot \omega| \le |\nabla_x u|$.

Let us consider u_2. We have,

$$\int_0^{+\infty} |u_2(t, x)|^2 \, dt \le \frac{1}{(4\pi)^2} \int_0^{+\infty} \int_{S^2} |f(x - t\omega)|^2 \, d\omega \, dt,$$

$$\le \frac{1}{(4\pi)^2} \int_0^{+\infty} \int_{S^2} \frac{|f(x - t\omega)|^2}{t^2} t^2 \, d\omega \, dt,$$

which, setting $y = t\omega$, gives the estimate

$$\int_0^{+\infty} |u_2(t, x)|^2 \, dt \le \frac{1}{(4\pi)^2} \int_{\mathbb{R}^3} \frac{|f(x - y)|^2}{|y|^2} \, dy. \tag{9.51}$$

2. For fixed x we can apply the Hardy inequality (see Problem 13) to the function $h_x(y) = f(x - y)$. We obtain

$$\int_{\mathbb{R}^3} \frac{|f(x - y)|^2}{|y|^2} \, dy \le 4 \sum_{i=1}^{3} \int_{\mathbb{R}^3} \left| \frac{\partial f}{\partial y_i}(x - y) \right|^2 \, dy.$$

To conclude we have just to set $z = x - y$ and to use (9.51).

Solution 50

Part 1

1. Setting $\partial = \partial_t$ or ∂_{x_j} we have $\partial v = e^u \partial u$, $\partial^2 v = e^u (\partial^2 u + (\partial u)^2)$. It follows that $\Box v = e^u \left(\Box u + (\partial_t u)^2 - \sum_{j=1}^{n} (\partial_{x_j} u)^2 \right) = 0$. Now, $v|_{t=0} = e^u|_{t=0} = 1$ and $\partial_t v|_{t=0} = e^u \partial_t u|_{t=0} = g$. Summing up we have,

$$\Box v = 0, \quad v|_{t=0} = 1, \quad \partial_t v|_{t=0} = g.$$

2. The above problem has a unique solution given by,

$$v(t, x) = 1 + w(t, x), \quad w(t, x) = \frac{t}{4\pi} \int_{S^2} g(x - t\omega) \, d\omega.$$

3. Since g tends to zero at infinity we can write,

$$g(x - t\omega) = -\int_t^{+\infty} \frac{d}{ds}[g(x - s\omega)]\,ds = \sum_{j=1}^3 \omega_j(\partial_{x_j}g)(x - t\omega)\,ds.$$

It follows that,

$$|w(t, x)| \le \frac{t}{4\pi} \sum_{j=1}^3 \int_t^{+\infty} \int_{\mathbf{S}^2} \frac{1}{s^2}|(\partial_{x_j}g)(x - t\omega)|s^2\,ds\,d\omega,$$

$$\le \frac{1}{4\pi t} \sum_{j=1}^3 \int_0^{+\infty} \int_{\mathbf{S}^2} |(\partial_{x_j}g)(x - t\omega)|s^2\,ds\,d\omega,$$

$$\le \frac{1}{4\pi t} \sum_{j=1}^3 \int_{\mathbf{R}^3} |(\partial_{x_j}g)(y)|\,dy.$$

4. a) We deduce from the above inequality that $|w(t, x)| < 1$ if $t > \frac{1}{4\pi}\|\nabla_x g\|_{L^1(\mathbf{R}^3)}$.
   b) If $t \le \frac{1}{4\pi}\|\nabla_x g\|_{L^1(\mathbf{R}^3)}$ we have, by definition of w,

$$|w(t, x)| \le \frac{1}{4\pi}\|\nabla_x g\|_{L^1(\mathbf{R}^3)}\|g\|_{L^\infty(\mathbf{R}^3)} \frac{1}{4\pi} \int_{\mathbf{S}^2} d\omega = \frac{1}{4\pi}\|\nabla_x g\|_{L^1(\mathbf{R}^3)}\|g\|_{L^\infty(\mathbf{R}^3)}.$$

5. If $\|\nabla_x g\|_{L^1(\mathbf{R}^3)}\|g\|_{L^\infty(\mathbf{R}^3)} < 4\pi$ we have $|w(t, x)| < 1$ so that $v(t, x) > 0$ for all $(t, x) \in \mathbf{R} \times \mathbf{R}^3$. Then $u(t, x) = \ln v(t, x)$ is the unique solution of problem (8.26).

Part 2

1. If $|x_0| \le \varepsilon\lambda$ and $1 < t_0 < (1 - \varepsilon)\lambda$ we have $|x_0 - t_0\omega| \le |x_0| + t_0 \le \lambda$. So $g(x_0 - t_0\omega) = -1$ and $w(t_0, x_0) = -\frac{t_0}{4\pi}\int_{\mathbf{S}^2} d\omega = -t_0$.
2. We have $\partial_{x_j}g(x) = -\frac{x_j}{|x|}\varphi_0'(|x|)$. Taking polar coordinates we obtain,

$$\int_{\mathbf{R}^3} |\partial_{x_j}g(x)|\,dx = \int_{\mathbf{S}^2} |\omega_j|\,d\omega \int_0^{+\infty} \varphi_0'(r)r^2\,dr.$$

Since $\sum_{j=1}^3 |\omega_j| \ge \left(\sum_{j=1}^3 \omega_j^2\right)^{\frac{1}{2}} = 1$, according to the form of φ_0 we have,

$$\|\nabla_x g\|_{L^1(\mathbf{R}^3)} \ge \frac{4\pi}{\delta} \int_1^{1+\delta} r^2\,dr = 4\pi(1 + \delta + \frac{\delta^2}{3}).$$

Part 3

1. Let $g(x) = -\varepsilon, \varepsilon > 0$. Then $w(t, x) = -\varepsilon t$ so that $v(t, x) = 1 - \varepsilon t < 0$ if $t > \frac{1}{\varepsilon}$.

Solution 51

Part 1

1. a) If $x \in A$ we have $|x|^2 - 4x_1 < 0$, $|x|^2 - 2x_1 \geq 0$, so $\frac{1}{4}|x|^2 \leq x_1 \leq \frac{1}{2}|x|^2$.

   b) In the spherical coordinates this reads $\frac{1}{4}\rho \leq \cos\theta_1 \leq \min(1, \frac{1}{2}\rho)$. This shows that $\cos\theta_1 \geq 0$ thus $\theta_1 \in (0, \frac{\pi}{2})$ and $\rho \leq 4$. Since the function arccos : $(-1, 1) \rightarrow (0, \pi)$ is decreasing this implies that $\arccos\min(1, (\frac{\rho}{2})) \leq \theta_1 \leq \arccos\left(\frac{\rho}{4}\right)$.

   c) In spherical coordinates one can write, $\int_{\mathbf{R}^3} g_A(x)^2 \, dx \leq J_1 + J_2$, where,

$$J_1 = \int_0^2 \int_0^{2\pi} \int_{\arccos(\frac{\rho}{2})}^{\arccos(\frac{\rho}{4})} \frac{\rho^2 \sin\theta_1}{\rho^4(1 + |\ln\rho|)^{2\alpha}} \, d\rho \, d\theta_1 \, d\theta_2,$$

$$J_2 = \int_2^4 \int_0^{2\pi} \int_0^{\arccos(\frac{\rho}{4})} \frac{\rho^2 \sin\theta_1}{\rho^4(1 + |\ln\rho|)^{2\alpha}} \, d\rho \, d\theta_1 \, d\theta_2.$$

We have,

$$J_1 \leq 2\pi \int_0^4 \frac{1}{\rho^2(1 + |\ln\rho|)^{2\alpha}} [-\cos\theta_1]_{\arccos(\frac{\rho}{2})}^{\arccos(\frac{\rho}{4})} \, d\rho$$

$$\leq \pi \int_0^4 \frac{1}{\rho(1 + |\ln\rho|)^{2\alpha}} \, d\rho.$$

   Setting $s = \ln\rho$, thus $ds = \frac{d\rho}{\rho}$ we get, $J_1 \leq \pi \int_{-\infty}^{2\ln 2} \frac{1}{(1+|s|)^{2\alpha}} \, ds < +\infty$, since $2\alpha > 1$. Moreover, it is clear that J_2 is a finite integral. This shows that $g \in L^2(\mathbf{R}^3)$.

2. a) Since $\cup_{k=0}^{+\infty} \{2^{-k+1} < |x| \leq 2^{-k+2}\} = \{0 < |x| \leq 4\}$ and $A_k \cap A_{k'} = \emptyset$ if $k \neq k'$ we have $A = \cup_{k=0}^{+\infty} A_k$ and $1_A(x) = \sum_{k=0}^{+\infty} 1_{A_k}(x)$. It follows from the Beppo–Levi theorem that,

$$u(t, x) = \frac{t}{4\pi} \sum_{k=0}^{+\infty} \int_{\mathbf{S}^2} g_{A_k}(x - t\omega) \, d\omega \leq +\infty.$$

   b) According to question 1a) the fact that $te_1 - t\omega \in A_k$ is equivalent to,

   (i) $\frac{1}{4}|te_1 - t\omega|^2 < t(1-\omega_1) \leq \frac{1}{2}|te_1 - t\omega|^2$ and (ii) $2^{-k+1} < |te_1 - t\omega| \leq 2^{-k+2}$.

Since for $\omega \in \mathbf{S}^2$ we have $|t e_1 - t\omega|^2 = 2t^2(1 - \omega_1)$ (i) is automatically satisfied if $1 < t < 2$ and (ii) is equivalent to $2^{-2k+1} < t^2(1-\omega_1) \leq 2^{-2k+3}$.

c) By question b) above we have,

$$I_k = \int_{B_k} d\omega$$

where $B_k = \left\{\omega \in \mathbf{S}^2 : 2^{-2k+1} < t^2(1 - \omega_1) \leq 2^{-2k+3}\right\}$.

In spherical coordinates we have $\omega_1 = \cos\theta_1$, $\omega_2 = \cos\theta_1 \sin\theta_2$, $\omega_3 = \cos\theta_1 \sin\theta_2$, $d\omega = \sin\theta_1 \, d\theta_1 \, d\theta_2$, $\theta_1 \in (0, \pi)$, $\theta_2 \in (0, 2\pi)$. If $\omega \in B_k$ we have then,

$$1 - \varepsilon_2 \leq \cos\theta_1 \leq 1 - \varepsilon_1, \quad \text{where} \quad \varepsilon_1 = t^{-2}2^{-2k+1}, \quad \varepsilon_2 = t^{-2}2^{-2k+3}.$$

If k is large $\cos\theta_1$ is close to one and θ_1 to zero so this is equivalent to,

$$\arccos(1 - \varepsilon_1) \leq \varphi \leq \arccos(1 - \varepsilon_2).$$

Therefore, we have,

$$I_k = \int_0^{2\pi} \int_{\arccos(1-\varepsilon_1)}^{\arccos(1-\varepsilon_2)} \sin\theta_1 \, d\theta_1 \, d\theta_2 = 2\pi \, [-\cos\theta_1]_{\arccos(1-\varepsilon_1)}^{\arccos(1-\varepsilon_2)} = 2\pi\,(\varepsilon_2 - \varepsilon_1).$$

We have just to notice that $\varepsilon_2 - \varepsilon_1 = 6t^{-2}2^{-2k}$.

d) On A_k for large k we have, $\dfrac{1}{|x|^2(1+|\ln|x||)^\alpha} \geq \dfrac{2^{2k-4}}{(1+|-k+2|\ln 2)^\alpha} \geq c\dfrac{2^{2k}}{(1+k)^\alpha}$.

According to question 2c) we have for large k,

$$m_k \geq c_0\,t^{-2}\,2^{-2k}c\frac{2^{2k}}{(1+k)^\alpha} \geq \frac{c_1 t^{-2}}{(1+k)^\alpha}.$$

3. We deduce from the previous questions that $u(t, te_1)$ is infinite since $u(t, te_1) \geq \sum_{k \geq k_0} m_k$ and $\alpha \leq 1$. It follows that (8.28) cannot hold. Indeed by question 1c) the left-hand side is finite while the left-hand side is infinite since,

$$\|u\|_{L^2((0,T),L^\infty(\mathbf{R}^3))} \geq \|u(\cdot, te_1)\|_{L^2(0,T)} = +\infty.$$

Part 2

1. a) By the Cauchy–Schwarz inequality we have,

$$|u_n(t, x) - u(t, x)|^2 \leq \frac{t^2}{16\pi^2}\mu(\mathbf{S}^2)\int_{\mathbf{S}^2} |(g_n - g)(x - t\omega)|^2 \, d\omega.$$

Integrating in t between 0 and $+\infty$ we obtain,

$$\|u_n(\cdot, x) - u(\cdot, x)\|^2_{L^2_t(\mathbf{R}^+)} \le C \int_0^{+\infty} \int_{S^2} (g_n - g)(x - t\omega)|^2 t^2 \, dt \, d\omega,$$

$$\le C \int_{\mathbf{R}^3} |(g_n - g)(x - y)|^2 \, dy = C\|g_n - g\|^2_{L^2(\mathbf{R}^3)}.$$

The right-hand side converges to zero when n goes to $+\infty$, therefore (u_n) converges to u in $L^\infty_x(\mathbf{R}^3, L^2_t(\mathbf{R}^+))$.

b) Since $L^2(\mathbf{R}^+, L^\infty(\mathbf{R}^3))$ is the dual of the Banach space $L^2(\mathbf{R}^+, L^1(\mathbf{R}^3))$ the Banach–Alaoglu theorem asserts that the closed ball $B(0, M)$ in $L^2(\mathbf{R}^+, L^\infty(\mathbf{R}^3))$ in weak-star compact. Therefore, there exists a subsequence $(u_{\sigma(n)})$ which converges to $v \in L^2(\mathbf{R}^+, L^\infty(\mathbf{R}^3))$ in the weak-star topology which means that,

$$\langle u_{\sigma(n)}, \varphi \rangle \to \langle v, \varphi \rangle, \quad \forall \varphi \in L^2(\mathbf{R}^+, L^1(\mathbf{R}^3)). \tag{9.52}$$

c) By question 1a) we know that (u_n) converges to u in $L^\infty_x(\mathbf{R}^3, L^2_t(\mathbf{R}^+))$. This implies that (u_n) converges to u in $\mathcal{D}'(\mathbf{R}^+ \times \mathbf{R}^3)$. Moreover, taking φ in $C_0^\infty(\mathbf{R}^+ \times \mathbf{R}^3)$ in (9.52) we see that $(u_{\sigma(n)})$ converges to v in $\mathcal{D}'(\mathbf{R}^+ \times \mathbf{R}^3)$. Therefore, $u = v \in L^2(\mathbf{R}^+, L^\infty(\mathbf{R}^3))$. This is impossible by Part 1 since we have proved there that the $L^2(\mathbf{R}^+, L^\infty(\mathbf{R}^3))$ norm of u was infinite.

Thus the sequence (u_n) is not bounded in $L^2(\mathbf{R}^+, L^\infty(\mathbf{R}^3))$.

d) This follows immediately from the last sentence above.

2. a) To prove the first claim we have just to set $x = \lambda y$ thus $dx = \lambda^3 \, dy$, when we compute the $L^2(\mathbf{R}^+)$ norm of $g^\lambda_{n_k}$. Now setting $s = \frac{t}{\lambda}$ we can write,

$$u^\lambda_{n_k}(t, x) = \frac{t}{4\pi} \lambda^{-\frac{3}{2}} \int_{S^2} g_{n_k}\left(\frac{x - t\omega}{\lambda}\right) d\omega = \lambda^{-\frac{1}{2}} \frac{s}{4\pi} \int_{S^2} g_{n_k}\left(\frac{x}{\lambda} - s\omega\right) d\omega,$$

$$= \lambda^{-\frac{1}{2}} u_{n_k}\left(s, \frac{x}{\lambda}\right) = \lambda^{-\frac{1}{2}} u_{n_k}\left(\frac{t}{\lambda}, \frac{x}{\lambda}\right).$$

It follows that $\|u^\lambda_{n_k}(t, \cdot)\|_{L^\infty(\mathbf{R}^3)} = \lambda^{-\frac{1}{2}} \|u_{n_k}\left(\frac{t}{\lambda}, \cdot\right)\|_{L^\infty(\mathbf{R}^3)}$ and,

$$\int_0^{+\infty} \|u^\lambda_{n_k}(t, \cdot)\|^2_{L^\infty(\mathbf{R}^3)} \, dt = \lambda^{-1} \int_0^{+\infty} \|u_{n_k}\left(\frac{t}{\lambda}, \cdot\right)\|^2_{L^\infty(\mathbf{R}^3)} \, dt$$

$$= \|u_{n_k}\|^2_{L^2(\mathbf{R}^+, L^\infty(\mathbf{R}^3))}.$$

To prove the last claim we notice that $\widehat{g^\lambda_{n_k}}(\xi) = \lambda^{\frac{3}{2}} \widehat{g_{n_k}}(\lambda\xi)$. Therefore,

$$\|g^\lambda_{n_k}\|^2_{H^s(\mathbf{R}^3)} = \lambda^3 \int_{\mathbf{R}^3} (1 + |\xi|^2)^s |\widehat{g_{n_k}}(\lambda\xi)|^2 \, d\xi = \int_{\mathbf{R}^3} \left(1 + |\frac{\eta}{\lambda}|^2\right)^s |\widehat{g_{n_k}}(\eta)|^2 \, d\eta.$$

It follows that,

$$\|g^\lambda_{n_k}\|^2_{H^s(\mathbf{R}^3)} - (2\pi)^3\|g_{n_k}\|^2_{L^2(\mathbf{R}^3)} = \int_{\mathbf{R}^3}\left[\left(1 + |\tfrac{\eta}{\lambda}|^2\right)^s - 1\right]|\widehat{g_{n_k}}(\eta)|^2\,d\eta.$$

Setting for $x \geq 0$, $f(x) = (1+x)^s$ we have $f(0) = 1$ and by the Taylor formula we get,

$$f(x) - 1 = sx\int_0^1 (1 + \mu x)^{s-1}\,d\mu.$$

Setting $x = |\tfrac{\eta}{\lambda}|^2$ and taking $\lambda \geq 1$ we obtain,

$$\left|\left(1 + |\tfrac{\eta}{\lambda}|^2\right)^s - 1\right| \leq s\left|\tfrac{\eta}{\lambda}\right|^2\int_0^1 (1 + \mu\left|\tfrac{\eta}{\lambda}\right|^2)^{s-1}\,d\mu \leq \tfrac{s}{\lambda^2}|\eta|^2\max(1, (1+|\eta|^2)^{s-1}).$$

It follows that,

$$\|g^\lambda_{n_k}\|^2_{H^s(\mathbf{R}^3)} - (2\pi)^3\|g_{n_k}\|^2_{L^2(\mathbf{R}^3)}$$

$$\leq \tfrac{s}{\lambda^2}\int_{\mathbf{R}^3}|\eta|^2\max(1, (1+|\eta|^2)^{s-1})|\widehat{g_{n_k}}(\eta)|^2\,d\eta,$$

and the right-hand side tends to zero when λ goes to $+\infty$ since $g_{n_k} \in \mathcal{S}(\mathbf{R}^3)$.

b) If (8.31) was true we would have with a fixed C and for all $\lambda \geq 1$,

$$\|u^\lambda_{n_k}\|_{L^2(\mathbf{R}^+, L^\infty(\mathbf{R}^3))} \leq C\|g^\lambda_{n_k}\|_{H^s(\mathbf{R}^3)}.$$

By the previous question this would imply,

$$\|u_{n_k}\|_{L^2(\mathbf{R}^+, L^\infty(\mathbf{R}^3))} \leq C\|g^\lambda_{n_k}\|_{H^s(\mathbf{R}^3)}.$$

Again by question a) above the right-hand side converges to $(2\pi)^3 C$ $\|g_{n_k}\|_{L^2(\mathbf{R}^3)}$ when λ goes to $+\infty$. Therefore, we would have,

$$\|u_{n_k}\|_{L^2(\mathbf{R}^+, L^\infty(\mathbf{R}^3))} \leq C'\|g_{n_k}\|_{L^2(\mathbf{R}^3)}.$$

But this is impossible since the right-hand side is bounded since (g_{n_k}) converges to g in $L^2(\mathbf{R}^3)$ while the left-hand side tends to $+\infty$ by question 1d).

Solution 52

1. We know (see the course reminder) that for $t \neq 0$,

$$\|S(t)u_0\|_{L^2(\mathbf{R}^d)} = \|u_0\|_{L^2(\mathbf{R}^d)}, \quad \forall u_0 \in L^2(\mathbf{R}^d),$$

$$\|S(t)u_0\|_{L^\infty(\mathbf{R}^d)} \leq \frac{1}{|4\pi t|^{\frac{d}{2}}}\|u_0\|_{L^1(\mathbf{R}^d)}, \quad \forall u_0 \in L^1(\mathbf{R}^d).$$

So $S(t)$ is continuous from $L^2(\mathbf{R}^d)$ to $L^2(\mathbf{R}^d)$ with norm $\leq (2\pi)^{-\frac{d}{2}}$ and from $L^1(\mathbf{R}^d)$ to $L^\infty(\mathbf{R}^d)$ with norm $\leq \frac{1}{|4\pi t|^{\frac{d}{2}}}$. By interpolation $S(t)$ is continuous from $L^p(\mathbf{R}^d)$ to $L^q(\mathbf{R}^d)$ where, for $0 < \theta < 1$, $\frac{1}{p} = \frac{1-\theta}{2} + \frac{\theta}{\infty}$, $\frac{1}{q} = \frac{1-\theta}{2} + \frac{\theta}{1}$. Then, $\frac{1}{p} + \frac{1}{q} = \frac{1-\theta}{2} + \frac{1-\theta}{2} + \theta = 1$. So $q = p'$. Moreover, its norm is bounded by $C[(2\pi)^{-\frac{d}{2}}]^{1-\theta}[|4\pi t|^{-\frac{d}{2}}]^{\theta} = C'|t|^{-d(\frac{1}{2}-\frac{1}{p})}$ since $\frac{\theta}{2} = \frac{1}{2} - \frac{1}{p}$.

2. Recall that the Sobolev embedding ensures that $H^1(\mathbf{R}^d)$ can be continuously embedded in $L^p(\mathbf{R}^d)$ for $\frac{1}{2} \geq \frac{1}{p} \geq \frac{1}{2} - \frac{1}{d} = \frac{d-2}{2d}$. Therefore, for $2 \leq p < \frac{2d}{d-2}$ there exists $C_0 > 0$ such that,

$$\|v\|_{L^p(\mathbf{R}^d)} \leq C_0\|v\|_{H^1(\mathbf{R}^d)}, \quad \forall v \in H^1(\mathbf{R}^d). \tag{9.53}$$

Let $u_0 \in H^1(\mathbf{R}^d)$ and $\varepsilon > 0$. There exists $\varphi \in C_0^\infty(\mathbf{R}^d)$ such that $\|u_0 - \varphi\|_{H^1(\mathbf{R}^d)} \leq \frac{\varepsilon}{2C_0}$. Then,

$$\|S(t)u_0 - S(t)\varphi\|_{H^1(\mathbf{R}^d)} = \|u_0 - \varphi\|_{H^1(\mathbf{R}^d)} \leq \frac{\varepsilon}{2C_0}.$$

From (9.53) for $2 \leq p < \frac{2d}{d-2}$ we have,

$$\|S(t)u_0 - S(t)\varphi\|_{L^p(\mathbf{R}^d)} \leq C_0\|S(t)u_0 - S(t)\varphi\|_{H^1(\mathbf{R}^d)} \leq \frac{\varepsilon}{2}. \tag{9.54}$$

Now from question 1 we have,

$$\|S(t)\varphi\|_{L^p(\mathbf{R}^d)} \leq C|t|^{-d(\frac{1}{2}-\frac{1}{p})}\|\varphi\|_{L^{p'}(\mathbf{R}^d)}.$$

Since $\frac{1}{2} - \frac{1}{p} > 0$ we have $\lim_{t \to +\infty} C|t|^{-d(\frac{1}{2}-\frac{1}{p})}\|\varphi\|_{L^{p'}(\mathbf{R}^d)} = 0$. So there exists $T > 0$ such that for $t \geq T$ we have,

$$\|S(t)\varphi\|_{L^p(\mathbf{R}^d)} \leq \frac{\varepsilon}{2}. \tag{9.55}$$

It follows from (9.54) and (9.55) that for $t \geq T$ we have $\|S(t)u_0\|_{L^p(\mathbf{R}^d)} \leq \varepsilon$.

Solution 53

In all what follows we set $L^2 = L^2(\mathbf{R}^d)$.

1. Set $F(t) = (f(t), g(t))_{L^2}$. Let $t_0 \in \mathbf{R}$. We have,

$$\frac{F(t_0 + h) - F(t_0)}{h} - (\partial_t f(t_0), g(t_0))_{L^2} - (f(t_0), \partial_t g(t_0))_{L^2} = I + II + III,$$

$$I = \left(\frac{f(t_0 + h - f(t_0)}{h} - \partial_t f(t_0), g(t_0 + h) \right)_{L^2},$$

$$II = (\partial_t f(t_0), g(t_0 + h) - g(t_0))_{L^2},$$

$$III = \left(f(t_0), \frac{g(t_0 + h) - g(t_0)}{h} - \partial_t g(t_0) \right)_{L^2}.$$

By the Cauchy–Schwarz inequality, since $f \in C^1(\mathbf{R}, L^2)$, we have for $|h| \leq 1$,

$$|I| \leq \left\| \frac{f(t_0 + h) - f(t_0)}{h} - \partial_t f(t_0) \right\|_{L^2} \sup_{|t - t_0| \leq 1} \|g(t)\|_{L^2} \to 0 \text{ when } h \to 0.$$

Similarly $III \to 0$. The term II tends to zero since $g \in C^0(\mathbf{R}, L^2)$.

2. By the previous question we have,

$$\partial_t (Bu(t), u(t))_{L^2} = (B\partial_t u(t), u(t))_{L^2} + (Bu(t), \partial_t u(t))_{L^2},$$

$$= (Bi \Delta u(t), u(t))_{L^2} + (Bu(t), i \Delta u(t))_{L^2},$$

$$= i (B\Delta u(t), u(t))_{L^2} - i (\Delta Bu(t), u(t))_{L^2}$$

$$= (i [B, \Delta]u(t), u(t))_{L^2}.$$

3. Integrating the previous inequality between 0 and T we get,

$$\int_0^T (i [B, \Delta]u(t), u(t))_{L^2} \, dt = (Bu(T), u(T))_{L^2} - (Bu(0), u(0))_{L^2}.$$

Since B is of order zero, therefore continuous from L^2 to L^2 and $\|u(t)\|_{L^2} = \|u_0\|_{L^2}$ for all $t \in \mathbf{R}$, there exists $C > 0$ such that,

$$\left| \int_0^T (i [B, \Delta]u(t), u(t))_{L^2} \, dt \right| \leq C\|u_0\|_{L^2}^2.$$

4. a) This follows from the fact that $\xi \in S^1$, $\int_0^x \frac{dy}{\langle y \rangle^{2\sigma}} \in S^0$, $\frac{1}{\langle \xi \rangle} \in S^{-1}$.

b) We have $b \in S^0$, $-\xi^2 \in S^2$. By the symbolic calculus the bracket $i\,[B, \Delta]$ is of order $0+2-1 = 1$ and, modulo a zero order term, its symbol is equal to the Poisson bracket $i \cdot \frac{1}{i}\{b, -\xi^2\} = \{\xi^2, b\} = 2\xi\,\partial_x b$. Since $\partial_x b = \frac{1}{2}\frac{1}{\langle x\rangle^{2\sigma}}\frac{\xi}{\langle\xi\rangle}$ we eventually have $i\,[B, \Delta] = \mathrm{Op}\left(\frac{\xi^2}{\langle\xi\rangle}\frac{1}{\langle x\rangle^{2\sigma}}\right) + R$, $R \in \mathrm{Op}(S^0)$. Setting $\Lambda = \mathrm{Op}(\langle\xi\rangle)$ we have $\Lambda^{-1} = \mathrm{Op}\left(\frac{1}{\langle\xi\rangle}\right)$ so that $i\,[B, \Delta] = -\Lambda^{-1}\partial_x^2\,\langle x\rangle^{-2\sigma} + R$, $R \in \mathrm{Op}(S^0)$.

c) Set $Q = \Lambda^{-\frac{1}{2}}\partial_x$. We have $Q^* = -Q$ and by Fourier transform, $Q^2 = \Lambda^{-1}\partial_x^2 \in \mathrm{Op}(S^1)$. Then,

$$(i\,[B, \Delta]u(t), u(t))_{L^2} = -\left(Q^2\,\langle x\rangle^{-2\sigma}\,u(t), u(t)\right)_{L^2} + (Ru(t), u(t))_{L^2}.$$
$$\tag{9.56}$$

We have $-\left(Q^2\,\langle x\rangle^{-2\sigma}\,u(t), u(t)\right)_{L^2} = (1) + (2)$, where,

$$(1) = -\left([Q^2, \langle x\rangle^{-\sigma}]\,\langle x\rangle^{-\sigma}\,u(t), u(t)\right)_{L^2},$$

$$(2) = \left(Q\,\langle x\rangle^{-\sigma}\,u(t), Q\,\langle x\rangle^{-\sigma}\,u(t)\right)_{L^2}.$$

Since $Q^2 \in \mathrm{Op}(S^1)$ and $\langle x\rangle^{-\sigma} \in \mathrm{Op}(S^0)$ the operator $[Q^2, \langle x\rangle^{-\sigma}]$ is of order zero. Therefore,

$$|(1)| \le C\|u(t)\|_{L^2}^2.$$

Now, $(2) = \|Q\,\langle x\rangle^{-\sigma}\,u(t)\|_{L^2}^2$. Using (9.56) and the fact that $R \in \mathrm{Op}(S^0)$ we see that there exists $C > 0$ such that,

$$\|Q\,\langle x\rangle^{-\sigma}\,u(t)\|_{L^2}^2 \le |(i\,[B, \Delta]u(t), u(t))_{L^2}| + C\|u(t)\|_{L^2}^2.$$

Integrating this inequality between 0 and T and using question 3 we obtain,

$$\int_0^T \|Q\,\langle x\rangle^{-\sigma}\,u(t)\|_{L^2}^2\,dt \le C(1 + T)\|u_0\|_{L^2}^2. \tag{9.57}$$

Now set $v = \langle x\rangle^{-\sigma}\,u$. We have for fixed t,

$$\|v(t)\|_{H^{\frac{1}{2}}}^2 = \int_{\mathbb{R}} \langle\xi\rangle\,|\widehat{v}(t, \xi)|^2\,d\xi$$

$$= \int_{|\xi|\le 1} \langle\xi\rangle\,|\widehat{v}(t, \xi)|^2\,d\xi + \int_{|\xi|\ge 1} \langle\xi\rangle\,|\widehat{v}(t, \xi)|^2\,d\xi,$$

$$\le \sqrt{2}\int_{|\xi|\le 1} |\widehat{v}(t, \xi)|^2\,d\xi + \int_{|\xi|\ge 1} \langle\xi\rangle\,|\widehat{v}(t, \xi)|^2\,d\xi = (1) + (2).$$

We have,

$$(1) \le C\|v(t)\|_{L^2}^2 = C\|\langle x \rangle^{-\sigma} u(t)\|_{L^2}^2 \le C\|u(t)\|_{L^2}^2 = C\|u_0\|_{L^2}^2.$$

For $|\xi| \ge 1$ we have $\langle \xi \rangle^2 \le 2\xi^2$ so that $\langle \xi \rangle \le 2\frac{\xi^2}{\langle \xi \rangle}$. It follows that,

$$(2) \le 2 \int_{\mathbf{R}} \frac{\xi^2}{\langle \xi \rangle} |\widehat{v}(t, \xi)|^2 \, d\xi \le C\|Q\langle x \rangle^{-\sigma} u(t)\|_{L^2}^2.$$

Integrating between 0 and T and using (9.57) we obtain eventually,

$$\int_0^T \|\langle x \rangle^{-\sigma} u(t)\|_{H^{\frac{1}{2}}}^2 \, dt \le C(1+T)\|u_0\|_{L^2}.$$

5. a) Applying the inequality obtained in question 4 to $u_j - u_k$ we see that $(\langle x \rangle^{-\sigma} u_k)$ is a Cauchy sequence in $L^2((0, T), H^{\frac{1}{2}})$. Since this space is complete we deduce that $(\langle x \rangle^{-\sigma} u_k)$ converges to v in $L^2((0, T), H^{\frac{1}{2}})$, therefore in $\mathcal{D}'((0, T) \times \mathbf{R}^d)$.

   b) We have,

$$\int_0^T \|u_k(t) - u(t)\|_{L^2}^2 \, dt = \int_0^T \|\mathcal{F}^{-1}(e^{-it\xi^2}\widehat{\varphi_k}) - \mathcal{F}^{-1}(e^{-it\xi^2}\widehat{\varphi})\|_{L^2}^2 \, dt,$$

$$= (2\pi)^{-1} \int_0^T \|e^{-it\xi^2}\widehat{\varphi_k} - e^{-it\xi^2}\widehat{\varphi}\|_{L^2}^2 \, dt,$$

$$= T\|\varphi_k - \varphi\|_{L^2}^2 \to 0.$$

   Now the convergence in $L^2((0, T) \times \mathbf{R}^d)$ implies that in $\mathcal{D}'((0, T) \times \mathbf{R}^d)$.

   c) Since the function $\langle x \rangle^{-\sigma}$ is C^∞, it follows from question b) that the sequence $(\langle x \rangle^{-\sigma} u_k)$ converges to $(\langle x \rangle^{-\sigma} u)$ in $\mathcal{D}'((0, T) \times \mathbf{R}^d)$. From question a) we have $\langle x \rangle^{-\sigma} u = v \in L^2((0, T), H^{\frac{1}{2}})$. We have just to pass to the limit in the estimate

$$\int_0^T \|\langle x \rangle^{-\sigma} u_k(t)\|_{H^{\frac{1}{2}}}^2 \, dt \le C(1+T)\|\varphi_k\|_{L^2}^2.$$

Solution 54

Part 1

1. First of all we have,

$$\|S(t)u_0\|_{H^s} = \|u_0\|_{H^s} = \frac{1}{2}R.$$

Now, using the estimate (ii) we obtain for $u \in B$ and $t \in (-T, T)$,

$$\left\| \int_0^t S(t - \sigma)[f(\sigma)], d\sigma \right\|_{H^s} \leq \int_0^t \|S(t - \sigma)[f(u(\sigma))]\|_{H^s} \, d\sigma,$$

$$\leq \int_0^t \|f(u(\sigma))\|_{H^s} \, d\sigma,$$

$$\leq \int_0^t \mathcal{F}(\|u(\sigma)\|_{H^s}) \|u(\sigma)\|_{H^s} \, d\sigma \leq \mathcal{F}(R)RT.$$

Taking T such that $\mathcal{F}(R)T \leq \frac{1}{2}$ we obtain $\|\mathcal{A}(u(t)\|_{H^s} \leq R$ for all $t \in (-T, T)$.

2. For $u_1, u_2 \in B$ we have,

$$\|\mathcal{A}(u_1) - \mathcal{A}(u_2)\|_{L^\infty(I, H^s)} \leq \int_{-T}^T \|f(u_1(\sigma)) - f(u_2(\sigma))\|_{H^s} \, d\sigma.$$

Since $s > \frac{d}{2}$, $H^s(\mathbf{R}^d)$ is an algebra and there exists $C > 0$ such that $\|uv\|_{H^s} \leq C\|u\|_{H^s}\|v\|_{H^s}$. Using the Taylor formula we deduce that,

$$\|\mathcal{A}(u_1) - \mathcal{A}(u_2)\|_{L^\infty(I, H^s)} \leq C \int_{-T}^T \int_0^1 \|df(\lambda u_1(\sigma) + (1 - \lambda)u_2(\sigma))\|_{H^s}$$

$$\times \|u_1(\sigma) - u_2(\sigma))\|_{H^s} \, d\lambda \, d\sigma,$$

where df is the differential of f. Since $u_1, u_2 \in B$ we have,

$$\|\lambda u_1(\sigma) + (1 - \lambda)u_2(\sigma)\|_{H^s} \leq \lambda \|u_1(\sigma)\|_{H^s} + (1 - \lambda)\|u_2(\sigma)\|_{H^s} \leq R.$$

It follows that there exists $\mathcal{F} : \mathbf{R}^+ \to \mathbf{R}^+$ increasing such that,

$$\|\mathcal{A}(u_1) - \mathcal{A}(u_2)\|_{L^\infty(I, H^s)} \leq C\mathcal{F}(R)2T \|u_1 - u_2\|_{L^\infty(I, H^s)}.$$

We have just to take $T > 0$ such that $2C\mathcal{F}(R)T < 1$.

3. The space $L^\infty(I, H^s) \cap C^0(I, H^s)$ endowed with the $L^\infty(I, H^s)$ norm is a Banach space. The set B is closed in this space. Therefore, it is a complete metric space. If T is small enough the map $\mathcal{A}$ sends B into B and it is contractive. By the fixed point theorem there exists a unique $u \in B$ such that $u = \mathcal{A}(u)$.

4. We have for $t \in (T, T)$,

$$u(t) = S(t)u_0 - i \int_0^t S(t - \sigma)[f(u(\sigma)] \, d\sigma.$$

Set $v(t) = S(t)u_0$. We have $v \in C^0(\mathbf{R}, H^s) \cap C^1(\mathbf{R}, H^{s-2})$ and,

$$(i\partial_t + \Delta)v(t) = 0, \quad v|_{t=0} = S(0)u_0 = u_0.$$

Now set $w(t) = -i \int_0^t S(t - \sigma)[f(u(\sigma)]\,d\sigma$. By the linear theory we have $w \in C^0(I, H^s) \cap C^1(I, H^{s-2})$ and,

$$(i\partial_t + \Delta)w = f(u(t)), \quad w|_{t=0} = 0.$$

Therefore, $u = w + w$ is the unique solution of the problem,

$$(i\partial_t + \Delta)u(t) = f(u(t)), \quad u|_{t=0} = u_0.$$

Part 2
1. We have $0 \in F$ since $u_1|_{t=0} = u_2|_{t=0} = u_0$. Moreover, F is closed since $u_1, u_2 \in C^0(I, H^s)$.
2. a) Let $t_0 \in F$. We have,

$$u_j(t + t_0) = S(t + t_0)u_0 - i \int_0^{t+t_0} S(t + t_0 - \sigma)[f(u_j(\sigma))]\,d\sigma = S(t + t_0)u_0$$

$$- i \int_0^{t_0} S(t + t_0 - \sigma)[f(u_j(\sigma))]\,d\sigma - i \int_{t_0}^{t+t_0} S(t + t_0 - \sigma)[f(u_j(\sigma))]\,d\sigma.$$

Using the Hint we can write,

$$S(t + t_0)u_0 - i \int_0^{t_0} S(t + t_0 - \sigma)[f(u_j(\sigma))]\,d\sigma$$

$$= S(t)\left[S(t_0)u_0 - i \int_0^{t_0} S(t_0 - \sigma)[f(u_j(\sigma))]\,d\sigma\right] = S(t)u_j(t_0).$$

On the other hand, setting $\sigma' = \sigma - t_0$ in the integral we obtain,

$$-i \int_{t_0}^{t+t_0} S(t + t_0 - \sigma)[f(u_j(\sigma))]\,d\sigma = -i \int_0^t S(t - \sigma')[f(u_j(\sigma' + t_0))]\,d\sigma',$$

$$= -i \int_0^t S(t - \sigma')[f(v_j(\sigma'))]\,d\sigma'.$$

Therefore, we have,

$$v_j(t) = S(t)u_j(t_0) - i \int_0^t S(t - \sigma')[f(v_j(\sigma'))]\,d\sigma'.$$

b) By hypothesis we have $t_0 \in F$ so $u_1(t_0) = u_2(t_0)$. Using the fact that $S(t-\sigma)$ is an isometry on H^s we have for $t \in (-\varepsilon, \varepsilon)$,

$$\|v_1(t) - v_2(t)\|_{H^s} \leq \int_0^t \|S(t-\sigma)[f(v_1(\sigma)) - f(v_2(\sigma))]\|_{H^s} \, d\sigma,$$

$$\leq \int_0^t \|f(v_1(\sigma)) - f(v_2(\sigma))\|_{H^s} \, d\sigma,$$

$$\leq \int_{-\varepsilon}^\varepsilon \|f(v_1(\sigma)) - f(v_2(\sigma))\|_{H^s} \, d\sigma.$$

c) Using the Taylor formula and the fact that H^s is an algebra we deduce that,

$$\|v_1(t) - v_2(t)\|_{H^s} \leq C \int_{-\varepsilon}^\varepsilon \int_0^1 \|df(\lambda v_1(\sigma) + (1-\lambda)v_2(\sigma))\|_{H^s}$$

$$\cdot \|v_1(\sigma) - v_2(\sigma)\|_{H^s} \, d\lambda \, d\sigma.$$

Therefore, we obtain,

$$\|v_1 - v_2\|_{L^\infty((-\varepsilon,\varepsilon),H^s)} \leq 2C\varepsilon \int_0^1 \|df(\lambda v_1 + (1-\lambda)v_2)\|_{L^\infty((-\varepsilon,\varepsilon),H^s)} \, d\lambda$$

$$\cdot \|v_1 - v_2\|_{L^\infty((-\varepsilon,\varepsilon),H^s)}.$$

If we choose $\varepsilon > 0$ such that $2C\varepsilon \int_0^1 \|df(\lambda v_1 + (1-\lambda)v_2)\|_{L^\infty((-\varepsilon,\varepsilon),H^s)} < 1$, we obtain that $v_1(t) = v_2(t)$ for $t \in (-\varepsilon, \varepsilon)$ which, coming back to the definition of v_j shows that there exists $\varepsilon > 0$ such that $u_1(t) = u_2(t)$ for $t \in (t_0 - \varepsilon, t_0 + \varepsilon)$, which proves that F is open.

By connectedness we have $F = (-T, T)$ that is $u_1 = u_2$ on $(-T, T)$.

Part 3

1. By construction we have $\|u\|_{L^\infty(I,H^s)} \leq R$, $\|v\|_{L^\infty(I,H^s)} \leq R$. Now we write,

$$u(t) - v(t) = S(t)(u_0 - v_0) - i \int_0^t S(t-\sigma)[f(u(\sigma)) - f(v(\sigma))] \, d\sigma.$$

Using the fact that $S(t)$ is an isometry on H^s we deduce,

$$\|u(t) - v(t)\|_{H^s} \leq \|u_0 - v_0\|_{H^s} + \int_{-T}^T \|f(u(\sigma)) - f(v(\sigma))\|_{H^s} \, d\sigma.$$

Using again the Taylor formula and the fact that $\|u\|_{L^\infty(I,H^s)} \leq R$, $\|v\|_{L^\infty(I,H^s)} \leq R$ we get,

$$\|u - v\|_{L^\infty(I,H^s)} \leq \|u_0 - v_0\|_{H^s} + C(R)T\|u - v\|_{L^\infty(I,H^s)}.$$

Taking T_0 such that $C(R)T_0 < 1$ and $T \leq T_0$ we absorb the second term in the right-hand side by the left-hand side and we obtain,

$$\|u - v\|_{L^\infty(I,H^s)} \leq C'(R, T_0)\|u_0 - v_0\|_{H^s}.$$

Solution 55

1. By definition $[A, B] = AB - BA$. Therefore

$$[P, L_j] = -2\,\partial_{x_j} + \left[\sum_{k=1}^{d} \partial^2_{x_k}, x_j\right] = -2\,\partial_{x_j} + 2\,\partial_{x_j} = 0.$$

It follows that $[i\partial_t + \Delta_x, L^\alpha] = 0$, for all $\alpha \in \mathbf{N}^d$.

2. If $g \in L^2$ the problem $i\partial_t u + \Delta_x u = 0$ with $u|_{t=0} = g$ has a unique solution $u \in C^0(\mathbf{R}, L^2(\mathbf{R}^d))$. Consider $v = L^\alpha u \in C^0(\mathbf{R}, S')$. It is easy to see that, $v_0 = x^\alpha u_0 = x^\alpha g \in L^2(\mathbf{R}^d)$. On the other hand, using question 1, we have

$$(i\partial_t + \Delta_x)\,v = [i\partial_t + \Delta_x, L^\alpha]u + L^\alpha(i\partial_t + \Delta_x)\,u = 0.$$

It follows that $L^\alpha u \in C^0(\mathbf{R}, L^2(\mathbf{R}^d))$ for all $\alpha \in \mathbf{N}^d$. Now,

$$(x + 2it\,\partial_x)^\alpha u = \sum_{|\beta| \leq |\alpha| - 1} a_{\alpha\beta}(t, x)\,\partial^\beta_x u + (2it)^{|\alpha|}\,\partial^\alpha_x u, \qquad (9.58)$$

where $a_{\alpha\beta}$ is a polynomial. Since the $C^\infty(\mathbf{R}^d)$ functions act on the spaces $H^s_{loc}(\mathbf{R}^d)$, an induction on $|\alpha|$, starting from $|\alpha| = 0$ and from the fact that $u \in C^0(\mathbf{R}, L^2(\mathbf{R}^d))$, shows that, for $t \neq 0$, we have $\partial^\alpha_x u(t, \cdot) \in L^2_{loc}(\mathbf{R}^d)$. Therefore, $u(t, \cdot) \in H^s_{loc}(\mathbf{R}^d)$ for all $s \in \mathbf{N}$ and $t \neq 0$. Since $\bigcap_{s \geq 0} H^s_{loc}(\mathbf{R}^d) \subset C^\infty(\mathbf{R}^d)$ we obtain the result.

Solution 56

1. We have,

$$\widehat{f}(x) = \int_{\mathbf{R}} e^{-ix\xi} e^{-|\xi|}\,d\xi = \int_{-\infty}^{0} e^{\xi(1-ix)}\,d\xi + \int_{0}^{+\infty} e^{-\xi(1+ix)}\,d\xi,$$

$$= \frac{1}{1 - ix} + \frac{1}{1 + ix} = \frac{2}{1 + x^2}.$$

2. We have $\widehat{g} = \widehat{g_1}$. On the other hand, by the Lebesgue differentiation of integrals theorem we can write,

$$\widehat{g_\lambda}(x) = \int e^{-ix\xi} |\xi| e^{-\lambda|\xi|} \, d\xi = -\frac{d}{d\lambda} \int e^{-ix\xi} e^{-\lambda|\xi|} \, d\xi$$

$$= -\frac{d}{d\lambda} \left\{ \frac{1}{\lambda} \int e^{-i\frac{x}{\lambda}\eta} e^{-|\eta|} \, d\xi \right\} = -\frac{d}{d\lambda} \left\{ \frac{2\lambda}{\lambda^2 + x^2} \right\}$$

$$= -\frac{2}{\lambda^2 + x^2} + \frac{4\lambda^2}{(\lambda^2 + x^2)^2}.$$

Taking $\lambda = 1$ we deduce that,

$$\mathcal{F}\left(|\xi| e^{-|\xi|} \right) = -\frac{2}{1 + x^2} + \frac{4}{(1 + x^2)^2}.$$

3. a) Using question 1 we can write, $e^{-|\xi|} = \mathcal{F}^{-1}\left(\frac{2}{1+x^2} \right) = \frac{2}{2\pi} \mathcal{F}\left(\frac{1}{1+x^2} \right)$ so,

$$\widehat{Q}(\xi) = \pi e^{-|\xi|},$$

which shows that $(1 + |\xi|^2)^{\frac{s}{2}} \widehat{Q}$ belongs to $L^2(\mathbf{R})$ for all $s \in \mathbf{R}$.

b) For all $t \in \mathbf{R}$ we have, $\widehat{|D_x|u}(t, \xi) = |\xi| \widehat{u}(t, \xi)$. On the other hand,

$$\widehat{u}(t, \xi) = 4c \int e^{-ix\xi} Q(c(x + ct)) \, dx = 4 \int e^{-i(\frac{y}{c} - ct)\xi} Q(y) \, dy,$$

$$= 4e^{ict\xi} \int e^{-iy\frac{\xi}{c}} Q(y) \, dy = 4e^{ict\xi} \widehat{Q}\left(\frac{\xi}{c} \right) = 4\pi e^{ict\xi} e^{-|\frac{\xi}{c}|},$$

so $\widehat{|D_x|u}(t, \xi) = 4\pi |\xi| e^{ict\xi} e^{-|\frac{\xi}{c}|}$.

Taking the inverse Fourier transform and using question 2 we can write,

$$|D_x|u(t, x) = \frac{1}{2\pi} \cdot 4\pi \int e^{ix\xi} |\xi| e^{ict\xi} e^{-|\frac{\xi}{c}|} \, d\xi = 2c^2 \int e^{i\eta(c^2 t + cx)} |\eta| e^{-|\eta|} \, d\eta,$$

$$= 2c^2 \left(-\frac{2}{1 + (c^2 t + cx)^2} + \frac{4}{(1 + (c^2 t + cx)^2)^2} \right),$$

$$= -cu(t, x) + \frac{1}{2} u(t, x)^2.$$

c) We deduce from the previous question that,

$$\partial_x |D_x|u = -c\partial_x u + \frac{1}{2}\partial_x(u^2).$$

To conclude we have just to notice that $\partial_t u = c\partial_x u$.

Solution 57

Part 1

1. Since $u \in L^2(\mathbf{R}^d \times \Omega)$, for almost all $x \in \mathbf{R}^d$ the function $\omega \to u(x, \omega)$ belongs to $L^2(\Omega)$ which is contained in $L^1(\Omega)$, since $\mu(\Omega) < +\infty$.
2. a) We have $|U(x)|^2 \le \mu(\Omega) \int_\Omega |u(x, \omega)|^2 \, d\mu(\omega) \in L^1(\mathbf{R}^d)$.
   b) We have for $\varphi \in \mathcal{S}(\mathbf{R}^d)$, $\langle \widehat{U}, \varphi \rangle = \langle U, \widehat{\varphi} \rangle = \int_{\mathbf{R}^d} \left(\int_\Omega u(x, \omega) \, d\mu(\omega) \right) \widehat{\varphi}(x) \, dx$. Since the function $(x, \omega) \to \widehat{\varphi}(x) u(x, \omega)$ belongs to $L^1(\mathbf{R}^d \times \Omega)$ we deduce from the Fubini theorem that,

$$\langle \widehat{U}, \varphi \rangle = \int_\Omega \left(\int_{\mathbf{R}^d} u(x, \omega) \widehat{\varphi}(x) \, dx \right) d\mu(\omega).$$

Since for almost all $\omega \in \Omega$ the function $x \to u(x, \omega)$ belongs to $L^2(\mathbf{R}^d)$ the Parseval theorem implies that,

$$\langle \widehat{U}, \varphi \rangle = \int_\Omega \left(\int_{\mathbf{R}^d} \widehat{u}(\xi, \omega) \varphi(\xi) \, d\xi \right) d\mu(\omega).$$

Since for almost all ω the function $\xi \to \widehat{u}(\xi, \omega) \varphi(\xi)$ belongs to $L^1(\mathbf{R}^d)$ the Fubini theorem implies that,

$$\langle \widehat{U}, \varphi \rangle = \int_{\mathbf{R}^d} \left(\int_\Omega \widehat{u}(\xi, \omega) \, d\mu(\omega) \right) \varphi(\xi) d\xi = \left\langle \int_\Omega \widehat{u}(\xi, \omega) \, d\mu(\omega), \varphi \right\rangle.$$

So $\widehat{U}(\xi) = \int_\Omega \widehat{u}(\xi, \omega) \, d\mu(\omega)$; the estimate is a consequence of the Cauchy–Schwarz inequality and the fact that $\mu(\Omega) < +\infty$.

3. Applying the Cauchy–Schwarz inequality we obtain,

$$|I_1(\xi)|^2 \le \mu \{ \omega \in \Omega : |p_\omega(\xi)| \le 1 \} \int_\Omega |\widehat{u}(\xi, \omega)|^2 \, d\mu(\omega).$$

Since p_ω is homogeneous of degree one in ξ we have,

$$\mu \{ \omega \in \Omega : |p_\omega(\xi)| \le 1 \} = \mu \left\{ \omega \in \Omega : \left| p_\omega \left(\frac{\xi}{|\xi|} \right) \right| \le \frac{1}{|\xi|} \right\} \le c_0 \frac{1}{|\xi|^{2\delta}},$$

using $(\star)$ with $a = 1/|\xi|$.

4. We have just to write,

$$I_2(\xi) = \int_{|p_\omega(\xi)| \ge 1} \frac{1}{|p_\omega(\xi)|} |p_\omega(\xi)| |\widehat{u}(\xi, \omega)| \, d\mu(\omega),$$

and then use the Cauchy–Schwarz inequality and the homogeneity of p_ω.

5. a) By the change of variables theorem and condition $(\star)$ we have,

$$\int_{\Omega} 1_{\{\omega:|p_{\omega}(\theta)|\le a\}}(\omega)d\mu(\omega) = \int_{\mathbf{R}^+} 1_{\{0\le t\le a\}}(t)\,d\nu(t) \le c_0\,a^{2\delta}.$$

So $\nu\{0\le t\le a\} \le c_0\,a^{2\delta}$.

b) The first equality is an immediate consequence of the change of variables theorem, and the second one follows from the Lebesgue dominated convergence theorem.

6. We have

$$J_N(\xi) \le \frac{1}{|\xi|^2}\sum_{k=1}^{N-1}\frac{|\xi|^2}{k^2}\nu\left\{\frac{k}{|\xi|}\le t\le\frac{k+1}{|\xi|}\right\},$$

$$\le \sum_{k=1}^{N-1}\frac{1}{k^2}\nu\left\{0\le t\le\frac{k+1}{|\xi|}\right\} - \sum_{k=1}^{N-1}\frac{1}{k^2}\nu\left\{0\le t<\frac{k}{|\xi|}\right\}.$$

In the first sum of the right-hand side set $k' = k+1$; we isolate the term $k' = N$ and in the second sum we isolate the term $k = 1$. We obtain

$$J_N(\xi) \le \frac{1}{(N-1)^2}\nu\left\{0\le t\le\frac{N}{|\xi|}\right\} + \sum_{k=2}^{N-1}\left(\frac{1}{(k-1)^2}-\frac{1}{k^2}\right)\nu\left\{0\le t\le\frac{k}{|\xi|}\right\}$$

$$- \nu\left\{0\le t\le\frac{1}{|\xi|}\right\} \le A+B-C.$$

Using $(\star)$ we obtain, since $2\delta \le 2$,

$$A \le \frac{1}{(N-1)^2}c_0\frac{N^{2\delta}}{|\xi|^{2\delta}} \le \frac{c_1}{|\xi|^{2\delta}}.$$

Now,

$$B \le \sum_{k=2}^{N-1}\frac{2k-1}{k^2(k-1)^2}k^{2\delta}\frac{c_0}{|\xi|^{2\delta}} \le \frac{c_2}{|\xi|^{2\delta}},$$

since $\delta < 1$. Eventually, $C \le \frac{c_0}{|\xi|^{2\delta}}$.

7. We deduce from the previous questions that $|I_2(\xi)|^2 \le \frac{c}{|\xi|^{2\delta}}$ and from question 2 that,

$$|\xi|^{2\delta}|\widehat{U}(\xi)|^2 \le C\left(\int_{\Omega}|\widehat{u}(\xi,\omega)|^2\,d\mu(\omega) + \int_{\Omega}|p_{\omega}(\xi)|^2|\widehat{u}(\xi,\omega)|^2\,d\mu(\omega)\right).$$

Since $|\widehat{U}(\xi)|^2 \leq C \int_\Omega |\widehat{u}(\xi, \omega)|^2 d\mu(\omega)$, integrating these inequalities with respect to ξ and using Parseval equality we obtain,

$$\|U\|_{H^\delta(\mathbf{R}^d)}^2 \leq C \left(\|u\|_{L^2(\mathbf{R}^d \times \Omega)}^2 + \|\operatorname{Op}(p_\omega)u\|_{L^2(\mathbf{R}^d \times \Omega)}^2 \right).$$

Part 2

1. Assume that there exists $\xi_0 \in \mathbf{R}^d$ such that $g(\xi_0) > 0$. Then there exists $\varepsilon > 0$ such that $g(\xi) > 0$ for $|\xi - \xi_0| \leq \varepsilon$. Let $\psi \in S(\mathbf{R}^d)$ be such that $\widehat{\psi} \in C_0^\infty(\mathbf{R}^d)$ with $\widehat{\psi}(\xi) = 1$ if $|\xi - \xi_0| \leq \frac{1}{2}\varepsilon$, $\widehat{\psi}(\xi) = 0$ if $|\xi - \xi_0| \geq \varepsilon$. Then,

$$\int_{\mathbf{R}^d} |\widehat{\psi}(\xi)|^2 g(\xi) \, d\xi = \int_{|\xi - \xi_0| \leq \varepsilon} |\widehat{\psi}(\xi)|^2 g(\xi) \, d\xi \geq \int_{|\xi - \xi_0| \leq \frac{1}{2}\varepsilon} g(\xi) \, d\xi > 0,$$

which contradicts our hypothesis.

2. Apply $(\star\star)$ to $u(x, \omega) = \psi(x)\varphi(\omega)$ where ψ belongs to $S(\mathbf{R}^d)$ and $\varphi \in L^2(\Omega)$. Using the fact that,

$$\|\operatorname{Op}(p_\omega)u\|_{L^2(\mathbf{R}^d \times \Omega)}^2 \leq C \|\widehat{\operatorname{Op}(p_\omega)}u\|_{L^2(\mathbf{R}^d \times \Omega)}^2,$$

$$\leq C \int_\Omega \int_{\mathbf{R}^d} |p_\omega(\xi)|^2 |\widehat{\psi}(\xi)|^2 |\varphi(\omega)|^2 \, d\xi \, d\omega,$$

we deduce that the quantity,

$$\int_{\mathbf{R}^d} |\widehat{\psi}(\xi)|^2 \left(|\xi|^{2\delta} \left| \int_\Omega \varphi(\omega) \, d\mu(\omega) \right|^2 - C \int_\Omega |\varphi(\omega)|^2 (1 + |p_\omega(\xi)|^2) \, d\mu(\omega) \right) d\xi$$

is negative. Since this holds for any function $\psi \in S(\mathbf{R}^d)$ and since the function inside the parentheses in the integral is continuous we deduce from the previous question that we have,

$$|\xi|^{2\delta} \left| \int_\Omega \varphi(\omega) \, d\mu(\omega) \right|^2 - C \int_\Omega |\varphi(\omega)|^2 (1 + |p_\omega(\xi)|^2) \, d\mu(\omega) \leq 0, \quad \forall \xi \in \mathbf{R}^d \tag{9.59}$$

for all $\varphi \in L^2(\mathbf{R}^d)$. Take $a > 0$, $\xi = \frac{\eta}{a}$ where $\eta \in S^{d-1}$ and $\varphi(\omega) = 1_{\{|p_\omega(\eta)| \leq a\}}$. We have

$$|\xi|^{2\delta} \left| \int_\Omega \varphi(\omega) \, d\mu(\omega) \right|^2 = \frac{1}{a^{2\delta}} (\mu \{\omega \in \Omega : |p_\omega(\eta)| \leq a\})^2,$$

$$\int_{\Omega} |\varphi(\omega)|^2 (1 + |p_\omega(\xi)|^2) \, d\mu(\omega) = \int_{\{|p_\omega(\eta)| \leq a\}} (1 + \frac{1}{a^2} |p_\omega(\eta)|^2) \, d\mu(\omega),$$

$$\leq 2\mu \{\omega \in \Omega : |p_\omega(\eta)| \leq a\}.$$

Applying the inequality (9.59) we obtain $(\star)$.

Solution 58

Part 1

1. We begin by a necessary condition which will give the uniqueness. Assume that f is a solution of (8.35). Consider the differential system,

$$\dot{t}(s) = 1, t(0) = 0, \quad \dot{x}(s) = v, x(0) = x_0 \in \mathbf{R}^d.$$

We have obviously $t(s) = s$ and $x(s) = x_0 + sv$. Then,

$$\frac{d}{ds}[f(s, x_0 + sv, v)] = \left(\frac{\partial f}{\partial t} + v \cdot \nabla_x f\right)(s, x_0 + sv, v) = 0.$$

Therefore, $f(s, x_0 + sv, v) = f(0, x_0, v) = f^0(x_0, v)$ which shows that for any $x \in \mathbf{R}^d$ we have $f(s, x, v) = f^0(x - sv, v)$.

On the other hand, if $f^0 \in C^1(\mathbf{R}_x^d \times \mathbf{R}_v^d)$, setting $f(t, x, v) = f^0(x - tv, v)$ we obtain easily a solution of problem (8.35).

2. According to question 1 we have,

$$\|f(t)\|_{L^p(\mathbf{R}_x^d \times \mathbf{R}_v^d)}^p = \int_{\mathbf{R}_x^d} \int_{\mathbf{R}_v^d} |f^0(x - tv, v)|^p \, dx \, dv.$$

We have just to perform the change of variables $y = x - tv$ in the integral in x to conclude.

3. According to the expression of the solution f we can write,

$$|\rho(t, x)| \le \int_{\mathbf{R}^d} |f^0(x - tv, v)| \, dv \le \int_{\mathbf{R}^d} \sup_{w \in \mathbf{R}^d} |f^0(x - tv, w)| \, dv.$$

Setting $v' = x - tv$ in the above integral we obtain $dv = \frac{1}{t^d} \, dv'$ so,

$$|\rho(t, x)| \le \frac{1}{t^d} \int_{\mathbf{R}^d} \sup_{w \in \mathbf{R}^d} |f^0(v', w)| \, dv'.$$

4. a) Set $A := \int_{\mathbf{R}} \int_{\mathbf{R}^d} |\rho(t, x)| |\Phi(t, x)| \, dx \, dt$. We have,

$$A \le \int_{\mathbf{R}} \int_{\mathbf{R}_x^d} \int_{\mathbf{R}_v^d} |f^0(x - tv, v)| |\Phi(t, x)| \, dv \, dx \, dt$$

$$= \int_{\mathbf{R}_y^d} \int_{\mathbf{R}_v^d} |f^0(y, v)| \left(\int_{\mathbf{R}} |\Phi(t, y + tv)| \, dt\right) dv \, dy,$$

$$\le \|f^0\|_{L^a(\mathbf{R}_y^d \times \mathbf{R}_v^d)} \left\|\int_{\mathbf{R}} |\Phi(t, y + tv)| \, dt\right\|_{L^{a'}(\mathbf{R}_y^d \times \mathbf{R}_v^d)}.$$

b) Recall that $a = \frac{2r}{r+1}$. Then $a' = \frac{2r}{r-1}$. Since $\frac{1}{r'} = 1 - \frac{1}{r} = \frac{r-1}{r}$ we have $a' = 2r'$. It follows that,

$$I = \left\| \int_{\mathbf{R}} |\Phi(t, x + tv)| \, dt \right\|^2_{L^{a'}(\mathbf{R}^d_x \times \mathbf{R}^d_v)} = \left\| \left(\int_{\mathbf{R}} |\Phi(t, x + tv)| \, dt \right)^2 \right\|_{L^{r'}(\mathbf{R}^d_x \times \mathbf{R}^d_v)}.$$

c) We have,

$$I = \left\| \int_{\mathbf{R}} \int_{\mathbf{R}} |\Phi(t, x + tv)||\Phi(s, x + sv)| \, ds \, dt \right\|_{L^{r'}(\mathbf{R}^d_x \times \mathbf{R}^d_v)},$$

$$= \left\| \int_{\mathbf{R}} \int_{\mathbf{R}} |\Phi(t, x + (t - s)v)||\Phi(s, x)| \, ds \, dt \right\|_{L^{r'}(\mathbf{R}^d_x \times \mathbf{R}^d_v)},$$

$$\leq \int_{\mathbf{R}} \int_{\mathbf{R}} \|\Phi(t, x + (t - s)v)\Phi(s, x)\|_{L^{r'}(\mathbf{R}^d_x \times \mathbf{R}^d_v)} \, ds \, dt.$$

Setting $x + (t - s)v = y$ in the integral in v we obtain,

$$I \leq \int_{\mathbf{R}} \int_{\mathbf{R}} \left(\int_{\mathbf{R}^d_x} \int_{\mathbf{R}^d_y} \frac{1}{|t - s|^d} |\Phi(t, y)|^{r'} |\Phi(s, x)|^{r'} \, dx \, dy \right)^{\frac{1}{r'}} \, ds \, dt,$$

$$\leq \int_{\mathbf{R}} \int_{\mathbf{R}} \frac{1}{|t - s|^{\frac{d}{r'}}} \|\Phi(t, \cdot)\|_{L^{r'}(\mathbf{R}^d)} \|\Phi(s, \cdot)\|_{L^{r'}(\mathbf{R}^d)} \, ds \, dt.$$

d) This follows from the Hölder inequality.

e) Since $1 < r < \frac{d}{d-1}$ we have, $0 < \frac{1}{r'} = 1 - \frac{1}{r} < \frac{1}{d}$. Thus $\frac{d}{r'} < 1$. Moreover, we have,

$$\frac{1}{q} - \frac{d}{r'} + 1 = \frac{1}{q} - d(1 - \frac{1}{r}) + 1 = \frac{1}{q} - \frac{2}{q} + 1 = -\frac{1}{q} + 1 = \frac{1}{q'}.$$

Thus $b = q'$. We can apply the Hardy–Littlewood–Sobolev inequality (see the inequality (3.21) in Chap. 2) to get the result.

f) This follows immediately from questions a) and d) above.

5. We have $q' < +\infty, r' < +\infty$. It follows that $C_0^\infty(\mathbf{R}_t \times \mathbf{R}^d_x)$ is dense in $L^{q'}(\mathbf{R}_t, L^{r'}(\mathbf{R}^d_x))$ and $L^q(\mathbf{R}_t, L^r(\mathbf{R}^d_x))$ is the dual of $L^{q'}(\mathbf{R}_t, L^{r'}(\mathbf{R}^d_x))$. Then,

$$\|\rho\|_{L^q(\mathbf{R}^+_t, L^r(\mathbf{R}^d_x))} = \sup_{\Phi \in C_0^\infty(\mathbf{R}_t \times \mathbf{R}^d_x)} \frac{\left| \int_{\mathbf{R}^+} \int_{\mathbf{R}^d} \rho(t, x) \Phi(t, x) \, dx \, dt \right|}{\|\Phi\|_{L^{q'}(\mathbf{R}_t, L^{r'}(\mathbf{R}^d_x))}},$$

$$\leq C(d) \|f^0\|_{L^a(\mathbf{R}^d_x \times \mathbf{R}^d_v)}$$

by the previous question.

Part 2

1. We have $w \in O(E^a_{\tau,\xi})$ if and only if $w = Ov$, $v \in E^a_{\tau,\xi}$. Then $|w| = |v|$ and $|\tau + \langle O^{-1}w, \xi \rangle| = |\tau + \langle w, O\xi \rangle| \leq a$. Moreover, setting $v = O^{-1}w$ in the integral we get,

$$\mu(E^a_{\tau,\xi}) = \int_{E^a_{\tau,\xi}} \chi(v)\, dv = \int_{O(E^a_{\tau,\xi})} \chi(O^{-1}w)\, dw.$$

2. For every $\xi \in \mathbf{R}^d \setminus \{0\}$ one can find an orthogonal matrix O such that $O\xi = (0, \ldots, 0, |\xi|)$ and we apply question 1.

3. We have obviously $E^a_{\tau,\xi} \subset B(0, M)$ so that,

$$\mu(E^a_{\tau,\xi}) \leq \mu(B(0, M)) \leq 2\mu(B(0, M))a,$$

since $a > \frac{1}{2}$.

4. a) If $|\xi| \leq \varepsilon_0 < 1$ we have $\tau^2 = 1 - |\xi|^2 \geq 1 - \varepsilon_0^2$ and,

$$|\tau + v \cdot \xi| \geq \sqrt{1 - \varepsilon_0^2} - |v||\xi| \geq \sqrt{1 - \varepsilon_0^2} - M\varepsilon_0.$$

The function $g(\varepsilon) = \sqrt{1 - \varepsilon^2} - M\varepsilon$ is such that $g(0) = 1$, so there exists $\varepsilon_0 \in (0, 1)$ depending only on M such that $g(\varepsilon_0) > \frac{1}{2} > a$.

b) If $|\xi| > \varepsilon_0$ we can write,

$$\{w : |\tau + |\xi|w_d| \leq a\} = \left\{w : \left|\frac{\tau}{|\xi|} + w_d\right| \leq \frac{a}{|\xi|}\right\} \subset \left\{w : \left|\frac{\tau}{|\xi|} + w_d\right| \leq \frac{a}{\varepsilon_0}\right\}$$

Using question 2 we have therefore,

$$\mu(E^a_{\tau,\xi}) \leq \mu_1 \left\{|w| \leq M : \left|\frac{\tau}{|\xi|} + w_d\right| \leq \frac{a}{\varepsilon_0}\right\}$$

$$\leq \mu_1 \left\{|w'| \leq M, \left|\frac{\tau}{|\xi|} + w_d\right| \leq \frac{a}{\varepsilon_0}\right\} \leq C(d, M)\frac{a}{\varepsilon_0}.$$

Here we have set $w = (w', w_d)$, $w' \in \mathbf{R}^{d-1}$.

5. This is an immediate application of the result proved in Part 1 of Problem 57 with $\delta = \frac{1}{2}$.

Solution 59

Part 1

1. This follows from the development of each determinant with respect to the ith line.
2. If $c_{ij} = \text{cof}(b_{ij})$ and $\text{adj}(B) = (d_{ij})$ we have $d_{ij} = c_{ji} = \text{cof}(b_{ji})$ so that,

$$\left(\frac{dB}{dt}(t)\,\text{adj}(B(t))\right)_{ij} = \sum_{k=1}^{n} \frac{db_{ik}}{dt}(t)\,d_{kj}(t) = \sum_{k=1}^{n} \frac{db_{ik}}{dt}(t)\,\text{cof}(b_{jk})(t).$$

Then,

$$\text{Tr}\left(\frac{dB}{dt}\,\text{adj}(B(t))\right) = \sum_{i=1}^{n}\sum_{k=1}^{n} \frac{db_{ik}}{dt}(t)\,\text{cof}(b_{ik}(t)) = \frac{d\Delta}{dt}(t).$$

3. Since $\frac{dB}{dt} = A(t)B(t)$ and $B(t)\,\text{adj}(B(t)) = (\det B(t))\,\text{Id} = \Delta(t)\text{Id}$ we deduce that,

$$\frac{d\Delta}{dt}(t) = \text{Tr}\left(A(t)B(t)\,\text{adj}(B(t))\right) = \Delta(t)\,\text{Tr}(A(t)).$$

Part 2

1. a) Assume that $X(t_0, y) = X(t_0, y')$. Set $E = \{t \in [0, T] : X(t, y) = X(t, y')\}$. By hypothesis $t_0 \in E$. On the other hand, by continuity E is closed. Let us show that it is open in $[0, T]$. Let $\bar{t} \in E$ and consider the differential systems (setting $\dot{f} = \frac{df}{dt}$)

$$\dot{Y}(t) = v(t, Y(t)), \quad Y(0) = X(\bar{t}, y), \tag{9.60}$$

$$\dot{Z}(t) = v(t, Z(t)), \quad Z(0) = X(\bar{t}, y'). \tag{9.61}$$

Since $\bar{t} \in E$ we have $X(\bar{t}, y) = X(\bar{t}, y')$. The uniqueness of the solution of this system implies that there exists $\varepsilon > 0$ such that $Y(t) = Z(t)$ for $|t - \bar{t}| \le \varepsilon$. But $X(t, y)$ is the unique solution of (9.60) and $X(t, y')$ the unique solution of (9.61). Therefore, we have $X(t, y) = X(t, y')$ for $|t - \bar{t}| \le \varepsilon$ which shows that $B(\bar{t}, \varepsilon) \subset E$. It follows that $E = [0, T]$ that is $0 \in E$, so $y = y'$.

   b) Using the continuity of the solution with respect to the initial data we see that the map $y \mapsto X(t, y)$ is C^1 and that we have,

$$\frac{d}{dt}\left(\frac{\partial X_j}{\partial y_k}\right)(t, y) = \sum_{\ell=1}^{d} \frac{\partial v_j}{\partial x_\ell}(t, X(t))\frac{\partial X_\ell}{\partial y_k}(t, y).$$

The matrix $B(t) = \left(\frac{\partial X_j}{\partial y_k}\right)$ is therefore solution of an equation of the form (8.36) with $A(t) = \left(\frac{\partial v_j}{\partial x_\ell}(t, X(t))\right)$ and $B(0) = \mathrm{Id}$. Using Part 1 we deduce that,

$$\Delta(t, y) = \det\left(\frac{\partial X_j}{\partial y_k}\right)(t, y) = \Delta(t_0) \exp\left(\int_{t_0}^t \mathrm{Tr}\,(A(s))\,ds\right),$$

$$= \exp\left(\int_0^t \mathrm{div}_x v(s, X(s, y))\,ds\right).$$

c) The map $y \mapsto \Phi_t(y) : \Omega_0 \to \Omega_t = \Phi_t(\Omega_0)$ is then bijective and its determinant is nonzero. The inverse function theorem implies that it is a C^1 diffeomorphism.

2. a) Set in the integral $x = \Phi_t(y)$, so $dx = |\Delta(t, y)|\,dy = \Delta(t, y)\,dy$ since $\Delta(t, y) > 0$. We obtain,

$$V(t) = \int_{\Omega_t} dx = \int_{\Omega_0} \Delta(t, y)\,dy.$$

b) Now Ω_0 is bounded and $t \to \Delta(t, y)$ is C^1. By the Lebesgue differentiation theorem we can differentiate inside the integral. Using Part 1 we obtain,

$$V'(t) = \int_{\Omega_0} \frac{d\Delta}{dt}(t, y)\,dy = \int_{\Omega_0} \Delta(t, y)(\mathrm{div}_x v)(t, X(t, y))\,dy.$$

c) It follows from question b), setting $x = \Phi_t(y)$, and from the hypothesis that,

$$V'(t) = \int_{\Omega_t} (\mathrm{div}\,v)(t, x)\,dx = 0, \quad \forall t \in [0, T).$$

Therefore, $V(t) = V(0)$ for all $t \in [0, T)$.

Part 3

1. We have seen in question 2b) of Part 1 that,

$$V'(t_0) = \int_{\Omega_0} (\mathrm{div}_x v)(t_0, X(t_0, y))\Delta(t_0, y)\,dy.$$

By the choice of Ω_0 and the fact that $\Delta(t_0, y) > 0$ we see that the function inside the integral is strictly positive. So $V'(t_0) > 0$. The Taylor formula shows then that for $t > t_0$, $|t - t_0|$ small enough, we have $V(t) > V(t_0)$.

Solution 60

1. We have just to differentiate the jth equation of (8.38) with respect to x_i then the ith equation with respect to x_j and make the difference.
2. We have,

$$\partial_i v_k \partial_k v_j - \partial_j v_k \partial_k v_i = \partial_i v_k (\partial_k v_j - \partial_j v_k) + \partial_j v_k (\partial_i v_k - \partial_k v_i)$$

$$= \partial_i v_k \omega_{kj} + \partial_j v_k \omega_{ki}.$$

3. It is the translation in terms of matrices of the result in question 1.
4. a) We have $\partial_s C(s) = [\partial_s A + (v \cdot \nabla_x) A](s, X(s, \cdot))$ so that this equality follows from the previous question.
   b) Integrating the above equation between 0 and s and taking the matrix norms of both members we obtain,

$$\|C(s)\| \leq \|C(0)\| + 2 \int_0^s \|B(\sigma, X(\sigma, \cdot))\| \|C(\sigma)\| \, d\sigma.$$

5. The Gronwall inequality implies that,

$$\|C(s)\| \leq \|C(0)\| \exp\left(2 \int_0^s \|B(\sigma, X(\sigma, \cdot))\| \, d\sigma\right).$$

We have $C(0) = A(0, y) = (\text{curl } v)(0, y) = (\text{curl } v_0)(y)$, so if $\text{curl } v_0 \equiv 0$, the above inequality shows that $A(s, X(s, y)) = 0$ for all $(s, y) \in [0, T) \times \mathbf{R}^d$ so $(\text{curl } v)(s, x) = 0$ for all $(s, x) \in [0, T) \times \mathbf{R}^d$.

Solution 61

1. Set for $1 \leq j \leq 3$, $w_j(\xi) = \frac{\xi_j}{i|\xi|^2} \widehat{f}(\xi)$. For all $s \in \mathbf{R}$ we have,

$$\int_{\mathbf{R}^3} \langle \xi \rangle^{2s} |w_j(\xi)|^2 \, d\xi \leq \int_{\{|\xi| \leq 1\}} \frac{\langle \xi \rangle^{2s} |\widehat{f}(\xi)|^2}{|\xi|^2} \, d\xi + \int_{\{|\xi| > 1\}} \langle \xi \rangle^{2s} |\widehat{f}(\xi)|^2 \, d\xi,$$

and the right-hand side is finite since $\frac{1}{|\xi|^2} \in L^1_{\text{loc}}(\mathbf{R}^3)$ and $\widehat{f} \in S(\mathbf{R}^3)$. Set $v_j = \mathcal{F}^{-1} w_j$. Then $v_j \in H^{+\infty}(\mathbf{R}^3)$. Let us show that $v = (v_j)$ solves the question. We have, $\sum_{j=1}^3 \widehat{\partial_j v_j}(\xi) = \sum_{j=1}^3 i \xi_j w_j(\xi) = \widehat{f}(\xi)$. So $\text{div } v = f$.
2. a) Choose $f = e^{-|x|^2}$. Then $\int_{\mathbf{R}^3} (\text{div } v)(x) \, dx = \int_{\mathbf{R}^3} e^{-|x|^2} \, dx \neq 0$.

b) Consider the function v described in question a). If we had $\partial_j v_j \in L^1(\mathbf{R}^3)$ for $j = 1, 2, 3$, we would have $\int_{\mathbf{R}^3}(\operatorname{div} v)(x)\,dx = \sum_{j=1}^3 \int_{\mathbf{R}^3} \partial_j v_j(x)\,dx = 0$ since, because $v_j \in H^{+\infty}(\mathbf{R}^3)$, we have $\lim_{|x|\to+\infty} v_j(x) = 0$.

3. Set $A = \| \langle\xi\rangle^{\sigma+1} \frac{1}{|\xi|}\widehat{u}\|^2_{L^2(\mathbf{R}^3)}$. Write,

$$A = \int_{|\xi|\leq 1} \frac{\langle\xi\rangle^{2\sigma+2}}{|\xi|^2}|\widehat{u}(\xi)|^2\,d\xi + \int_{|\xi|>1} \frac{\langle\xi\rangle^{2\sigma+2}}{|\xi|^2}|\widehat{u}(\xi)|^2\,d\xi = A_1 + A_2.$$

We have,

$$A_1 \leq C\|\widehat{u}\|^2_{L^\infty(\mathbf{R}^3)} \int_{|\xi|\leq 1} \frac{d\xi}{|\xi|^2} \leq C'\|u\|^2_{L^1(\mathbf{R}^3)}.$$

$$A_2 \leq C \int_{\mathbf{R}^3} \langle\xi\rangle^{2\sigma}|\widehat{u}(\xi)|^2\,d\xi = C\|u\|^2_{H^\sigma(\mathbf{R}^3)}.$$

Solution 62

1. If $f(x) = C \in \mathbf{C}$ we have $\widehat{f}(k) = 0$ if $k \neq 0$ and $\widehat{f}(0) = C(2\pi)^2$ so that $\|f\|_{H^s(\mathbf{T}^2)} = |C|(2\pi)^2$. If $f(x) = \sin nx_1 = \frac{1}{2i}(e^{inx_1} - e^{-inx_1})$ we have,

$$\widehat{f}(k) = \frac{1}{2i}\left(\int_0^{2\pi} e^{i(n-k_1)x_1}\,dx_1 - \int_0^{2\pi} e^{-i(n+k_1)x_1}\,dx_1\right)\int_0^{2\pi} e^{-ik_2x_2}\,dx_2.$$

The right-hand side vanishes for $k_2 \neq 0$. If $k_2 = 0$ it vanishes also if $k_1 \neq n$ and $k_1 \neq -n$. Eventually $\widehat{f}(-n, 0) = -\frac{2\pi^2}{i}$, $\widehat{f}(n, 0) = \frac{2\pi^2}{i}$. It follows that,

$$\|f\|_{H^s(\mathbf{T}^2)} = 2\sqrt{2}\pi^2 \langle n\rangle^s.$$

2. a) First of all, since u_1 does not depend on x_1 and u_2 on x_2, we have $A = \operatorname{div} U_{n,\omega} = 0$. Now,

$$\partial_t U_{n,\omega} = \left(\omega n^{-s}\sin(nx_2 - \omega t),\ \omega n^{-s}\sin(nx_1 - \omega t)\right)$$

$$(U_{n,\omega} \cdot \nabla_x) U_{n,\omega} = \left(-(\omega n^{-1} + n^{-s}\cos(nx_1 - \omega t))n^{-s+1}\sin(nx_2 - \omega t),\right.$$

$$\left. -\left(\omega n^{-1} + n^{-s}\cos(nx_2 - \omega t)\right)n^{-s+1}\sin(nx_1 - \omega t)\right).$$

It follows that

$$B = \partial_t U_{n,\omega} + \left(U_{n,\omega} \cdot \nabla_x\right) U_{n,\omega}$$

$$= \left(-n^{-2s+1} \cos(nx_1 - \omega t) \sin(nx_2 - \omega t), \right. \tag{9.62}$$

$$\left. -n^{-2s+1} \cos(nx_2 - \omega t) \sin(nx_1 - \omega t)\right).$$

Then we have,

$$C = \operatorname{div}\left(U_{n,\omega} \cdot \nabla_x\right) U_{n,\omega} = 2n^{-2s+2} \sin(nx_1 - \omega t) \sin(nx_2 - \omega t). \tag{9.63}$$

b) We have,

$$\Delta \left(\sin(nx_1 - \omega t) \sin(nx_2 - \omega t)\right) = -2n^2 \sin(nx_1 - \omega t) \sin(nx_2 - \omega t)$$

so $\Delta \left(\frac{1}{2n^2} w_n\right) = -w_n$.

c) This follows from b).

d) Since $R(x) = \frac{1}{(2\pi)^2} \sum_{k \in \mathbf{Z}^2} \widehat{R}(k) e^{i \langle k, x \rangle}$, we have $\sum_{k \in \mathbf{Z}^2} -|k|^2 \widehat{R}(k) e^{i \langle k, x \rangle} = 0$. This implies that $\widehat{R}(k) = 0$ if $k \neq 0$ so $R(x) = \frac{1}{(2\pi)^2} \widehat{R}(0)$.

e) We have,

$$\nabla_x \Delta^{-1} \operatorname{div}\left(\left(U_{n,\omega} \cdot \nabla_x\right) U_{n,\omega}\right) = \left(-n^{-2s+1} \cos(nx_1 - \omega t) \sin(nx_2 - \omega t), \right.$$

$$\left. -n^{-2s+1} \sin(nx_1 - \omega t) \cos(nx_2 - \omega t)\right).$$
$$\qquad \cdot \quad (9.64)$$

f) This follows from (9.62) and (9.64).

3.

$$U_n^0 = U_{n,1}^0, \quad V_n^0 = U_{n,-1}^0.$$

Then $U_n^0 - V_n^0 = (\frac{2}{n}, \frac{2}{n})$ so that $\|U_n^0 - V_n^0\|_{(H^s(\mathbf{T}^2))^2} = \frac{8\sqrt{2}\pi^2}{n}$. Therefore,

$$\lim_{n \to +\infty} \|U_n^0 - V_n^0\|_{(H^s(\mathbf{T}^2))^2} = 0. \tag{9.65}$$

4. Denote by U_n and V_n the solutions of the Euler system corresponding to these data. By (8.43) we have,

$$U_n(t, x) = \left(n^{-1} + n^{-s} \cos(nx_2 - t), \quad n^{-1} + n^{-s} \cos(nx_1 - t)\right),$$

$$V_n(t, x) = \left(-n^{-1} + n^{-s} \cos(nx_2 + t), \quad -n^{-1} + n^{-s} \cos(n\dot{x}_1 + t)\right).$$

Then,

$$U_n(t,x) - V_n(t,x) = \left(\frac{2}{n} + n^{-s}(\cos(nx_2 - t) - \cos(nx_2 + t)),\right.$$

$$\left.\frac{2}{n} + n^{-s}(\cos(nx_1 - t) - \cos(nx_1 + t))\right).$$

Using the formula $\cos p - \cos q = 2\sin(\frac{q+p}{2})\sin(\frac{q-p}{2})$ we can write,

$$U_n(t,x) - V_n(t,x) = \left(\frac{2}{n} + 2n^{-s}\sin(nx_2)\sin(t), \ \frac{2}{n} + 2n^{-s}\sin(nx_1)\sin(t)\right).$$

If $f(x) = \sin(nx_2)$ or $f(x) = \sin(nx_1)$ we have, from question 1 $\|f\|_{H^s(\mathbf{T}^2)} = C\langle n\rangle^s$. Therefore, the $(H^s(\mathbf{T}^2))^2$ norm of each component of $U_n - V_n$ is bounded below by $C_1|\sin(t)| - \frac{C_2}{n}$. It follows that,

$$\|U_n(t,\cdot) - V_n(t,\cdot)\|_{(H^s(\mathbf{T}^2))^2} \geq C_3|\sin(t)| - \frac{C_4}{n}.$$

Let $t > 0$ be arbitrary small. This construction contradicts the uniform continuity of the map $U_0 \mapsto U(t)$ from $H^s(\mathbf{T}^2)$ to $H^s(\mathbf{T}^2)$. Indeed denying the uniform continuity reads,

$$\exists \varepsilon_0 > 0 : \forall \delta > 0 \ \exists U_0, V_0 \in (H^s(\mathbf{T}^2))^2 : \|U_0 - V_0\|_{(H^s(\mathbf{T}^2))^2} \leq \delta \text{ and}$$

$$\|U(t) - V(t)\|_{(H^s(\mathbf{T}^2))^2} \geq \varepsilon_0.$$

We have just to take $\varepsilon_0 = \frac{C_3}{2}|\sin t|$, $\delta \leq \frac{8\sqrt{2}\pi^2}{n}$, n large enough.

Solution 63

Part 1
1. a) It follows from (8.45) that, $|\partial_t \rho(t,x_0)| \leq M\sum_{j=1}^3 |v_j(t,x_0)| + 3M|\rho(t,x_0)|$. We deduce from (8.46) if $\rho(t,x_0) \neq 0$, (dividing by $\rho(t,x_0)$) and from (C_2) that,

$$|\partial_t v_k(t,x_0)| \leq M\sum_{j=1}^3 |v_j(t,x_0)| + 3M|\rho(t,x_0)|, \quad \text{if } \rho(t,x_0) \neq 0,$$

$$|\partial_t v_k(t,x_0)| \leq M\sum_{j=1}^3 |v_j(t,x_0)| \quad \text{if } \rho(t,x_0) = 0.$$

b) We have,

$$|\partial_t u(t, x_0)| \le 2 \left(|\partial_t \rho(t, x_0)||\rho(t, x_0)| + \sum_{j=1}^{3} |\partial_t v_j(t, x_0)||v_j(t, x_0)| \right).$$

$$\le 2 \left(|\partial_t \rho(t, x_0)| + \sum_{j=1}^{3} |\partial_t v_j(t, x_0)| \right) \left(|\rho(t, x_0)| + \sum_{j=1}^{3} |v_j(t, x_0)| \right),$$

We have then just to use question a) and the Cauchy Schwarz inequality to conclude.

2. For $t \in [0, T_0]$ we deduce from question 1b) that, $u(t) \le u(0) + C \int_0^t u(s)\,ds$. The Gronwall inequality implies that $u(t) \le u(0)e^{CT_0}$. Since $|x_0| > R$ and supp $\rho_0 \cup$ supp $v_0 \subset B(0, R)$, we have $u(0) = \rho_0(x_0)^2 + \sum_{j=1}^3 |v_{j,0}(x_0)|^2 = 0$. It follows that $u(t) = 0$ so $\rho(t, x_0) = v_j(t, x_0) = 0$ for all $t \in [0, T_0]$ and $1 \le j \le 3$. In other words $\rho(t, \cdot)$ and $v(t, \cdot)$ vanish identically when $|x| > R$ and $t \in [0, T)$; this proves our claim.

Part 2

1. The first identity follows immediately from (8.45). To obtain the second one we have just to multiply (8.45) by v_k and to add it to (8.46). Denote by A the left-hand side of the third identity. We have,

$$A = |v|^2 \partial_t \rho + 2\rho \sum_{k=1}^{3} v_k \partial_t v_k + 3\rho^2 \partial_t \rho + \rho|v|^2 (\text{div } v) + 3\rho^3 (\text{div } v)$$

$$+ |v|^2 (v \cdot \nabla)\rho + 2\rho \sum_{j=1}^{3}\sum_{k=1}^{3} v_j v_k \partial_j v_k + 9\rho^2 (v \cdot \nabla)\rho = \sum_{\ell=1}^{8} A_\ell.$$

From (8.45) we get $A_1 + A_4 = -|v|^2 (v \cdot \nabla)\rho$. Now multiplying (8.46) by v_k and summing we obtain, $\rho \sum_{k=1}^{3} v_k \partial_t v_k + \rho \sum_{k=1}^{3} v_k (v \cdot \nabla)v_k + 3\rho^2 (v \cdot \nabla)\rho = 0$. It follows that $A_2 + A_7 = -6\rho^2 (v \cdot \nabla)\rho$. Using (8.45) we obtain $A_3 = -3\rho^2 (v \cdot \nabla)\rho - 3\rho^3 (\text{div } v)$. Now we have $A_5 = 3\rho^3 (\text{div } v)$, $A_6 = |v|^2 (v \cdot \nabla)\rho$, $A_8 = 9\rho^2 (v \cdot \nabla)\rho$. Therefore,

$$\sum_{\ell=1}^{8} A_\ell = -|v|^2 (v \cdot \nabla)\rho - 6\rho^2 (v \cdot \nabla)\rho - 3\rho^2 (v \cdot \nabla)\rho - 3\rho^3 (\text{div } v)$$

$$+ 3\rho^3 (\text{div } v) + |v|^2 (v \cdot \nabla)\rho + 9\rho^2 (v \cdot \nabla)\rho = 0.$$

2. By hypothesis there exists $R > 0$ such that $\mathrm{supp}(\rho_0, v_0) \subset B(0, R)$. From Part 1 we have $\mathrm{supp}(\rho(t, \cdot), v(t, \cdot)) \subset B(0, R)$. So,

$$E(t) = \int_{B(0,R)} \left(\rho(t, x)|v(t, x)|^2 + \rho(t, x)^3 \right) dx.$$

The functions inside the integral being C^1 in t we can, by the Lebesgue differentiation theorem, differentiate inside the integral. Using the previous question we obtain,

$$\frac{\partial E}{\partial t}(t) = \int \frac{\partial}{\partial t} \left[\rho(t, x)|v(t, x)|^2 + \rho(t, x)^3 \right] dx$$

$$= -\sum_{j=1}^{3} \int \frac{\partial}{\partial x_j} \left[\rho(t, x)|v(t, x)|^2 + 3\rho(t, x)^3 \right] dx = 0, \quad 0 \le t < T,$$

since the functions inside the bracket have compact support in x.

Therefore, we have $E = \int \left(\rho_0(x)|v_0(x)|^2 + \rho_0(x)^3 \right) dx > 0$ since ρ_0 is ≥ 0 and $\rho_0 \ne 0$.

3. As above we can differentiate inside the integral. Using the first relation in (8.48) we obtain,

$$\frac{\partial}{\partial t} \int \rho(t, x) \, dx = \int \frac{\partial \rho}{\partial t}(t, x) \, dx = -\sum_{j=1}^{3} \int \frac{\partial}{\partial x_j} (\rho(t, x)v_j(t, x)) \, dx = 0,$$

since the functions ρ and v have compact support x.

4. We have,

$$H(t) = \int_{B(0,R)} |x|^2 \rho(t, x) \, dx \le R^2 \int \rho(t, x) \, dx$$

$$= R^2 \int \rho_0(x) \, dx = M < +\infty.$$

5. The fact that these integrals are C^1 has been considered in question 2. So we have, using the first relation in (8.48),

$$H'(t) = \int |x|^2 \frac{\partial \rho}{\partial t}(t, x) \, dx = -\sum_{j=1}^{3} \int |x|^2 \frac{\partial}{\partial x_j} (\rho(t, x)v_j(t, x)) \, dx.$$

Integrating by parts and using the fact that ρ, v have compact support in x we get,

$$H'(t) = 2 \sum_{j=1}^{3} \int \rho(t, x) x_j v_j(t, x) \, dx.$$

6. a) By the same argument as in 2. we see that the right-hand side of $H'(t)$ is a C^1 function in t and that we have,

$$H''(t) = \sum_{k=1}^{3} \int x_k \partial_t \, (\rho v_k) \, (t, x) \, dx.$$

We use the second equality in (8.48). We obtain,

$$H''(t) = -\sum_{j=1}^{3}\sum_{k=1}^{3} \int x_k \frac{\partial}{\partial x_j}(\rho v_j v_k)(t, x) \, dx - \sum_{k=1}^{3} \int x_k \frac{\partial}{\partial x_k}\left[\rho(t, x)^3\right] dx.$$

Integrating by parts we see that the first integral vanishes if $j \neq k$ so,

$$H''(t) = \sum_{k=1}^{3} \int \rho(t, x) v_k(t, x)^2 \, dx + \int \rho(t, x)^3 \, dx.$$

b) This follows from the fact that $\rho \geq 0$.
7. By the Taylor formula with integral reminder and the previous question we have,

$$H(t) = H(0) + t H'(0) + t^2 \int_0^1 (1 - \lambda) H''(\lambda t) \, d\lambda \geq H(0) + t H'(0) + \frac{1}{2} E t^2.$$

8. From question 4 we have $H(t) \leq M$. So, $H(0) + t H'(0) + \frac{1}{2} E t^2 \leq M$ for all $t < T$. Since from question 2 we have $E > 0$ this inequality shows that T must be bounded.

Solution 64

1. Recall that,

$$W^{1+\alpha,\infty}(\mathbf{R}^3, \mathbf{R}^3) = \left\{ u \in W^{1,\infty}(\mathbf{R}^3, \mathbf{R}^3) : \sup_{x \neq y} \frac{|\nabla_x u(x) - \nabla_x u(y)|}{|x - y|^\alpha} < +\infty \right\}.$$

First of all we have, $\|u(t, \cdot)\|_{L^\infty} \leq C(\|f\|_{L^\infty} + \|h\|_{L^\infty})$. Then $\|\partial_{x_1} u(t, \cdot)\|_{L^\infty} \leq$
$\|h'\|_{L^\infty}$ and $\|\partial_{x_2} u(t, \cdot)\|_{L^\infty} \leq C(\|f'\|_{L^\infty} + T\|h'\|_{L^\infty}\|f'\|_{L^\infty})$.
Now,

$$\partial_{x_1} u(t, x) - \partial_{x_1} u(t, y) = (0, 0, h'(x_1 - tf(x_2)) - h'(y_1 - tf(y_2))).$$

Since $f, h \in W^{1+\alpha,\infty}(\mathbf{R})$ we have $h' \in W^{\alpha,\infty}$, $f' \in L^\infty$. Therefore, we have,

$$|\partial_{x_1} u(t, x) - \partial_{x_1} u(t, y)| \leq \|h'\|_{W^{\alpha,\infty}} |x_1 - y_1 - t(f(x_2) - f(y_2))|^\alpha,$$
$$\leq \|h'\|_{W^{\alpha,\infty}} \left(|x_1 - y_1| + T\|f'\|_{L^\infty}|x_2 - y_2|\right)^\alpha$$
$$\leq C(f, h, T)|x - y|^\alpha.$$

Eventually we have,

$$\partial_{x_2} u(t, x) - \partial_{x_2} u(t, y) = (f'(x_2) - f'(y_2), \, 0, \, -tf'(x_2)h'(x_1 - tf(x_2))$$
$$+ tf'(y_2)h'(y_1 - tf(y_2)).$$

Set $(1) := f'(y_2)h'(y_1 - tf(y_2)) - f'(x_2)h'(x_1 - tf(x_2))$. Then $(1) = A + B$
where,

$$A = (f'(y_2) - f'(x_2))h'(y_1 - f(y_2)),$$
$$B = f'(x_2)(h'(y_1 - tf(y_2)) - h'(x_1 - tf(x_2))).$$

We obtain,

$$|(1)| \leq \|h'\|_{L^\infty}\|f'\|_{W^{\alpha,\infty}}|x_2 - y_2|^\alpha$$
$$+ \|f'\|_{L^\infty}\|h'\|_{W^{\alpha,\infty}}|y_1 - x_1 + t(f(x_2) - f(y_2))|^\alpha \leq C(f, h, T)|x - y|^\alpha.$$

It follows that,

$$|\partial_{x_2} u(t, x) - \partial_{x_2} u(t, y)| \leq C(f, h, T)|x - y|^\alpha.$$

This shows that $u \in W^{1+\alpha,\infty}$. The computation is exactly the same for v.
2. We have, $u_0 = (f(x_2), 0, h(x_1))$. Then div $u_0 = \partial_{x_1} u_{01} + \partial_{x_2} u_{02} + \partial_{x_3} u_{03} = 0$.
On the other hand, since $u_1(t, x) = f(x_2), u_2(t, x) = 0, u_3(t, x) = h(x_1 - tf(x_2))$ we have, $u \cdot \nabla_x = f(x_2)\partial_{x_1} + h(x_1 - tf(x_2))\partial_{x_3}$. It follows that,

$$\partial_t u_1 + (u \cdot \nabla_x)u_1 = 0, \quad \partial_t u_2 + (u \cdot \nabla_x)u_2 = 0,$$
$$\partial_t u_3 + (u \cdot \nabla_x)u_3 = -f(x_2)h'(x_1 - tf(x_2)) + f(x_2)h'(x_1 - tf(x_2)) = 0.$$

Therefore, u satisfies the system (8.49) with $P = 1$.

3. We have $u_0 - v_0 = (f(x_2) - g(x_2), 0, 0)$. Let $g \in W^{1+\alpha,\infty}(\mathbf{R})$. Let $\delta > 0$. Set $f(x) = g(x) + \delta$. Then $f(x) \neq g(x)$ for all $x \in \mathbf{R}$ and,

$$\|u_0 - v_0\|_{W^{1+\alpha,\infty}} = \|f - g\|_{W^{1+\alpha,\infty}} = \delta\|1\|_{W^{1+\alpha,\infty}} = \delta.$$

4. Keeping the notation in the statement we have,

$$A(t) \geq \|\partial_{x_1}(u(t,\cdot) - v(t,\cdot))\|_{W^{\alpha,\infty}} = \|h'(\cdot - tf(\cdot)) - h'(\cdot - tg(\cdot))\|_{W^{\alpha,\infty}}.$$

5. Since $\{(x,y) : x_1 \neq y_1, x, y \in [-a,a]^2\} \subset \{(x,y) : x \neq y\}$ we have,

$$\|U\|_{W^{\alpha,\infty}} \geq \sup_{\substack{x_1 \neq y_1 \\ x,y \in [-a,a]^2}} \frac{|U(x) - U(y)|}{|x_1 - x_2|^\alpha}.$$

Now if $|x_1| \leq a$ we have $|x_1 - tf(x_2)| \leq a + T\|f\|_{L^\infty} \leq 2a$. Therefore, by our choice of h' we have $h'(x_1 - tf(x_2)) = |x_1 - tf(x_2)|^\alpha$. This gives the desired lower bound for $A(t)$.

6. When $x_2 = y_2 = c$ we obtain from the previous question that $A(t)$ is greater than,

$$\sup_{\substack{x_1 \neq y_1 \\ x_1, y_1 \in [-a,a]}} \frac{\big||x_1 - tf(c)|^\alpha - |x_1 - tg(c)|^\alpha - |y_1 - tf(c)|^\alpha + |y_1 - tg(c)|^\alpha\big|}{|x_1 - y_1|^\alpha}.$$

Then for $t \in (0,T]$ one can take $x_1 = tg(c), y_1 = tf(c)$ since for $t > 0$, $x_1 = tg(c) \neq tf(c) = y_1$, and $|x_1| \leq T\|g\|_{L^\infty} \leq a, |y_1| \leq T\|f\|_{L^\infty} \leq a$. We obtain,

$$A(t) \geq 2\frac{(t|g(c) - f(c)|)^\alpha}{(t|g(c) - f(c)|)^\alpha} = 2.$$

7. Fix $v_0 = (g(x_2), 0, 0)$. We have proved that there exists $c_0(= 2)$ such that, for all $\delta > 0$ there exists $u_0 \in W^{1+\alpha,\infty}$ such that,

$$\|u_0 - v_0\|_{W^{1+\alpha,\infty}} \leq \delta \text{ and } \sup_{t \in [0,T]} \|u(t,\cdot) - v(t,\cdot)\|_{W^{1+\alpha,\infty}} \geq c_0.$$

This contradicts the continuity of the map $u_0 \mapsto u$ at the point v_0 from $W^{1+\alpha,\infty}$ to $L^\infty([0,T], W^{1+\alpha,\infty})$.

Solution 65

Part 1

1. By using the equation, we have

$$\frac{d}{ds}[u(s, x(s))] = (\partial_s u + \dot{x}(s)\partial_x u)(s, x(s)) = (\partial_s u + u\,\partial_x u)(s, x(s)) = 0.$$

Therefore, $u(s, x(s)) = u(0, x(0)) = u_0(y)$. Coming back to the equation of the characteristic we find that $\dot{x}(s) = u_0(y)$ so $x(s) = y + su_0(y)$ for all $s \in [0, +\infty)$.

2. Since $u(s, x(s)) = u_0(y)$ we deduce from question 1 that,

$$u(s, y + u_0(y)) = u_0(y).$$

Part 2

1. We have, $\frac{\partial F}{\partial y}(t, y) = 1+tu_0'(y)$.If $\inf_{\mathbf{R}} u_0' \geq 0$ we have $\frac{\partial F}{\partial y}(t, y) \geq 1$ for all $t \geq 0$ and if $\inf_{\mathbf{R}} u_0' = -\alpha$, with $0 < \alpha < +\infty$, we have $1 + tu_0'(y) \geq 1 - t\alpha > 0$ if $t < \frac{1}{\alpha}$.

2. a) For $t \in [0, T^*)$ the map $y \rightarrow F(t, y)$ from $\mathbf{R} \rightarrow \mathbf{R}$ is then strictly increasing. It is therefore invertible and there exists a map $x \rightarrow k(t, x)$ such that $F(t, y) = x \iff y = \kappa(t, x)$. The fact that the map κ is C^1 follows from the inverse function theorem.

   b) Obvious.

3. According to question 2 Part 1 set $u(t, x) = u_0(\kappa(t, x))$. Then u is a C^1 function on $[0, T^*) \times \mathbf{R}$. Differentiating the equality $\kappa(t, y + tu_0(y)) = y$ with respect to t and setting $y + tu_0(y) = x$ we find,

$$\partial_t \kappa(t, x) + u_0(\kappa(t, x))\partial_x \kappa(t, x) = 0. \tag{9.66}$$

It follows from the form of u and (9.66) that we have,

$$\partial_t u(t, x) = u_0'(\kappa(t, x))\partial_t \kappa(t, x) = -u_0'(\kappa(t, x))u_0(\kappa(t, x))\partial_x \kappa(t, x),$$

$$= -u(t, x)u_0'(\kappa(t, x))\partial_x \kappa(t, x),$$

$$\partial_x u(t, x) = u_0'(\kappa(t, x))\partial_x \kappa(t, x).$$

Therefore, $\partial_t u + u\,\partial_x u = 0$ in $[0, T^*) \times \mathbf{R}$. Moreover, since $\kappa(0, x) = x$ we have $u(0, x) = u_0(x)$. This proves the existence of a solution. The uniqueness follows from the necessary condition proved in question 2 Part 1.

4. Let $\mathrm{supp}\, u_0 \subset [a, b]$. We show by contradiction that there exists $x_0 \in \mathbf{R}$ such that $u_0'(x_0) < 0$. Indeed assume that $u_0'(x) \geq 0$ for every $x \in \mathbf{R}$. Then u_0 is increasing so $u_0(x) \geq u_0(a) = 0$ for $x \geq a$ and since $u_0 \not\equiv 0$ there exists $x_1 > a$

such that $u_0(x_1) > 0$. Then for $x \geq x_1$ we have,

$$u_0(x) - u_0(x_1) = \int_0^1 u_0'(tx + (1-t)x_1)(x - x_1)\, dt \geq 0.$$

Therefore, $u_0(x) \geq u_0(x_1) > 0$ which contradicts the fact that $u_0 = 0$ for $x \geq b$.

5. We have by definition $\kappa(t, y + tu_0(y)) = y$. Differentiating this equality with respect to y and setting $y + tu_0(y) = x$ that is $y = \kappa(t, x)$ we obtain the expression of $\partial_x \kappa$. Differentiating the above equality with respect to t we obtain the expression of $\partial_t \kappa$.

Part 3

1. a) Since $q(t) = (\partial_x u)(t, x_0 + tu_0(x_0))$ we have,

$$q'(t) = \partial_t(\partial_x u)(t, x_0 + tu_0(x_0)) + u_0(x_0)(\partial_x^2 u)(t, x_0 + tu_0(x_0)). \quad (9.67)$$

Differentiating the equation (8.50) with respect to t and setting $x = x_0 + tu_0(x_0)$ we obtain,

$$\partial_t(\partial_x u)(t, x_0 + tu_0(x_0)) + u(t, x_0 + tu_0(x_0))(\partial_x^2 u)(t, x_0 + tu_0(x_0))$$

$$+ (\partial_x u)^2(t, x_0 + tu_0(x_0)) = 0.$$

Since $u(t, x_0 + tu_0(x_0)) = u_0(x_0)$ we deduce from the above equation that,

$$\partial_t(\partial_x u)(t, y) + u_0(x_0)(\partial_x^2 u)(t, y) = -(\partial_x u)^2(t, y)$$

where $y = x_0 + tu_0(x_0)$.

Using (9.67) we obtain $q'(t) = -q(t)^2$. Moreover, $q(0) = (\partial_x u)(0, x_0) = u_0'(x_0) < 0$ by the choice of x_0. It follows that q is decreasing so, since $q(0) < 0$ we have $q(t) < 0$ for $0 \leq t < T_{x_0}^*$. Therefore, the equation on q is equivalent to,

$$\frac{q'(t)}{q(t)^2} = -\frac{d}{dt}\left(\frac{1}{q(t)}\right) = -1.$$

Integrating between 0 and $t > 0$ we obtain $\frac{1}{q(t)} = t + \frac{1}{u_0'(x_0)}$. Since $q < 0$ this equality shows that q exists only for $0 \leq t < -\frac{1}{u_0'(x_0)}$. Moreover, in this interval we have,

$$q(t) = \frac{1}{t + \frac{1}{u_0'(x_0)}} = \frac{1}{t - T_{x_0}^*}.$$

b) This equality shows in particular that $\lim_{t \to T^*_{x_0}} = -\infty$. Now if x_0 changes we see that our solution can exist as a C^1 function for t smaller than the smallest $T^*_{x_0}$ that is for,

$$t < \inf_{x_0 \in \mathbf{R}} T^*_{x_0} = \inf_{x_0 \in \mathbf{R}} \left(-\frac{1}{u'_0(x_0)} \right) = -\sup_{x_0 \in \mathbf{R}} \left(\frac{1}{u'_0(x_0)} \right) = \frac{1}{-\inf_{x_0 \in \mathbf{R}} u'_0(x_0)} = T^*.$$

Part 4

1. This follows from the fact that if $u_0 \in H^2$ then $u'_0 \in H^1 \subset C^0 \cap L^\infty$ and that we have proved in question 2b) Part 2 that if $T \|u'_0\|_{L^\infty} < 1$ then $T < T^*$.
2. Setting $\kappa(t, x) = y \Leftrightarrow x = y + tu_0(y)$ in the integral we obtain,

$$\|u(t, \cdot)\|^2_{L^2} = \int |u_0(\kappa(t, x))|^2 \, dx = \int |u_0(y)|^2 |1 + tu'_0(y)| \, dy,$$

$$\leq (1 + T \|u'_0\|_{L^\infty(\mathbf{R})}) \|u_0\|^2_{L^2}.$$

Now we have, $\partial_x u(t, x) = u'_0(\kappa(t, x)) \partial_x \kappa(t, x)$. Using question 6 Part 2 we have, $|\partial_x \kappa(t, x)| \leq \frac{1}{1 - T \|u'_0\|_{L^\infty}}$. Therefore, the same computation as above gives,

$$\|\partial_x u(t, \cdot)\|^2_{L^2} \leq \frac{1 + T \|u'_0\|_{L^\infty}}{(1 - T \|u'_0\|_{L^\infty})^2} \|u'_0\|^2_{L^2}.$$

3. We have $\partial^2_x u(t, x) = (1) + (2)$ where,

$$(1) = ((\partial_x \kappa)(t, x))^2 u''_0(\kappa(t, x)), \quad (2) = (\partial^2_x \kappa)(t, x) u'_0(\kappa(t, x)).$$

The term (1) can be estimated by the same method as that used in the previous question. We find,

$$\|(1)\|_{L^2} \leq \frac{(1 + T \|u'_0\|_{L^\infty})^{\frac{1}{2}}}{(1 - T \|u'_0\|_{L^\infty})^2} \|u''_0\|_{L^2}.$$

Let us estimate (2). Differentiating the equality in question 6 Part 2 with respect to x we find,

$$(\partial^2_x \kappa)(t, x) = -\frac{tu''_0(\kappa(t, x)) \partial_x \kappa(t, x)}{(1 + tu'_0(\kappa(t, x)))^2} = -\frac{tu''_0(\kappa(t, x))}{(1 + tu'_0(\kappa(t, x)))^3}. \tag{9.68}$$

It follows that,

$$|(2)| \leq \frac{T \|u'_0\|_{L^\infty(\mathbf{R})}}{(1 - T \|u'_0\|_{L^\infty(\mathbf{R})})^3} |u''_0(\kappa(t, x))|.$$

We may therefore apply the same method as above to prove that

$$\|(2)\|_{L^2} \le \frac{T\|u_0'\|_{L^\infty}(1 + T\|u_0'\|_{L^\infty})^{\frac{1}{2}}}{(1 - T\|u_0'\|_{L^\infty})^3}\|u_0''\|_{L^2}.$$

4. a) Let $|x| \ge R$. Set $y_n = \kappa(t_n, x) \Leftrightarrow x = y_n + t_n u_0(y_n)$. and $y_0 = \kappa(t_0, x) \Leftrightarrow x = y_0 + t_0 u_0(y_0)$. We have,

$$|\kappa(t_n, x)| = |y_n| \ge |x| - (t_0 + 1)\|u_0\|_{L^\infty(\mathbf{R})} \ge R_0,$$

$$|\kappa(t_0, x)| = |y_0| \ge |x| - t_0\|u_0\|_{L^\infty(\mathbf{R})} \ge R_0,$$

so $\varphi(\kappa(t_n, x)) = \varphi(\kappa(t_0, x)) = 0$.
b) For $n \ge n_0$, using question 6 Part 2 we can write,

$$I_n \le |B_R|\|\varphi'\|_{L^\infty}|\kappa(t_n, x) - \kappa(t_0, x)| \le |B_R|\|\varphi'\|_{L^\infty} \sup_{t\in[0,T]} |\frac{\partial\kappa}{\partial t}(t, x)||t_n - t_0|$$

$$\le |B_R|\,\|\varphi'\|_{L^\infty}\frac{\|u_0\|_{L^\infty(\mathbf{R})}}{1 - T\|u_0'\|_{L^\infty}}|t_n - t_0|.$$

Therefore, $\lim_{n\to+\infty} I_n = 0$.
5. Let $v \in L^2$. There exists $(\varphi_k) \subset C_0^\infty$ such that $\varphi_k \to v$ in L^2. We write,

$$J_n \le C\|v(\kappa(t_n, \cdot)) - \varphi_k(\kappa(t_n, \cdot))\|_{L^2}^2 + C\|\varphi_k(\kappa(t_n, \cdot)) - \varphi_k(\kappa(t_0, \cdot))\|_{L^2}^2$$

$$+ C\|\varphi_k(\kappa(t_0, \cdot)) - v(\kappa(t_0, \cdot))\|_{L^2}^2 = (1) + (2) + (3).$$

Setting successively $y = \kappa(t_n, x)$ then $y = \kappa(t_0, x)$ we see that,

$$(1) + (3) \le 2C\int_{\mathbf{R}} |v(y) - \varphi_k(y)|^2 \, dy \, (1 + (t_0 + 1))\|u_0'\|_{L^\infty}).$$

Let $\varepsilon > 0$. We fix k so large that $(1) + (3) \le \frac{2\varepsilon}{3}$. Since $\varphi_k \in C_0^\infty(\mathbf{R})$ we may use the previous question to infer that for n large enough we have $(2) \le \frac{\varepsilon}{3}$. Therefore, $J_n \le \varepsilon$.
6. We write,

$$w(t_n, x) - w(t_0, x) = f(t_n, x)\,(v(\kappa(t_n, x)) - v(\kappa(t_0, x)))$$

$$+ (f(t_n, x) - f(t_0, x))\,v(\kappa(t_0, x)).$$

The first term tends to zero in L^2 using question 5 above and the fact that f is uniformly bounded. The second term tends to zero in L^2 by the dominated convergence theorem. Indeed, for fixed x the quantity $|(f(t_n, x) - f(t_0, x)) v(\kappa(t_0, x))|^2$ tends to zero and it is bounded by,

$$4\|f\|^2_{L^\infty((0,T)\times\mathbf{R})}|v(\kappa(t_0, x))|^2,$$

which is integrable.

7. Recall that $u(t, x) = u_0(\kappa(t, x))$ where $u_0 \in H^2$. The fact that $u \in C^0([0, T), L^2)$ follows immediately from question 5 taking $v = u$. Now we have,

$$\partial_x u(t, x) = \partial_x \kappa(t, x) u_0'(\kappa(t, x)). \tag{9.69}$$

Since $\partial_x \kappa(t, x)$ is continuous, bounded, and $u_0' \in L^2$ it follows from question 6 above that $\partial_x u \in C^0([0, T), L^2)$. We are left with the second derivative. We have,

$$\partial_x^2 u(t, x) = (\partial_x \kappa(t, x))^2 u_0''(\kappa(t, x)) + \partial_x^2 \kappa(t, x) u_0'(\kappa(t, x)). \tag{9.70}$$

Therefore, by (9.68) we have,

$$\partial_x^2 u(t, x) = (\partial_x \kappa(t, x))^2 u_0''(\kappa(t, x)) - t(\partial_x \kappa(t, x))^3 u_0''(\kappa(t, x)) u_0'(\kappa(t, x)).$$

We use question 6 above. The first term in the right-hand side belongs to $C^0([0, T), L^2)$ since $(\partial_x \kappa(t, x))^2$ is continuous, bounded, and $u_0'' \in L^2$. The same is true for the second term since $t(\partial_x \kappa(t, x))^3$ is continuous, bounded, and $u_0' u_0'' \in L^2$ since $u_0' \in H^1 \subset L^\infty$ and $u_0'' \in L^2$.

Part 5

1. The uniform continuity of this map reads: for all $\varepsilon > 0$ there exists $\delta > 0$ such that,

$$\left(u_0, v_0 \in H^2 \text{ and } \|u_0 - v_0\|_{H^2} \le \delta\right) \Rightarrow \sup_{t\in[0,T]} \|u(t, \cdot) - v(t, \cdot)\|_{H^2} \le \varepsilon.$$

Therefore, the nonuniform continuity will be proved if: there exists $\varepsilon_0 > 0$ such that,

$$\forall n \in \mathbf{N} \, \exists u_n^0, v_n^0 \in H^2 : \|u_n^0 - v_n^0\|_{H^2} \le \frac{1}{n} \text{ but } \sup_{t\in[0,T]} \|u_n(t, \cdot) - v_n(t, \cdot)\|_{H^2} > \varepsilon_0.$$

This proves our claim.

2. a) This follows from the equalities, $(u_n^0)'(x) = \lambda_n^{-\frac{1}{2}} \chi'(\lambda_n x)$ and $(v_n^0)'(x) = (u_n^0)'(x) = \varepsilon_n \chi'(x)$.

   b) This follows from the discussion made in Part 2.

3. We have $1 - T\|(u_n^0)'\|_{L^\infty} \leq |1 + t(u_n^0)'(\kappa(t,x))| \leq 1 + T\|(u_n^0)'\|_{L^\infty}$. Since by hypothesis $T\|(u_n^0)'\|_{L^\infty} \leq \frac{1}{2}$ we deduce that, $\frac{1}{2} \leq |1 + t(u^0)'(\kappa(t,x))| \leq \frac{3}{2}$. The same holds for v_n^0. The first estimate follows from question 6 Part 2. Now again by question 6 Part 2 we have,

$$\frac{\partial^2 \kappa_1}{\partial x^2}(t,x) = -\frac{t(u_n^0)''(\kappa(t,x))\partial_x\kappa_1(t,x)}{(1 + t(u_n^0)'(\kappa_1(t,x)))^2}.$$

Now by (8.52) we have $(u_n^0)''(x) = \lambda_n^{\frac{1}{2}}\chi''(\lambda_n x)$. By the above estimates we deduce that,

$$\left|\frac{\partial^2 \kappa_1}{\partial x^2}(t,x)\right| \leq Ct\lambda_n^{\frac{1}{2}}|\chi''(\lambda_n\kappa_1(t,x))|.$$

Since $v_n^0(x) = u_n^0(x) + \varepsilon_n\chi(x)$ we have in the same way,

$$\left|\frac{\partial^2 \kappa_2}{\partial x^2}(t,x)\right| \leq Ct(\lambda_n^{\frac{1}{2}}|\chi''(\lambda_n\kappa_2(t,x))| + \varepsilon_n).$$

4. Setting $\kappa_1(t,x) = y \Leftrightarrow x = y + tu_n^0(y)$ in the integral we obtain,

$$\|(u_n^0)''(\kappa_1(t,\cdot))\|_{L^2}^2 = \lambda_n\int |\chi(\lambda_n y)|^2|1 + t(u_n^0)'(y)|\,dy.$$

By question 2a) we have fixed T such that $T\|(u_n^0)'\|_{L^\infty} \leq \frac{1}{2}$. It follows that,

$$\|(u_n^0)''(\kappa_1(t,\cdot))\|_{L^2}^2 \geq \frac{1}{2}\lambda_n\int |\chi(\lambda_n y)|^2\,dy = \frac{1}{2}\int |\chi(y)|^2\,dy.$$

5. Set $I_j(t) = \|(\partial_x^j A)(t,\cdot)\|_{L^2}$, $j = 0,1,2$. Setting $\kappa_2(t,x) = y$ in the integral and using the fact that $T\|(v_n^0)'\|_{L^\infty} \leq \frac{1}{2}$ we obtain,

$$I_0(t) \leq \varepsilon_n\left(\int |\chi(y)|^2|1 + t(v_n^0)'(y)|\,dy\right)^{\frac{1}{2}} \leq C\varepsilon_n.$$

The estimate of $I_1(t)$ is analogue since by (8.54) we have $\left|\frac{\partial\kappa_2}{\partial x}(t,x)\right| \leq 2$. Eventually,

$$I_2(t) \leq \varepsilon_n\left(\|(\partial_x\kappa_2(t,\cdot))^2\,\chi'(\kappa_2(t,\cdot))\|_{L^2} + \|\partial_x^2\kappa_2(t,\cdot)\chi''(\kappa_2(t,\cdot))\|_{L^2}\right).$$

The first term in the right-hand side is estimated as I_1. For the second term, we use question 3. Since, $|\partial_x^2\kappa_2(t,x)| \leq Ct(\lambda_n^{\frac{1}{2}}|\chi''(\lambda_n\kappa_2(t,x))| + 1)$, its square can be bounded by,

$$C\varepsilon_n^2(1 + T\|(v_n^0)'\|_{L^\infty(\mathbf{R})})T^2 \int (\lambda_n|\chi''(\lambda_n y)|^2 + 1)|\chi''(y)|^2 \, dy \le C(T)\varepsilon_n^2.$$

Therefore, we have proved that $\|A_n(t, \cdot)\|_{H^2} \le C(T)\varepsilon_n$ which gives the conclusion.

6. Setting as before $\kappa_j(t, x) = y$ in the integral we obtain,

$$\|u_n^0(\kappa_j(t, \cdot))\|_{L^2}^2 = \lambda^{-3} \int |\chi(\lambda_n\kappa_j(t, x))| \, dx \le C\lambda_n^{-3} \int |\chi(\lambda_n y)|^2 \, dy \to 0$$

when $n \to +\infty$. Using the same method and question 3 we can write,

$$\|\partial_x[u_n^0(\kappa_j(t, \cdot))]\|_{L^2}^2 \le C\lambda_n^{-1} \int |\chi'(\lambda_n y)|^2 \, dy \to 0, \quad n \to +\infty.$$

7. a) Using question 3 and setting as usual $y = \kappa_j(t, x)$ in the integral we can write,

$$\left\|(u_n^0)'(\kappa_j(t, \cdot)) \frac{\partial^2 \kappa_j}{\partial x^2}(t, \cdot)\right\|_{L^2}^2 \le C\lambda_n^{-1} \int |\chi'(\lambda_n y)|^2 t^2 (\lambda_n|\chi''(\lambda_n y)|^2 + 1) \, dy.$$

Setting $z = \lambda_n y$ we see that the right-hand side tends to zero when $n \to +\infty$.

b) This follows from (8.56), questions 5, 6, 7a) and the fact that,

$$\partial_x^2 v_n(t, x) = (u_n^0)''(\kappa_2(t, x)) \left(\frac{\partial \kappa_2}{\partial x}(t, x)\right)^2 + (u_n^0)'(\kappa_2(t, x)) \left(\frac{\partial^2 \kappa_2}{\partial x^2}(t, x)\right)^2.$$

c) First of all we have, $\kappa_2(t, y + tv_n^0(y)) = y$. Moreover, $y + tu_n^0(y) = x \Leftrightarrow y = \kappa_1(t, x)$. Thus if $|\lambda_n\kappa_1(t, x)| \le 1$ we have $\lambda_n|y| \le 1$ so $\chi(y) = 1$. Since $u_n^0(y) - v_n^0(y) = -\varepsilon_n\chi(y)$ we can write $\kappa_2(t, y + tu_n^0(y)) = y - t\varepsilon_n\frac{\partial\kappa_2}{\partial x}(t, x^*)$. Setting $y + tu_n^0(y) = x$ we deduce from question 3 that,

$$|\lambda_n\kappa_2(t, x)| \ge \lambda_n\varepsilon_n t \left|\frac{\partial\kappa_2}{\partial x}(t, x^*)\right| - \lambda_n|y| \ge C\lambda_n\varepsilon_n t - 1$$

Therefore, if n is large enough we obtain, $|\lambda_n\kappa_2(t, x)| \ge 2$ since $\lambda_n\varepsilon_n \to +\infty$ and $t > 0$.

d) This follows from the equality $(\alpha - \beta)^2 = \alpha^2 + \beta^2 - 2\alpha\beta$ and the fact that $\alpha\beta = 0$ since they have disjoint supports. We may apply the previous question to $\alpha = f, \beta = g$.

e) This follows from the previous question and questions 3 and 4. where we have proved that $|\partial_x\kappa_1(t, x^*)|$ is bounded from below and

$$\|(u_n^0)''(\kappa_1(t, \cdot))\|_{L^2} \ge c > 0.$$

Chapter 10
Classical Results

In this chapter we recall some results which are used in the problems.

10.1 Some Classical Formulas

10.1.1 The Leibniz Formula

- Let N be an integer $N \geq 1$, Ω be an open subset of $\mathbf{R}^d$, and u, v be two C^N functions on Ω. Then for $\alpha \in \mathbf{N}^d$, $|\alpha| \leq N$ we have,

$$\partial^\alpha(uv) = \sum_{\beta \leq \alpha} \binom{\alpha}{\beta} \partial^\beta u \, \partial^{\alpha-\beta} v.$$

Here $\beta \leq \alpha$ means $\beta_j \leq \alpha_j, 1 \leq j \leq d$; $\binom{\alpha}{\beta} = \frac{\alpha!}{\beta!(\alpha-\beta)!}$, $\alpha! = \alpha_1! \ldots \alpha_d!$.

10.1.2 The Taylor Formula with Integral Reminder

- Let Ω be an open subset of $\mathbf{R}^d$, N be an integer, $N \geq 1$, and $u \in C^N(\Omega)$. Let x, y be two points in Ω and assume that the segment that joins them is contained in Ω. Then we have,

$$u(x) = \sum_{|\alpha| \leq N-1} \partial^\alpha u(y)(x-y)^\alpha$$

$$+ N \sum_{|\alpha|=N} \frac{(x-y)^\alpha}{\alpha!} \int_0^1 (1-t)^{N-1} \partial^\alpha u(tx + (1-t)y) \, dt.$$

© The Editor(s) (if applicable) and The Author(s), under exclusive license
to Springer Nature Switzerland AG 2020
T. Alazard, C. Zuily, *Tools and Problems in Partial Differential Equations*,
Universitext, https://doi.org/10.1007/978-3-030-50284-3_10

10.1.3   The Faa–di-Bruno Formula

- Let Ω be an open subset of $\mathbf{R}_x^d$, $k, N \in \mathbf{N}$, $N \geq 1, k \geq 1, U = (u_1, \ldots, u_k) \in \left(C^N(\Omega, \mathbf{R})\right)^k$ and $F \in C^N(\mathbf{R}_y^k)$. Then for every $\alpha \in \mathbf{N}^d$, $1 \leq |\alpha| \leq N$, we have,

$$\partial_x^\alpha \left(F(U(x))\right) = \sum_{1 \leq |\beta| \leq |\alpha|} c_{\alpha,\beta} \, (\partial_y^\beta F)(U(x)) \prod_{j=1}^s \left(\partial_x^{\ell_j} u(x)\right)^{k_j},$$

where $c_{\alpha,\beta}$ are constants and in the sum we have,

$$s \geq 1, \quad |k_j| \geq 1, \quad |\ell_j| \geq 1, \quad \sum_{j=1}^s k_j = \beta, \quad \sum_{j=1}^s |k_j| \ell_j = \alpha.$$

- In dimension $d = 1$ we can give the expression of the real coefficients which occur in the above sum. Indeed we have the following formula.
  Set $D = \frac{d}{dx}$. Let $n \in \mathbf{N}^*$ and let F, g be two C^n functions on $\mathbf{R}$. Then we have,

$$D^n[F(g(x))] = \sum \frac{n!}{k_1! \cdots k_n!} (D^k F)(g(x)) \left(\frac{Dg(x)}{1!}\right)^{k_1} \cdots \left(\frac{D^n g(x)}{n!}\right)^{k_n},$$

where, in the sum, we have $k_1 + 2k_2 + \cdots + nk_n = n$ and $k = k_1 + \cdots + k_n$.

10.2   Elements of Integration

In this section we recall several important results in integration theory.

10.2.1   Convergence Theorems

Let $(X, \mathcal{T}, \mu)$ be a measured space. In what follows a.e. will stand for *almost everywhere* with respect to μ.

- The Beppo–Levi theorem.
  Let $(f_n)_{N \in \mathbf{N}}$ be an increasing sequence of real integrable functions. Then

$$\lim_{n \to +\infty} \int_X f_n \, d\mu = \int_X (\lim_{n \to +\infty} f_n) \, d\mu \leq +\infty.$$

Notice that we have the same result for a decreasing sequence.

- The Fatou lemma.

  Let $(f_n)_{n\in\mathbf{N}}$ be a sequence of positive measurable functions on X. Then

$$\int_X \liminf_{n\to+\infty} f_n \, d\mu \le \liminf_{n\to+\infty} \int_X f_n \, d\mu \le +\infty.$$

- The Lebesgue dominated convergence theorem.

  Let $(f_n)_{n\in\mathbf{N}}$ be a sequence of integrable functions such that,

  (i) the sequence $(f_n(x))_{n\in\mathbf{N}}$ converges a.e. in $\mathbf{C}$,

  (ii) there exists g positive integrable such that $|f_n(x)| \le g(x)$, a.e.$\forall n \in \mathbf{N}$.

  Then there exists f integrable such that $\lim_{n\to+\infty} f_n(x) = f(x)$ a.e. and,

$$\lim_{n\to+\infty} \int_X f_n \, d\mu = \int_X f(x) \, d\mu.$$

10.2.2   Change of Variables in $\mathbf{R}^d$

Let $\Omega \subset \mathbf{R}_x^d$ and $O \subset \mathbf{R}_\xi^d$ be two open subsets such that there exists a C^1 diffeomorphism $\varphi : O \to \Omega$. Recall that for $\xi \in O$ the Jacobian $J_\varphi(\xi)$ of φ is the determinant of the differential of φ at ξ.

- A function $f : \Omega \to \mathbf{C}$ is integrable on Ω if and only if the function $(f \circ \varphi)|J_\varphi| : O \to \mathbf{C}$ is integrable on O and we have,

$$\int_\Omega f(x) \, dx = \int_O f(\varphi(\xi))|J_\varphi(\xi)| \, d\xi.$$

10.2.3   Polar Coordinates in $\mathbf{R}^d$

- They consist to use the coordinates (r, ω) where $r \in (0, +\infty)$ and $\omega \in \mathbf{S}^{d-1}$ where $\mathbf{S}^{d-1}$ is the unit sphere in $\mathbf{R}^d$. If we call $d\omega$ the Lebesgue measure on $\mathbf{S}^{d-1}$ then we have,

$$\int_{\mathbf{R}^d} f(x) \, dx = \int_0^{+\infty} \int_{\mathbf{S}^{d-1}} f(r\omega) r^{d-1} \, d\omega \, dr.$$

In particular if f is a radial function, this means that $f(x) = F(|x|)$, we have,

$$\int_{\mathbf{R}^d} f(x)\,dx = |\mathbf{S}^{d-1}| \int_0^{+\infty} F(r) r^{d-1}\,dr,$$

where $|\mathbf{S}^{d-1}|$ is the Lebesgue measure of the unit sphere.

The diffeomorphism related to these coordinates can be written explicitly. Set,

$$x_1 = r\cos\theta_1,\ x_2 = r\sin\theta_1\cos\theta_2,\dots,$$
$$x_{d-1} = r\sin\theta_1\sin\theta_2\cdots\sin\theta_{d-2}\cos\theta_{d-1}$$
$$x_d = r\sin\theta_1\sin\theta_2\cdots\sin\theta_{d-2}\sin\theta_{d-1}.$$

Then we obtain a diffeomorphism from the open set

$$\left\{ r \in (0, +\infty), \theta_j \in (0, \pi), 1 \le j \le d-2, \theta_{d-1} \in (0, 2\pi) \right\},$$

to $\mathbf{R}^d \setminus A$ where A has Lebesgue measure zero. Then the Jacobian of this diffeomorphism is,

$$r^{d-1}(\sin\theta_1)^{d-2}(\sin\theta_2)^{d-3}\cdots(\sin\theta_{d-2}).$$

- Notice that in the polar coordinates (r, ω) in $\mathbf{R}^d$ the Laplacian takes the form,

$$\Delta = \frac{\partial^2}{\partial r^2} + \frac{d-1}{r}\frac{\partial}{\partial r} + \frac{1}{r^2}\Delta_\omega,$$

where Δ_ω is a second order operator in $\omega \in \mathbf{S}^{d-1}$ called the "Laplace–Beltrami" operator.

10.2.4   The Gauss–Green Formula

This is a formula of integration by parts in more than one dimension. It is sometimes called *the divergence theorem*,

Let Ω be an open subset of $\mathbf{R}^d$ and $\partial\Omega$ be its boundary. Assume that one has,

$$\Omega = \left\{ x \in \mathbf{R}^d : \rho(x) < 0 \right\}, \quad \partial\Omega = \left\{ x \in \mathbf{R}^d : \rho(x) = 0 \right\}, \tag{10.1}$$

where ρ is a real C^1 function on $\mathbf{R}^d$ such that grad $\rho(x) \ne 0$ for all $x \in \partial\Omega$.

For instance the unit ball B in $\mathbf{R}^d$ can be written as $B = \left\{ x \in \mathbf{R}^d : |x|^2 - 1 < 0 \right\}$ and $\rho(x) = |x|^2 - 1$ satisfies the above condition. Its boundary is $\mathbf{S}^{d-1}$. Of course the exterior of the ball B can also be written as (10.1) by taking $\rho(x) = 1 - |x|^2$.

Let $x \in \partial\Omega$. Then the vector $n(x) = \frac{\mathrm{grad}\,\rho(x)}{\|\mathrm{grad}\,\rho(x)\|}$ is called the *unit exterior normal* to the boundary at the point x.

Recall that for $k \in \mathbf{N}$, $C^k(\overline{\Omega})$ is the space of restrictions to Ω of elements of $C^k(\mathbf{R}^d)$. If Ω is not bounded we shall set $C_0^k(\overline{\Omega}) = C^k(\overline{\Omega}) \cap C_0^k(\mathbf{R}^d)$ which means that we consider C^k functions on $\overline{\Omega}$ having compact support in $\mathbf{R}^d$. However, $C_0^k(\overline{\Omega}) = C^k(\overline{\Omega})$ if Ω is bounded.

Then we have the following result.

- There exists a positive measure $d\sigma$ on the boundary $\partial\Omega$ such that for every functions $\varphi \in C_0^1(\overline{\Omega})$ and $f_1, \ldots, f_d \in C^1(\overline{\Omega})$ we have,

$$\int_\Omega \mathrm{div}\, F(x)\,\varphi(x)\,dx = -\int_\Omega F(x) \cdot \mathrm{grad}\,\varphi(x)\,dx + \int_{\partial\Omega} \varphi(x)\,n(x) \cdot F(x)\,d\sigma,$$

$$(10.2)$$

where $F(x) = (f_1(x), \ldots, f_d(x))$, $\mathrm{div}\, F = \sum_{j=1}^d \frac{\partial f_j}{\partial x_j}$ and $X \cdot Y = \sum_{j=1}^d X_j Y_j$. Moreover, the same formula is true if $\varphi \in C^1(\overline{\Omega})$ and $f_1, \ldots, f_d \in C_0^1(\overline{\Omega})$.

In particular if $\varphi \equiv 1$ and $F \in C_0^1(\overline{\Omega})$ we have,

$$\int_\Omega \mathrm{div}\, F(x)\,dx = \int_{\partial\Omega} n(x) \cdot F(x)\,d\sigma.$$

Taking $F = \mathrm{grad}\, u$ where $u \in C_0^2(\overline{\Omega})$ and $\varphi \in C^2(\overline{\Omega})$ we deduce from (10.2) the so-called *Green formula* for the Laplacian,

$$\int_\Omega \Delta u(x)\varphi(x)\,dx = \int_\Omega u(x)\Delta\varphi(x)\,dx + \int_{\partial\Omega} \left(\frac{\partial u}{\partial n}(x)\varphi(x) - u(x)\frac{\partial\varphi}{\partial n}(x)\right)d\sigma,$$

where $\Delta = \sum_{j=1}^d \frac{\partial^2}{\partial x_j^2}$ is the Laplacian and $\frac{\partial}{\partial n}$ is the exterior normal derivative to the boundary.

10.2.5 Integration on a Graph

- Let O be an open subset of $\mathbf{R}^{d-1}$ and let S be a graph over O given by,

$$S = \left\{x = (x', x_d) : x' \in O : x_d = \psi(x')\right\}.$$

Then the Lebesgue measure $d\sigma$ on S is given for $\varphi \in C^0(S)$ by,

$$\int_S \varphi(x)\,d\sigma = \int_O \varphi(x', \psi(x'))\sqrt{1 + |\nabla_{x'}\psi(x')|^2}\,dx'.$$

10.3   Elements of Differential Calculus

- **The inverse function theorem.** Let E, F be two Banach spaces. Let Ω be an open subset of E and $x_0 \in E$. Let f be a C^1 map from Ω to F. Assume that $df(x_0)$ (the differential of f at x_0) is an isomorphism from E to F.

  Then there exists V neighborhood of x_0 and W neighborhood of $f(x_0)$ such that f is a C^1 diffeomorphism from V to W.

- **The implicit function theorem.** Let E, F, G be three Banach spaces and let Ω be a neighborhood of $(x_0, y_0) \in E \times F$. Let $f : \Omega \to G$ be a C^1 map. Assume that there exists a linear and continuous map $L : G \to F$ such that $f'_y(x_0, y_0) \circ L = Id_G$.

  Then there exists a C^1 map g from a neighborhood of x_0 to F such that $f(x, g(x)) = f(x_0, y_0)$. Moreover, if $f'_y(x_0, y_0)$ is bijective then g is unique and $f(x, y) = f(x_0, y_0)$ is equivalent to $y = g(x)$ for (x, y) close to (x_0, y_0).

- **The Hadamard theorem.** Recall first that a map $f : \mathbf{R}^d \to \mathbf{R}^d$ is said to be *proper* if the inverse image of a compact set is compact. This is equivalent to say that $\lim_{|x| \to +\infty} |f(x)| = +\infty$.

  Let $f : \mathbf{R}^d \to \mathbf{R}^d$ be a C^1 map. Then the two following conditions are equivalent.

  (i) f is a diffeomorphism from $\mathbf{R}^d$ to $\mathbf{R}^d$.

  (ii) f is proper and its Jacobian is different from zero everywhere.

10.4   Elements of Differential Equations

We recall in this section some basic facts about systems of differential equations.

10.4.1   The Precise Cauchy–Lipschitz Theorem

- Let a, b be two strictly positive real numbers and $(t_0, x_0) \in \mathbf{R} \times \mathbf{R}^d$. Set

$$Q = \left\{ (t, x) \in \mathbf{R} \times \mathbf{R}^d : |t - t_0| \leq a, |x - x_0| \leq b \right\}.$$

Let $f : Q \to \mathbf{R}^d$ be continuous and let $M > 0$ be such that,

$$|f(t, x)| \leq M, \quad \forall (t, x) \in Q. \tag{10.3}$$

Assume moreover that,

$$\exists C > 0 : |f(t, x) - f(t, y)| \le C|x - y|, \quad \forall (t, x), (t, y) \in Q. \tag{10.4}$$

Then the Cauchy problem,

$$\frac{dx}{dt}(t) = f(t, x(t)), \quad x(t_0) = x_0, \tag{10.5}$$

has a unique solution $x(t)$ defined for $t \in J = [t_0 - T, t_0 + T]$, where $T = \min(a, \frac{b}{M})$, such that $(s, x(s)) \in Q$ for all $s \in J$.

- Notice that, in this version, the time T of existence of the solution does not depend on the constant C in (10.4).
- Differentiability with respect to the initial conditions.

  Assume conditions (10.3) and (10.4) satisfied. Then there exists a neighborhood V of (t_0, x_0) such that for every $(s, y) \in V$ the problem,

$$\frac{dx}{dt}(t) = f(t, x(t)), \quad x(s) = y, \tag{10.6}$$

has a unique solution defined for $t \in J = [t_0 - T, t_0 + T]$ where $T = \min(a, \frac{b}{M})$ and this solution denoted by $x(t; s, y)$ is continuous with respect to (s, y) in V.

  Moreover, if $f \in C^k(Q)$ for $k \in \mathbf{N}, k \ge 1$ then the solution is C^k with respect to (s, y) and C^{k+1} with respect to t.

10.4.2 The Cauchy–Arzela–Péano Theorem

- We keep the notations above. Let $f : Q \to \mathbf{R}^d$ be a continuous function satisfying (10.3). Then the problem (10.5) has a solution defined on J.
- Notice that in this case we lose the uniqueness of the solution.

  Moreover, we have the same differentiability results with respect to the initial conditions.

10.4.3 Global Theory

- Let x_1 (resp. x_2) be a solution of (10.5) on an interval J_1 (resp. J_2.) We say that x_2 extends x_1 if $J_1 \subset J_2$ and $x_1 = x_2$ on J_1. A *maximal* solution of (10.5) is a solution which has no nontrivial extension.
- Let $I \subset \mathbf{R}$ be an interval, $\Omega \subset \mathbf{R}^d$ be an open set, $(t_0, x_0) \in I \times \mathbf{R}^d$ and let $f : I \times \Omega \to \mathbf{R}^d$ be a continuous function.

Then the problem (10.5) has a maximal solution defined on an open interval $J = (T_*, T^*) \subset I$.

Moreover, if f is locally Lipschitz then this solution is unique.

- Set $I = (a, b)$ where $-\infty \leq a < b \leq +\infty$ and assume $\Omega = \mathbf{R}^d$. Let x be a maximal solution of (10.5) defined on $(T_*, T^*) \subset (a, b)$. Then,

  (i)   either $T^* = b$ or $T^* < b$ and $\lim_{t \to T^*} |x(t)| = +\infty$,

  (ii)   either $T_* = a$ or $T_* > a$ and $\lim_{t \to T_*} |x(t)| = +\infty$.

- Notice that in the above results we have a true limit and not only $\overline{\lim}$.

10.4.4   The Gronwall Inequality

Here is a very useful lemma.

- Let φ, k be two nonnegative and continuous functions on an interval $[a, b] \subset \mathbf{R}$. Assume that there exist two constants $A \geq 0$, $B \geq 0$ such that,

$$\varphi(t) \leq A + B \int_a^t k(s)\varphi(s)\,ds, \quad \forall t \in [a, b].$$

Then,

$$\varphi(t) \leq A \exp\left(B \int_a^t k(s)\,ds \right), \quad \forall t \in [a, b].$$

10.5   Elements of Holomorphic Functions

In this section we recall some basic facts of the theory of holomorphic functions in one variable.

In $\mathbf{C}$, the variable is denoted by $z = x + iy$ with $x, y \in \mathbf{R}$. Moreover, in all what follows Ω will denote an open subset of $\mathbf{C}$.

- **The Cauchy–Riemann operator.** It is defined by

$$\bar{\partial} = \frac{1}{2}\left(\frac{\partial}{\partial x} + i\frac{\partial}{\partial y} \right).$$

- **Holomorphic functions.** A function $f : \Omega \to \mathbf{C}$ is said to be holomorphic if,

    (i) f is a C^1 function with respect to x, y,   (ii) $\bar\partial f = 0$ in Ω.

  The vector space of holomorphic functions on Ω is denoted by $\mathcal{H}(\Omega)$.
- **The maximum modulus principle.** Assume Ω is connected and let $f \in \mathcal{H}(\Omega)$. If there exists $z_0 \in \Omega$ such that $|f(z_0)| \geq |f(z)|$ for all $z \in \Omega$, then f is constant on Ω.
- **The principle of isolated zeroes.** Assume Ω is connected and let $f, g \in \mathcal{H}(\Omega)$. Let S be a subset of Ω containing an accumulation point. If $f(z) = g(z)$ for $z \in S$ then $f = g$.
- **The index.** Let γ be a closed path in Ω and $z_0 \notin \gamma$. The index of z_0 with respect to γ is defined by,

$$\mathrm{Ind}(\gamma, z_0) = \frac{1}{2i\pi} \int_\gamma \frac{d\zeta}{\zeta - z_0}.$$

  It is an integer. For instance if γ is the boundary of the disc $D = \{z : |z| < R\}$ then $\mathrm{Ind}(\gamma, z_0) = 1$ if z_0 is inside D and $\mathrm{Ind}(\gamma, z_0) = 0$ if it is outside.
- **Simply connected domain.** In $\mathbf{C}$ this is a connected open set Ω such that Ω^c (its complementary) has no bounded connected component.

    Star shaped domains are simply connected. An annulus is not simply connected.
- **The Cauchy integral formula.** Assume Ω is simply connected and let $f \in \mathcal{H}(\Omega)$. Let γ be a closed path. Then,

    (i) $\displaystyle \int_\gamma f(z)\,dz = 0,$

    (ii) $\forall z_0 \in \Omega,\ z_0 \notin \gamma,$ $\displaystyle \mathrm{Ind}(\gamma, z_0) f(z_0) = \frac{1}{2i\pi} \int_\gamma \frac{f(z)}{z - z_0}\,dz.$

- **Characterization of the holomorphy.** The two following conditions are equivalent:

    (i) $f \in \mathcal{H}(\Omega)$,

    (ii) $\forall z_0 \in \Omega,\ \exists r > 0 : f(z) = \displaystyle\sum_{n=0}^{+\infty} a_n (z - z_0)^n,\ \forall z \in \{z \in \mathbf{C} : |z - z_0| < r\},$

  where the series is absolutely convergent for $|z - z_0| < r$.
- **Convergence of sequences of holomorphic functions.** Let $(f_n)_{n \in \mathbf{N}} \subset \mathcal{H}(\Omega)$. Assume that for each compact subset K of Ω the sequence converges uniformly on K to a function f. Then f belongs to $\mathcal{H}(\Omega)$.

- **Laurent series. Residues. Poles.** A *Laurent* series is a series of the form $f(z) = \sum_{n=-\infty}^{+\infty} a_n(z - z_0)^n$. It is said to be absolutely (resp. uniformly) convergent in a set $A \subset \mathbf{C}$ if the series $\sum_{n \geq 0} a_n(z - z_0)^n$ and $\sum_{n < 0} a_n(z - z_0)^n$ converge absolutely (resp. uniformly) on A.

  Let Ω be an open subset of $\mathbf{C}$ and $z_0 \in \Omega$. If f is holomorphic in $\Omega \setminus \{z_0\}$ then f has a Laurent expansion $f(z) = \sum_{n=-\infty}^{+\infty} a_n(z - z_0)^n$ which converges absolutely and uniformly in the set $\varepsilon \leq |z - z_0| \leq r$ for some $\varepsilon > 0, r > 0$.

  The *residue* of f at z_0 is defined by, Res $(f, z_0) = a_{-1}$.

  If f is holomorphic in $\Omega \setminus \{z_0\}$ we shall say that f has a *pole of order* $m \geq 1$ if the function $(z - z_0)^m f(z)$ is holomorphic in Ω and does not vanish at z_0.

  If f has a pole of order $m \geq 1$ at z_0 then,

$$\text{Res}\,(f, z_0) = \frac{1}{(m-1)!} \lim_{z \to z_0} \left(\frac{d}{dz}\right)^{m-1} \left((z - z_0)^m f(z)\right).$$

- **The Residue formula.** Let Ω be a simply connected domain and $A = \{z_1, \ldots, z_n\}$ be a subset of Ω. Let γ be a closed path in Ω such that $z_j \notin \gamma, j = 1 \ldots n$. Let $f \in \mathcal{H}(\Omega \setminus A)$. Then,

$$\int_\gamma f(z)\,dz = 2i\pi \sum_{a \in A} \text{Res}\,(f, z_j)\,\text{Ind}(\gamma, z_j).$$

- **Meromorphic functions.** Let Ω be an open subset of $\mathbf{C}$. A function f is said to be *meromorphic* in Ω if there exists $A \subset \Omega$ with no accumulation point such that,

$$(i) \quad f \in \mathcal{H}(\Omega \setminus A), \quad (ii) \quad \text{every point of } A \text{ is a pole of } f.$$

  Meromorphic functions are quotient of two holomorphic functions.

- **Interior of a path.** Let Ω be an open subset of $\mathbf{C}$ and let γ be a closed path in Ω. We say that γ has an interior if Ind $(\gamma, a) = 0$ or 1 for every $a \notin \gamma$. Then the set $\{a \in \Omega : \text{Ind}\,(\gamma, a) = 1\}$ will be called the *interior* of γ.

  This notion of interior coincides obviously with the usual one if for instance γ is the boundary of a disc.

  Assume Ω simply connected and let $\gamma \subset \Omega$ be a closed path. Let f be a meromorphic function in Ω with only a finite number of zeros and poles $z_1, \ldots, z_n$ such that $z_j \notin \gamma, j = 1, \ldots, n$. Then,

$$\frac{1}{2i\pi} \int_\gamma \frac{f'(z)}{f(z)}\,dz = N - P,$$

  where N (resp. P) is the number of zeros (resp. poles) of f (counted with their multiplicity) in the interior of γ.

References

1. Alazard, T.: Analyse et équations aux dérivées partielles. Lectures Notes. Available at http://talazard.perso.math.cnrs.fr (2020)
2. Alinhac, S.: Hyperbolic Partial Differential Equations. Universitext. Springer, Dordrecht (2009)
3. Alinhac, S., Gérard, P.: Pseudo-Differential Operators and the Nash-Moser Theorem. Graduate Studies in Mathematics, vol. 82. American Mathematical Society, Providence, RI (2007) (Translated from the 1991 French original by Stephen S. Wilson)
4. Ambrosio, L., Caffarelli, L.A., Brenier, Y., Buttazzo, G.M., Villani, C.: In: Caffarelli, L.A., Salsa, S. (eds.) Optimal Transportation and Applications. Lectures from the C.I.M.E. Summer School held in Martina Franca, September 2–8, 2001. Lecture Notes in Mathematics, vol. 1813. Springer/Centro Internazionale Matematico Estivo (C.I.M.E.), Berlin/Florence (2003)
5. Bahouri, H., Chemin, J.-Y., Danchin, R.: Fourier Analysis and Nonlinear Partial Differential Equations. Grundlehren der Mathematischen Wissenschaften (Fundamental Principles of Mathematical Sciences), vol. 343. Springer, Heidelberg (2011)
6. Bergh, J., Löfström, J.: Interpolation Spaces. An Introduction. Grundlehren der Mathematischen Wissenschaften, vol. 223. Springer, Berlin (1976)
7. Brezis, H.: Functional Analysis, Sobolev Spaces and Partial Differential Equations. Universitext. Springer, New York (2011)
8. Carleman, T.: Sur un problème d'unicité pur les systèmes d'équations aux dérivées partielles à deux variables indépendantes. Ark. Mat., Astr. Fys. **26**(17), 9 (1939)
9. Evans, L.C.: Partial Differential Equations. Graduate Studies in Mathematics, vol. 19, 2nd edn. American Mathematical Society, Providence, RI (2010)
10. Gérard, P.: Microlocal defect measures. Commun. Partial Differ. Equ. **16**(11), 1761–1794 (1991)
11. Gilbarg, D., Trudinger N.S.: Elliptic Partial Differential Equations of Second Order. Grundlehren der Mathematischen Wissenschaften (Fundamental Principles of Mathematical Sciences), vol. 224, 2nd edn. Springer, Berlin (1983)
12. Himonas, A.A., Misiołek, G.: Non-uniform dependence on initial data of solutions to the Euler equations of hydrodynamics. Commun. Math. Phys. **296**(1), 285–301 (2010)
13. Hörmander, L.: Hypoelliptic second order differential equations. Acta Math. **119**, 147–171 (1967)
14. Hörmander, L.: The Analysis of Linear Partial Differential Operators. I. Distribution Theory and Fourier Analysis. Classics in Mathematics. Springer, Berlin (2003) (Reprint of the second (1990) edition [Springer, Berlin; MR1065993 (91m:35001a)].)

© The Editor(s) (if applicable) and The Author(s), under exclusive license
to Springer Nature Switzerland AG 2020
T. Alazard, C. Zuily, *Tools and Problems in Partial Differential Equations*,
Universitext, https://doi.org/10.1007/978-3-030-50284-3

15. Hörmander, L.: The Analysis of Linear Partial Differential Operators. IV. Fourier Integral Operators. Classics in Mathematics. Springer, Berlin (2009) (Reprint of the 1994 edition)
16. John, F.: Partial Differential Equations. Applied Mathematical Sciences, vol. 1, 4th edn. Springer, New York (1982)
17. Kato, T.: The Cauchy problem for quasi-linear symmetric hyperbolic systems. Arch. Ration. Mech. Anal. **58**(3), 181–205 (1975)
18. Lerner, N.: Carleman Inequalities. An Introduction and More. Grundlehren der Mathematischen Wissenschaften (Fundamental Principles of Mathematical Sciences), vol. 353. Springer, Cham (2019)
19. Lieb, E.H., Loss, M.: Analysis. Graduate Studies in Mathematics, vol. 14, 2nd edn. American Mathematical Society, Providence, RI (2001)
20. Makino, T., Ukai, S., Kawashima, S.: Sur la solution à support compact de l'équations d'Euler compressible. Jpn. J. Appl. Math. **3**(2), 249–257 (1986)
21. Métivier, G.: Para-Differential Calculus and Applications to the Cauchy Problem for Nonlinear Systems. Centro di Ricerca Matematica Ennio De Giorgi (CRM) Series, vol. 5. Edizioni della Normale, Pisa (2008)
22. Misiołek, G., Yoneda, T.: Continuity of the solution map of the Euler equations in Hölder spaces and weak norm inflation in Besov spaces. Trans. Am. Math. Soc. **370**(7), 4709–4730 (2018)
23. Nishida, T.: A note on a theorem of Nirenberg. J. Differ. Geom. **12**(4), 629–633 (1978), 1977.
24. Rudin, W.: Functional Analysis. International Series in Pure and Applied Mathematics, 2nd edn. McGraw-Hill, Inc., New York (1991)
25. Smoller, J.: Shock Waves and Reaction-Diffusion Equations. Grundlehren der Mathematischen Wissenschaften (Fundamental Principles of Mathematical Sciences), vol. 258, 2nd edn. Springer, New York (1994)
26. Sogge, C.D.: Lectures on Non-Linear Wave Equations, 2nd edn. International Press, Boston, MA (2008)
27. Sogge, C.D.: Hangzhou Lectures on Eigenfunctions of the Laplacian. Annals of Mathematics Studies, vol. 188. Princeton University Press, Princeton, NJ (2014)
28. Stein, E.M.: Harmonic Analysis: Real-Variable Methods, Orthogonality, and Oscillatory Integrals. Princeton Mathematical Series, vol. 43. Princeton University Press, Princeton, NJ (1993) (With the assistance of Timothy S. Murphy, Monographs in Harmonic Analysis, III.)
29. Tao, T.: Nonlinear Dispersive Equations. Local and Global Analysis. CBMS Regional Conference Series in Mathematics, vol. 106. American Mathematical Society, Providence, RI (2006) (Published for the Conference Board of the Mathematical Sciences, Washington, DC)
30. Taylor, M.E.: Partial Differential Equations II. Qualitative Studies of Linear Equations. Applied Mathematical Sciences, vol. 116, 2nd edn. Springer, New York (2011)
31. Taylor, M.E.: Partial Differential Equations III. Nonlinear Equations. Applied Mathematical Sciences, vol. 117, 2nd edn. Springer, New York (2011)
32. Zuily, C.: Problems in Distributions and Partial Differential Equations. North-Holland Mathematics Studies, vol. 143. North-Holland Publishing Co./Hermann, Amsterdam/Paris (1988) (Translated from the French.)
33. Zuily, C.: Distributions et équations aux dérivées partielles. Dunod, Paris (2002)
34. Zworski, M.: Semiclassical Analysis. Graduate Studies in Mathematics, vol. 138. American Mathematical Society, Providence, RI (2012)

Index

Printed in the United States
By Bookmasters